Analysis of Variance (ANOVA)

Dawn Iacobucci

Nashville, TN: Earlie Lite Books, Inc.

Readers and bookstores: please order this book directly from Amazon.

Notes to the Reader

I am writing this book on "Analysis of Variance (ANOVA)" primarily to the Ph.D. students I've taught over the years at Kellogg and Wharton, as something they might share with their students. The book grew from the compilation of the case packet notes I used when teaching their statistics methods Ph.D. seminars. (The homework assignments, and solutions, are also included!)

The seminar covered analysis of variance, some experimental design, and a little matrix algebra to get into multivariate analysis variance. These topics are essential for conducting behavioral research in consumer psychology and the micro side of organizational behavior.

Statistical and methodological topics like measurement and factor analysis, structural equations models, logits, multidimensional scaling, cluster analysis, etc., were covered in a companion Ph.D. seminar on multivariate statistics. The essence of those topics are addressed in my *Marketing Models: Multivariate Statistics and Marketing Analytics* book (look for the most recent edition on Amazon, currently the 2nd ed., ©2015).

As with the models book, my hope is that you will find the tone of the writing in this ANOVA book to be as if I'm sitting down with you, explaining the techniques to you, and coaching you. And a rare author's offer: if you find some material confusing or incomplete (or, eek, if you find a typo!), please email me and I'll try to address your issues.

The use of ANOVA is typically that an experiment is being planned or has been run, and the analysis of variance is an extraordinarily useful and powerful technique that answers the research questions posed by the experiment(er). We manipulate a set of factors, and we measure their potential impact on some outcome or dependent variables.

The analysis of variance is related to the regression model (run with dummy variables), as both are exemplars of the general linear model. However, the focus in the analyses in either case is slightly different, and expectations regarding the models' usage and reporting vary with journals for which the research is written and targeted. This book does not touch upon the general linear model or much about regression. Per the title, it's 100%, full-on, 24/7, analysis of variance.

Finally, a brief word about notation: independent variables (or predictors or factors) are labeled A, B, etc. The dependent variable is called X, and the data matrix, **X**. Means are denoted $\bar{x}$ for the overall grand mean (but not with double-bars $\bar{\bar{x}}$ or dots $\bar{x}_{..}$), and with subscripts, e.g., $\bar{x}_i$ for the ith group mean. Equations in the text do not end with periods, because they look too much like the dots sometimes used to depict means.

The real trick in interpreting any model is to watch the subscripts: it doesn't matter whether a term in a model is depicted as μ or β or φⅎι$\Im$oF$\mathfrak{U}\mu$ (sound it out). If we're told that the raw data point X_{ijk} stands for a rating on brand "i" at time "j" for person "k" then if we see a mean or a model term $\bar{x}_{ij}$ or β_{ij} then we'll know we have aggregated (summed or averaged) over the missing subscript "k" all people in the sample, and we'll have a $\bar{x}_{ij}$ and β_{ij} for every brand "i" at every time point "j." If the model term is τ_i then we'll have averaged over time points "j" and people "k" and we'll have a tau for each brand "i."

Good luck! Have fun!

Table of Contents

Section I: Analysis of Variance (ANOVA) and Experimental Design

Section II: Related Topics

Section III: Multivariate Analysis of Variance (MANOVA)

The seminars covered the topics in these weekly units (this table is a rough syllabus):

Week	Learning Objectives	Chs.	HWs
1	Analysis of variance and experimental design • Review of basic inferential statistics • One-way ANOVA Related topic: SAS intro	1, 2, 11	ch.2 (due next class)
2	Factorials (2- and 3-way), effect sizes • Two-way ANOVA, w replications (equal n's) • Two-way ANOVA. without replications • Three-way ANOVA and higher-order ANOVA (equal n's) • Omega squared	3, 4	ch.3
3	• Contrasts and comparisons	5	ch.5
4	Experimental design • Fixed vs. random effects, computation of Expected Mean Squares • Factorial designs, randomized blocks, Latin Squares, nested designs, fractional factorials, split-plot designs	6, 7	ch.6
5	More on designs, repeated measures, ANCOVA • Repeated measures, within subjects designs • Analysis of covariance	8, 9	ch.8
6	Unbalanced designs (Unequal n's) Related topic: matrix algebra (needed to go multivariate)	10, 12	ch.12
7	Multivariate Analysis of Variance • Overview, multivariate normal distribution, T^2, assumptions and robustness • MANOVA intro, test statistics, assumptions, robustness, power	13, 14	ch.14
8	Finish MANOVA • Contrasts and follow-up testing • Application to repeated measure	15, 16	ch.15
9, 10	• Advanced issues • Assorted topics		

Regarding the homework assignments... There are 8 homework assignments that I use to torture my Ph.D. students over the course of a 10-week quarter.

In the class packet, as in this book, the solutions were also provided. My philosophy is that for many topics, like stats, immediate feedback is more useful than feedback that is a week old (upon my returning their graded papers). If students believe they are solving a homework problem correctly, but in fact are doing something incorrectly, it's better that they can check their answers and see immediately upon completion that something is wrong, so they can re-check their work and see where they were led astray. By the time the students handed in their homeworks to me, they should have been perfect. I graded their homeworks pass/fail (with happy cartoon stickers!), and it never looked like a student took advantage of the solutions being present—these were Ph.D. students after all, smart, and knowledgeable in that trying to do the work would be the only way they'd learn the material. I might mention that they were motivated as well, given that they could see how important this course material would be to a successful career in research (and their professors let them know this as well). I also encouraged students to write questions throughout their homeworks, like, "I don't really understand this," or even, "I saw that I did the problem incorrectly the first time, but I still don't quite understand why this or that is so." I encouraged questions in class of course, but sometimes certain questions don't arise during the seminar, occurring only while trying to complete a problem on one's own.

The content covered in each homework is roughly as follows:

- HW from Ch.2: demonstrating that a property holds in relation to SS_{total}, $SS_{between}$, and SS_{within} (or SS_{error}), calculating the elements of two separate one-factor ANOVAs on two tiny datasets (to get a "feel" for what an ANOVA is doing with a dataset by hand, and to come to an appreciation for computers to be used in subsequent HWs), and a treasure hunt finding critical values (no longer by looking them up in the tables in the back of stats books, rather, deriving them from functions in Excel; we are so contemporary).
- HW from Ch.3 (a long HW assignment, need to torture students early in the term to get them committed): defining what an "interaction" is (they're so important), getting a feel for results by staring at tables of means and trying to determine by eye whether a main effect for factor A or B or their interaction is present in the data, running a small dataset (3×3 factorial, with n=3) through SAS, getting a feel for data in a 3-factor factorial, and running another dataset through SAS, extending the ANOVA model to what it would look like for 4 factors, filling in an ANOVA table by understanding that df and SS sum.
- HW from Ch.5 (a short HW assignment): a study is described along with conceptual comparisons that the researcher wants to make and the task is to create the contrast coefficients to logically test each hypothesis, and two questions converge on testing for the orthogonality of contrast coefficients.
- HW from Ch.6: two scenarios are presented, each of which has at least one random factor in the experimental design, hence students must derive the Expected Mean Squares (EMS) so that they could calculate the proper F-tests, the SAS code for each is then requested, and an optional (extra credit) question is posed to extend the logic further to yet another design.
- HW from Ch.8 (very short assignment, the term is a little over mid-way, students are exhausted, with assignments due in other classes, this week is a break for them): specifies a particular repeated measures experimental design and asks that the ANOVA table be delineated.

  - HW from Ch.12 on matrix algebra (a little long, what the heck, students had last week "off"): what is a matrix, basic matrix manipulations (additions, subtractions, scalar multiplications, vectors), playing with a small 'data set,' definitions, converting a covariance matrix into a correlation matrix, obtaining a covariance matrix.
  - HW from Ch.14 (looks like a short HW but is a little hairy): gets students familiar with manipulating the four major MANOVA test statistics.
- HW from Ch.15: prepares students for the take home final, presents a small data set 3×2 with n=8 observations per cell, and p=2 dependent variables measured on each study participant. Students are to analyze the heck out of the data set, starting with basic descriptives (means, standard deviations, correlations, within cells, across factors, really getting a feel for the data), then ANOVAs and a MANOVA, with contrasts where appropriate.

CHAPTER 1

BRIEF REVIEW OF BASIC INFERENTIAL STATISTICS

Questions to guide your learning:

1. *This chapter is a refresher to make sure you recall these terms:*
    a. *population and sample*
    b. *mean and standard deviation*
    c. *null and alternative hypotheses*
    d. *one-tailed vs. two-tailed tests and directional vs. non-directional hypotheses*
    e. *Type I and Type II errors in hypothesis testing.*
2. *Make sure you're confident that you can compute a z-test or t-test for a one-sample and two-sample test.*

This chapter provides a brief review of statistics. We'll start simply, first with just a single sample from a single population.

We define some target population. Statistical theory will assume the population distribution will follow (approximately) a normal curve like in Figure 1.

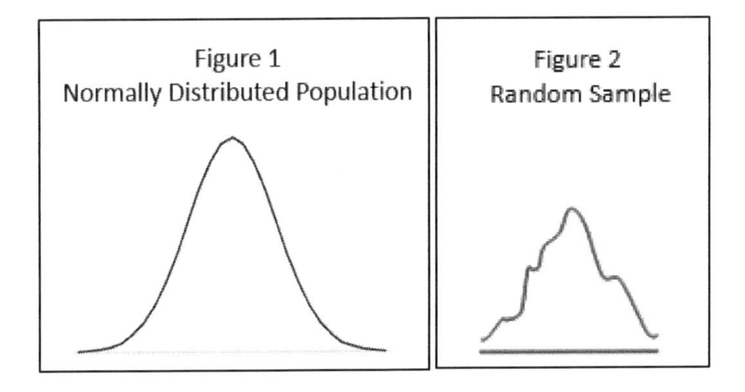
Figure 1
Normally Distributed Population

Figure 2
Random Sample

From that gorgeous population distribution, we draw a random sample of size "n." It will look a little rougher, like Figure 2.

In that sample, we collect our data, such as a survey full of questions about respondents' attitudes toward something such as an advertisement, a brand, their health care plan, a presidential hopeful, or we can measure their height, etc. The survey items can tap the cognitive or emotional elements of a person's perceptions, their motivations, or likelihoods to engage in some behavior of interest, etc. An amazing number of things we measure follow a normal curve like that in Figure 1, at least roughly.[1]

---

[1] When we create scales by averaging multiple items, the result is even more likely to be normal in appearance, in kind of a "more data are better than fewer data" corollary of the Central Limit Theorem (based purely on sample size). We won't talk much about robustness in this book (the extent to which our analyses are likely to lead to the correct conclusion regarding rejecting the null hypothesis or not

For the purposes of continuing in this statistical adventure, we'll focus for the moment on just one measurement or variable, i.e., a single attitude item "x" from the survey of our n respondents. Call those observations: $x_1, x_2, x_3, \ldots, x_n$.

One Sample Descriptive Statistics

The population has a mean, μ. We'll estimate the population mean using the sample mean based on the n data points, the x_i's (i stands for the ith respondent, i ranges from 1 to n):

$$\hat{\mu} = \bar{x} = \frac{1}{n}\sum_{i=1}^{n} x_i$$

The population and sample also both have other measures of central tendency. Basic stats courses also present the median (the 50th percentile; as many scores observed above the median as below it) and mode (the most frequently occurring value in the distribution). For a normal, bell-shaped curve like that in Figure 1, the mean, median, and mode are all the same value. For a sample drawn from it, like that in Figure 2, the mean, median, and mode are likely to be approximately similar values. The distribution in Figure 2 doesn't look skewed—with a lump of data at the left (low values) and few observations toward the right (high values) that makes it look like a tail at the right, indicative of a positive skew, or the reverse with the majority of values toward the right and a long tail with only a few observations at the left (negative skew). In skewed distributions, a mean will be more sensitive to extreme values (so it will be higher in positive skews, and lower in negative skews). With a distribution like that in Figure 2, which doesn't look very skewed, even if the mean is pulled slightly above or below the mode and median, it probably won't be far away.

Back to the statistics we care more about... In addition to a mean, the population and sample also have measures of dispersion. These measures capture how far apart the observed scores are; if the observations are clustered tightly together, we would say there seems to be a good deal of consensus in the perceptions of the study respondents. For samples with observations spread further apart, we'd say there were more individual differences or heterogeneity. The two main indicators of dispersion that we tend to care about are the variance, σ_x^2, and standard deviation, σ_x. Just as the notation implies, the standard deviation is the square root of the variance. The standard deviation is measured on the same units as the raw data (e.g., height in inches or attitudinal ratings on a 7-point scale), whereas the variance is measured on those units squared (e.g., inches squared), so the standard deviation is often more intuitive when we're trying to understand the data. Nevertheless, we'll need both, in different roles.

For a standard deviation, the sample estimator is called $\hat{\sigma}_x$ or simply, s_x. There are three ways to compute and understand the essence of a standard deviation.

1. First, it is standard to begin with the "definitional" formula, which depicts most clearly the numerator as the squared sum of differences between each data point, the x_i's, and the sample's overall mean:

---

even if model assumptions are violated), in part because we don't need to—the analysis of variance model is extraordinarily robust. In later chapters, we'll look at assumptions and robustness more carefully in the realm of multivariate statistics because there, the models can be a little touchier.

$$\hat{\sigma}_x = s_x = \sqrt{\frac{\sum_{i=1}^{n}(x_i - \bar{x})^2}{n-1}} \quad (1)$$

That equation shows how the standard deviation captures the notion of variability: if all x_i scores were the same, they'd all equal the mean as well, and all the differences would be zero. Hence, to the extent that the x_i's vary, the standard deviation grows, $s_x > 0$. (Given the numerator is a squared term, the smallest it can be is 0 if there is no variability, and it cannot be negative.)

2. Second, while that first equation shows how a standard deviation captures the variability of the data points from the overall mean, there is an easier "computational" formula:

$$\hat{\sigma}_x = s_x = \sqrt{\frac{\sum x_i^2 - \left(\frac{1}{n}(\sum x_i)^2\right)}{n-1}} \quad (2)$$

The first term in the numerator adds up the squared x_i scores, whereas the second term adds up the x_i scores and then squares that sum. This formula is referred to as computational because if we had to compute a standard deviation by hand, this one is easier to do than that in equation (1). If we were stranded on a deserted island and yet had to calculate a standard deviation by hand, this formula is certainly the one we'd use. However, we will assume that we have access to a computer, so we'll rely more frequently on definitional formulae. Just by looking at the difference term in parentheses in equation (1), we get an immediate sense of what we're comparing—the raw data points x_i to the mean $\bar{x}$. That intuition is somewhat clouded in equation (2).

Also note the denominator in equations (1) and (2). When the variance is defined with "n" in the denominator, it is a "maximum likelihood estimator" (which is a good thing in math stats) but it tends to underestimate very slightly the population variance, that is, it's biased (which is a quality we don't want when we're analyzing data and trying to test hypotheses about the population). By comparison, the variance defined with "n-1" in the denominator is unbiased. Hence we proceed with "n-1."

3. Third, the definitional formula (1) can be written conceptually in the short-hand form,

$$\hat{\sigma}_x = s_x = \sqrt{\frac{SS_x}{df_x}} \quad (3)$$

That is, the standard deviation is the (square-root of a) "sum of squares" (SS) divided by a "degrees of freedom" (df). The sum of squares term is easy to understand—the deviations of the observed x_i's from the mean are squared and then summed. The degrees of freedom will be explained in more detail shortly.

Hypothesis Testing for One Population

What's next? We'll use the sample information, $\bar{x}$ and s_x to infer something about the nature of the population, hence, we're entering the zone of "inferential" statistics, compared to simply "descriptive" statistics characterizing the sample.

To do so, we posit a "model," a simplified representation of reality. The simplest model we'll see takes the form:

$$x_i = \mu + \epsilon_i \quad (4)$$

which basically says that we believe each of our sample's data points, each x_i, to be a function of the population mean, μ, and some noise, ϵ_i. If we had to guess a sample respondent's score, we'd guess μ, and we know that our estimate will be off by some amount of error, ϵ_i. Note the subscript of "i" on the error term; the amount of error or adjustment varies for each sample score (each respondent or observation in the sample). Some observed scores will be higher than the mean, some lower, some large, some small. We typically assume that the errors, big and small, negative and positive, will average out to zero (due to the bell-shaped normal distribution assumption). (We'll get more specific about other assumptions when we need them.)

In the model (4), μ is referred to as "fixed" (or even "structural"), and it's the part of the model we're trying to estimate. That is, we're trying to answer the question, "What is the mean in the population?" Our best guess is obviously going to be the sample mean. In model (4), the error term ϵ_i is said to be "random," and it's what makes the data point x_i random.

Given the model in (4), we pose a hypothesis. The null hypothesis states that we expect the population mean to equal some constant.[2] We'll call the constant "c" and it can be whatever makes sense for the context (zero, last year's mean, the midpoint on a scale, etc.):

$$H_0\colon\ \mu = c \quad (5a)$$

In contrast, the alternative hypothesis takes the form:

$$H_A\colon\ \mu \neq c \quad (5b)$$

Recall from basic stats, we could also have "directional" hypotheses (e.g., $H_0\colon\ \mu < c$ vs. $H_A\colon\ \mu \geq c$). However, for the analysis of variance, we'll focus on the non-directional hypotheses as posited in (5a) and (5b). In the null hypothesis stated in (5a), we're investigating a point estimate, and in the alternative hypothesis stated in (5b), we're basically saying we'd be interested in knowing if our hypothesis (5a) is wrong, in either direction—too high or too low. By comparison, if we had a directional hypothesis of say, $H_0\colon\ \mu < 100$, the test would only inform us if the sample mean was significantly greater than 100, but it would not tell us if the sample mean was far smaller than 100.

The accompanying jargon is that we'll focus on 2-tailed tests. That is, we'll place half of the test's rejection region, .05/2, in the extreme left tail and the other half at the extreme right, and we'll reject H_0 if our mean is significantly less than or greater than the value "c." (By comparison, recall that if the hypotheses are directional, then the entire rejection region is placed in one tail, on the side of the alternative hypothesis, e.g., for $H_A\colon\ \mu \geq c$, the tail would be at the right.)

The Philosophy of Hypothesis Testing

Before proceeding to the mechanics of testing hypotheses, let's first have a brief word about what we're doing. The logic of hypothesis testing is so fundamental to research and the

---

[2] The "null" gets its name from applications where we're comparing one sample mean to another, and H_0 would say there is no difference, that is, the groups are the same, they come from the same population. Or, we might compare an experimentally treated group to a control group and H_0 would say the treatment yielded no difference and so it was ineffective. So "null" means essentially, "no differences," or "nothing's going on in the data." Naturally we are more often interested in when groups differ or when experimental manipulations are effective, thus we usually hope to reject the null in favor of the alternative hypothesis.

logic of decision-making, that once you understand it well, you will lead a more fulfilling and enriched life. Honest!

Our goal is to reject H_0. Statistically and logically, we cannot prove a relationship or hypothesis such as H_0 to be "true." There always exists another possible explanation, if we are clever enough. And there will always exist data that suggest the relationship is not as simple as what we're considering, if the variables included in the study and analyses are encompassing enough. So the only thing we can do is evaluate the current explanation or hypothesis as more or less plausible or implausible. It is easier statistically and logically to demonstrate that a hypothesis is implausible. When we cannot demonstrate that a hypothesis is implausible (i.e., we cannot reject the null hypothesis), we are not demonstrating that the null hypothesis is true. It is simply plausible until further data come along. All models are "false" given that they are always simplifications of the world. When comparing how well different models explain some data, a model has utility if it explains the data (in the current dataset and that known in the literature) better than the other models, while being parsimonious.

When we reject a null hypothesis, we say the finding is statistically significant. Significance means that, if the null hypothesis were true, our result would be exceedingly rare (our result is significantly different from what the null hypothesis posits); hence, we'll conclude that the null hypothesis is not likely true, and that the alternative hypothesis is more likely.

One Sample Inferential Statistics—Testing Hypotheses

To test the (non-directional) null hypothesis in (5a), we compute the sample mean:

$$\bar{x} = \frac{1}{n}\sum_{i=1}^{n} x_i = \frac{1}{n}(x_1 + x_2 + x_3 + \cdots + x_n) \tag{6}$$

and the sample variance:

$$s_x^2 = \sum_{i=1}^{n} \frac{(x_i - \bar{x})^2}{n-1} \tag{7}$$

from which we derive the "standard error" which adjusts the variability for the sample size of the study:

$$SE = s_{\bar{x}} = \frac{s_x}{\sqrt{n}} \tag{8}$$

and then we can calculate the z-statistic:

$$z = \frac{(\bar{x} - \mu)}{SE} \tag{9}$$

The z-statistic is what will determine whether we reject the null hypothesis or not. Note that it serves us well if the standard error is small (because the standard error is in the denominator, so if it is small, then z will be large), and while we typically have little control over the size of s_x, we can make our samples as large as possible, i.e., to ensure n is large, and the SE in (8) is small.

The logic underlying rejecting null hypotheses is based on the notion of what a distribution would look like if we drew many many samples of size n from the population, and in each sample, calculate a mean, and collect the means over all those samples. (Of course, we don't really use an empirical sampling distribution of means, but an analytically derived theoretical sampling distribution.) The mean of the population is μ or call it μ_X. The mean of the "sampling distribution of means" is $\mu_{\bar{x}}$. These two means are identical, $\mu_x = \mu_{\bar{x}}$ and given that our sample mean $\bar{x}$ is an unbiased estimator of μ_x, it follows that $\bar{x}$ is also an unbiased estimator of the mean of the sampling distribution: $\bar{x} = \hat{\mu}_x = \hat{\mu}_{\bar{x}}$. The standard deviation of the sample, $s_x = \hat{\sigma}_x$, is used in the equation for the standard error, $\sigma_{\bar{x}} = \frac{\sigma_x}{\sqrt{n}}$ with estimators $\hat{\sigma}_{\bar{x}} = s_{\bar{x}} = \frac{s_x}{\sqrt{n}}$. The shape of the sampling distribution of means is like Figure 1 but the bell shape is narrower, because the variability of the sampling distribution is smaller than the standard deviation of the population, by that factor of $\frac{1}{\sqrt{n}}$.

In Figure 3, we see the sampling distribution of means with two marks, corresponding to $\pm(1.96)(s_{\bar{x}})$. (A z-score of 1.96 corresponds to a rejection region of 5% of the area under the curve.) If the null hypothesis H_0 is true, we'd expect to see a mean in the range of $-(1.96)(s_{\bar{x}})$ to $+(1.96)(s_{\bar{x}})$ pretty often, indeed 95% of the time. By comparison, we'd expect to see $\bar{x}$'s in the extremes, in the tails, less than $-(1.96)(s_{\bar{x}})$ or greater than $+(1.96)(s_{\bar{x}})$ less frequently, only 5% of the time. We know we can make an error (reject the null when it's true if we happen to get a very small or very large mean), but we're willing to take that 5% risk.

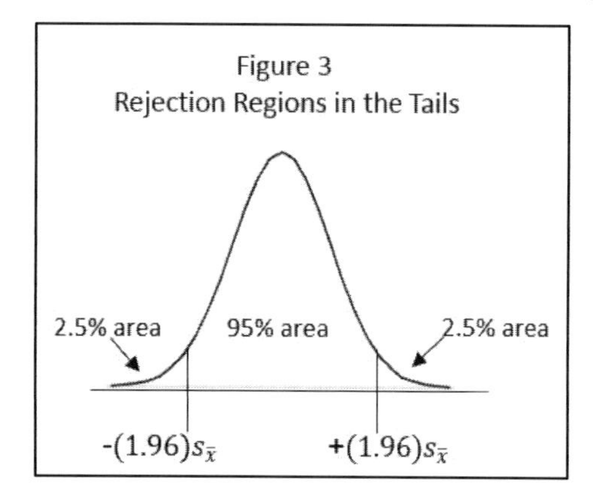

Thing is... no one could possibly create or store tables of critical values for all possible values of μ and σ, so we translate or "standardize" our values to follow a "standard normal curve," the z-distribution, which has a mean $\mu = 0$ and a standard deviation $\sigma = 1$. We achieve $\mu = 0$ with the numerator of the z-test, $(\bar{x} - \mu)$, and $\sigma = 1$ with the the division of $s_{\bar{x}}$.

Procedurally, we follow these steps:

a) First, do we know σ? If we do, we calculate a z-test: $z = \frac{(\bar{x}-\mu)}{\sigma_{\bar{x}}}$.

b) If we do not know σ, we ask whether the sample size n is 30 or larger. If $n \geq 30$, then again, we calculate a z-test: $z = \frac{(\bar{x}-\mu)}{s_{\bar{x}}}$.

c) If we do not know σ, and we have a small sample ($n < 30$), then we calculate the same statistic but it follows a t-distribution with (n-1) degrees of freedom: $t = \frac{(\bar{x}-\mu)}{s_{\bar{x}}}$.

The difference between options (a) and (b) is whether the standard deviation is known or estimated (it's rarely known), and the difference between those options and (c) is whether we compare the statistic we've just calculated to a "z" or "t" distribution to find the "critical values" that determine whether or not to reject the null.

The probability density function (pdf, or the equation for) a univariate normal curve, like the z-distribution, is:

$$\frac{1}{\sigma\sqrt{2\pi}} exp^{-\left\{\frac{(\bar{x}-\mu)^2}{2\sigma_{\bar{x}}^2}\right\}}$$

When we've standardized the data by subtracting the mean and dividing by the standard deviation, the new mean is 0 and the new standard deviation is 1, and the pdf for z-scores in the standard normal curve simplifies to: $\frac{1}{\sqrt{2\pi}} exp^{-\left\{\frac{z^2}{2}\right\}}$.

Figure 4 shows how the standard normal curve z-distribution compares to the t-distribution. The t-distribution isn't quite as tall and has more of its area in heavier tails. As a result, our statistic has to be larger in magnitude in order to reject the null. For example, where a 95% confidence level z-test uses the critical values $\pm(1.96)$, the 95% confidence t-test on, say, 10 degrees of freedom, would use the critical values $\pm(2.23)$. The idea is that in the t-distribution, the cutoff is a little bit more stringent (bigger) to compensate for smaller, possibly quirky samples. It makes the t-test a little more "conservative" (harder to reject the null), a good thing in research because we can be more confident that a significant result isn't attributable to just random sampling variability.

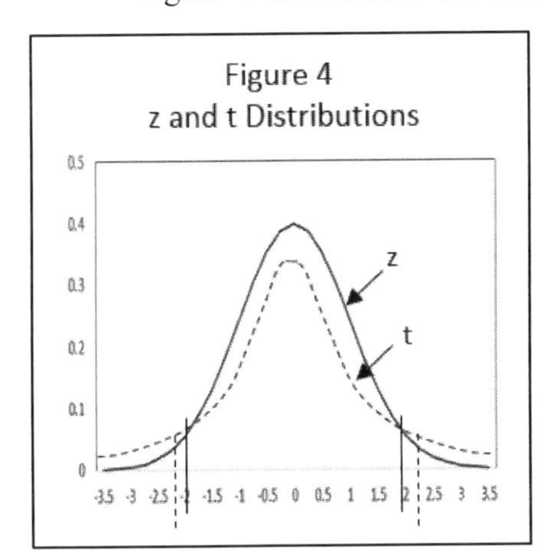

With these statistical hypotheses, statistical inference puts us into one of two situations:

(1) If our observed z-statistic equals or exceeds the critical value, that is, $z \geq 1.96$ or ≤ -1.96 (i.e., the magnitude $|z| \geq 1.96$), then we reject H_0. That means that if H_0 were true, our data or results would be unlikely. Therefore, we conclude that H_0 is unlikely, which means that H_A is more plausible.

(2) If the magnitude of our observed z-statistic < the critical value, then we cannot reject H_0. That means that if H_0 were true, our results would be likely. Therefore we'd conclude that H_0 is plausible.

Because the logic of statistical testing allows us only to reject the null hypothesis, we want to phrase our research questions in such a way that the statement we would like to support is included in the alternative hypothesis. For example, we might pose a null hypothesis that is a tentative statement that the population parameter equals some value (e.g., $\mu = 50$). The (non-directional, two-tailed) alternative hypothesis is that the population parameter does not equal the hypothesized value (i.e., $\mu \neq 50$). Or, in comparing two groups, the null hypothesis would be that the groups are not different. The alternative hypothesis would be that the groups do differ.

Errors—Ruh roh.

It can happen (α % of the time) that the H_0 might indeed be true (e.g., $\mu = 50$), and yet in our sample we observe some extreme value for our statistic z. We'd reject H_0 by the decision-rule criterion. Yet we'd be making a logical error. Oops.

Figure 5 shows there are two good scenarios and two scenarios that depict different statistical errors. If H_0 is actually true and we do not reject it (cell "d"), that's proper. If the H_0 is actually false and we do reject it ("a"), that's proper. If H_0 is true but we reject it ("b"), that's a "Type I" error (or a "false positive"). If H_0 is false but we fail to reject it because we obtain a small z-statistic ("c"), that's a "Type II" error (or a false negative).

Figure 5
Hypothesis Testing: Correct Decisions and Errors

	State of nature about μ	
Test stat says	H_0 false	H_0 true
Reject H_0	a) OK	b) Type I
Do not reject H_0	c) Type II	d) OK

A Type I error, the error of rejecting H_0 when it's true (in "b") occurs with probability α. When H_0 is true and we do not reject it, that's good (in "d"), and that occurs with probability $(1-\alpha)$. Conditional on the state that the null is true, these outcomes are probabilistically opposites, hence the α and $(1-\alpha)$. The value $(1-\alpha)$ is our "confidence" level; e.g., if $\alpha = 0.05$, then $(1-\alpha) = 0.95$.

In contrast, if H_0 is false, and we do not reject it ("c"), a Type II error occurs, and it does so with probability β. If H_0 is false, and we properly reject it ("a"), that's good, and that occurs with probability (l-β). Thus, conditional on the state that the null is false, these outcomes are probabilistically opposites, β and (l-β). The probability value (l-β) is referred to as "power" or the likelihood that we'll identify a significant result when we should.

As an example, consider Figure 6. The solid normal curve to the left is associated with H_0: $\mu = 0$, the null we're testing. In that distribution, the left and right tails comprise the area under the curve of $\alpha = 0.05$, or 5%, equally split to the left and right so that's $\alpha/2 = 0.025$, or 2.5%. If our data yielded a z-statistic ≤ -1.96 or ≥ 1.96, we'd reject the null, when in fact the null was true, so 5% of the time, we'd be making a Type I error. The area under the curve between the two rejection regions is 95% and that's the likelihood that we would not reject the null if the null were true, e.g., if we obtain a z-statistic of -0.3 or +1.2, for example.

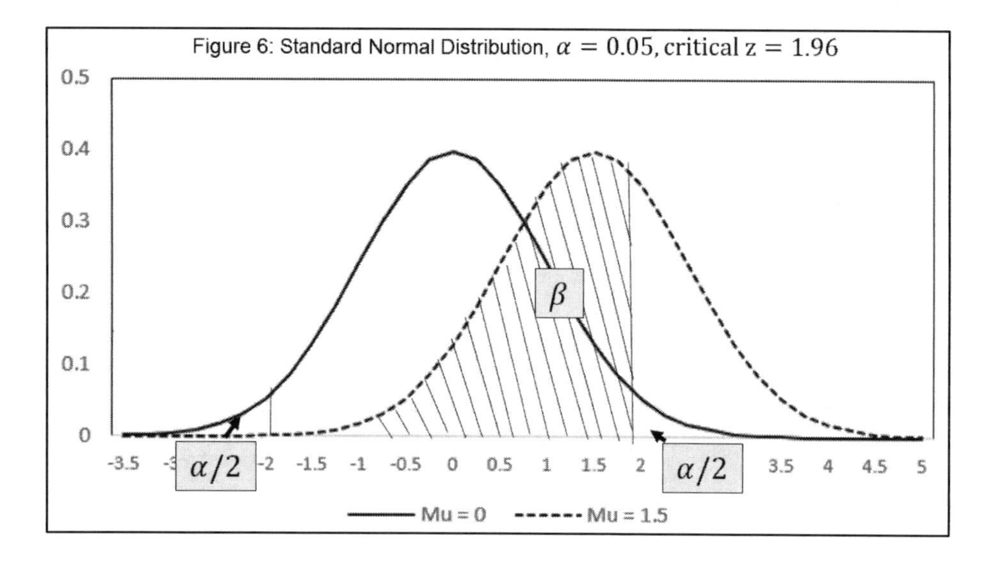

Now imagine that the null hypothesis H_0: $\mu = 0$ does not hold, and instead, the real population is skootched to the right, with $\mu = 1.5$ (the dashed bell-shape distribution to the right, still in Figure 6). We don't know that "truth" of course, all we know is that we're testing a certain null hypothesis, in this case, we believe that $\mu = 0$, and we use the rules about rejecting that null when the z-statistic exceeds 1.96 in magnitude. But if, unbeknown to us, μ actually is 1.5, that is, the truth is depicted by the normal curve to the right, then any time we did not reject the null, i.e., our calculated |z| was less than 1.96, we'd be making a Type II error, and the probability that we'd be doing that is β, depicted in the shaded area. (The area depicting power, (1-β), is under the right dashed curve to the right of the shaded β lines.)

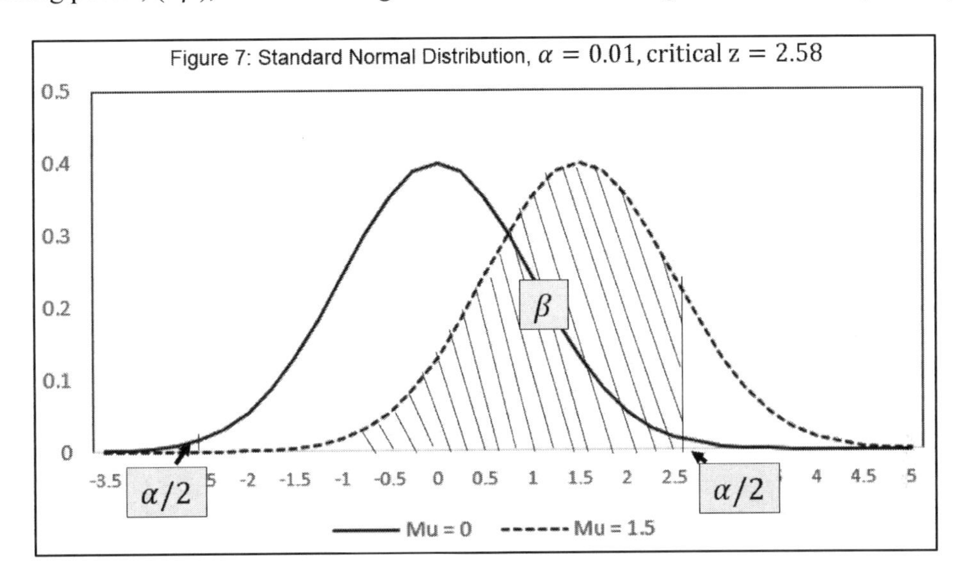

In sum regarding errors…statistical parameter estimation and testing is contained within the logic of probabilities, so we can make errors. The errors in our statistical inferences can be 1 of 2 types. A "Type I" error is committed when we are led to reject the null hypothesis when it is in fact true. The probability that we commit this error is α (alpha). Because α is the level of significance of the test, we can to some extent control the probability that we make this error by simply making α smaller. Reducing α (e.g., from .05 to .01) makes it more difficult to reject H_0 (whether the null hypothesis is true or not), because the observed test statistic needs to be more extreme in order to obtain significance, thus it will reduce Type I errors (if the null is true) but also power (if the null is false). Figure 7 shows that if we select $\alpha = 0.01$ rather than 0.05, the rejection region is reduced (from 5% to 1% of the area under the curve, corresponding to critical z-values of ± 2.58). Thus, our data have to yield an even larger z-statistic to reject the null (whether the null is true or not).

A "Type II" error is committed when we are unable to reject the null hypothesis when in fact it is false. This second error is made with probability β (beta), and (1-β) is the "power" of the statistical test. The power of the test is the probability that we will reject the null hypothesis when it is false (i.e., when it should be rejected). Figure 7 shows that reducing α from 0.05 to 0.01 increases the area labeled β, the likelihood of making a Type II error. Thus, reducing α reduces the likelihood of making a Type I error if the null is true, and it increases the likelihood of making a Type II error if the null is not true, or, same thing, it reduces power or the likelihood of finding a significant result if the null is not true. As α increases, β decreases; as the likelihood of making a Type I error increases, the likelihood of making a Type II error decreases. Conversely, as $(1-\alpha)$ decreases, $(1-\beta)$ increases; as the likelihood of correctly not rejecting the null decreases, power or the likelihood of rejecting the null when the null does not hold increases.

The Power of "n"

Power sounds like a good thing—a greater likelihood of rejecting a null hypothesis and finding some result to be significant when we should legitimately do so. We had seen briefly when discussing the standard error in equation (8) that a larger sample size can be helpful in this regard. Let's see that in greater detail.

In the z-test or the t-test, take the t… start with:

$$t = \frac{(\bar{x} - \mu)}{s_{\bar{x}}} \tag{10}$$

Let's elaborate the standard error in the denominator:

$$t = \frac{(\bar{x} - \mu)}{\left(\frac{s_x}{\sqrt{n}}\right)} \tag{11}$$

Now, rearrange the terms to take the denominator of the denominator up to the numerator:

$$t = \frac{\sqrt{n}(\bar{x} - \mu)}{s_x} \tag{12}$$

Equation (12) shows us that regardless of how different $\bar{x}$ might be from μ, that is, for a fixed nonzero effect size $(\bar{x} - \mu)$, and for a fixed s_x, then the magnitude of $|t|$ increases as $\sqrt{n}$ (or n) increases. So if we increase our sample size, n →we get more power →which means a greater likelihood of rejecting H_0, which is typically our goal. We'll see the tremendous advantage of big n again (and again!)

Hypothesis Testing for Two Populations

Now that we're pros with one population, next, let's imagine drawing random samples from two separate, independent populations. (We'll talk about what to do if we're dealing with matched samples when we talk about "repeated measures" in Chapter 8.) For example, we might wonder whether the mean rating for men is the same as for women, or whether users and non-users of our brand like an ad equally or purchase similar quantities.

In this scenario, per Figure 8, we draw samples from the two populations of sizes n_1 and n_2. In the first sample, we'll collect the data: $x_{11}, x_{12}, x_{13}, \ldots, x_{1n1}$ (note the first subscript is a "1" for group 1) and in the second sample, we'll collect the data: $x_{21}, x_{22}, x_{23}, \ldots, x_{2n2}$ (the first subscript is a "2" for group 2).

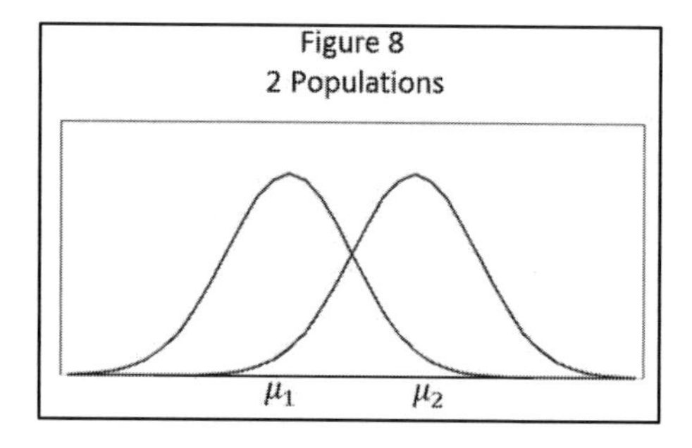

We calculate means and standard deviations in each sample to serve as estimates of their respective population parameters. We'll denote them as $\bar{x}_1$ and s_{x1} and $\bar{x}_2$ and s_{x2}.

We posit a null hypothesis usually of the form:

$$H_0\colon \mu_1 = \mu_2 \text{ or } H_0\colon (\mu_1 - \mu_2) = 0 \quad (13a)$$

$$\text{vs. } H_A\colon \mu_1 \neq \mu_2 \text{ or } H_A\colon (\mu_1 - \mu_2) \neq 0 \quad (13b)$$

We could entertain a different hypothesis, such as $H_0\colon \mu_1 = 5\,\mu_2$ or $H_0\colon \mu_1 = 10 - \mu_2$ but the simple form in (13a) is the most frequently considered.

Before crunching, let's see some examples of real null hypotheses:

a) H_0: there are no differences between groups of users and non-users of "Tide" in terms of how consumers rate the detergent on its ability to clean.
b) H_0: a grocery purchase simulation is to be conducted online, and a marketing variable such as package color will have no effect; e.g., the red package will attract the same number of purchasers as the blue package.
c) H_0: the status quo holds; there is no change from last year or last quarter; e.g., our new product attracts no more customers than our old product, there is no difference or improvement and the new sales will be the same as the old sales, or say, the relative market share of a toy maker observed during this most recent Christmas season will resemble the relative market share of the toymaker from the previous Christmas season.

If the null is stated "H_0: the groups are not different" then the alternative hypothesis is formed by simply making it mutually exclusive to the null:

H_A: "the H_0 is not true"; i.e., there are group differences.

Or, for example, if "H_0: advertising had no effect," then "H_A: advertising had some effect."

How do we proceed to test H_0? If we know both variances or standard deviations, σ_{x1} and σ_{x2}, then we calculate z as:

$$z = \frac{(\bar{x}_1 - \bar{x}_2) - (\mu_1 - \mu_2)}{\sigma_{(\bar{x}_1 - \bar{x}_2)}} \quad (14)$$

where

$$\sigma_{(\bar{x}_1 - \bar{x}_2)} = \sqrt{\sigma_{\bar{x}_1}^2 + \sigma_{\bar{x}_2}^2} = \sqrt{\frac{\sigma_{x_1}^2}{n_1} + \frac{\sigma_{x_2}^2}{n_2}} \quad (15)$$

If we don't know both σ_{x1} and σ_{x2} (and usually we do not), then we estimate them and calculate t:

$$t = \frac{(\bar{x}_1 - \bar{x}_2) - (\mu_1 - \mu_2)}{s_{(\bar{x}_1 - \bar{x}_2)}} \quad (16)$$

where

$$s_{(\bar{x}_1 - \bar{x}_2)} = \sqrt{\frac{s_{x_1}^2}{n_1} + \frac{s_{x_2}^2}{n_2}} \quad (17)$$

Even better, we typically assume "homogeneity of variances." That is, even if we wonder whether the group means (μ's) are different, per the hypotheses, we assume that the variability within each group is the same; i.e., $\sigma_{x_1}^2 = \sigma_{x_2}^2$. That assumption allows us to create a better standard error using a "pooled" estimate:

$$s_p^2 = \frac{\sum(x_{1i} - \bar{x}_1)^2 + \sum(x_{2i} - \bar{x}_2)^2}{(n_1 - 1) + (n_2 - 1)} \tag{18}$$

and replace it in the equation (17):

$$s_{(\bar{x}_1 - \bar{x}_2)} = \sqrt{s_p^2\left(\frac{1}{n_1} + \frac{1}{n_2}\right)} \tag{19}$$

The pooled estimate in (19) would then serve as the denominator in the t-test in equation (16). This two-sample t-statistic has $(n_1 - 1) + (n_2 - 1) = n_1 + n_2 - 2$ degrees of freedom.

In general, like in equation (16), for many test statistics, we'll find this general form will hold true:

$$observed\ value\ of\ test\ stat = \frac{observed\ sample\ stat - hypothesized\ population\ value}{std.error}$$

And the standard decision rules apply, specifically:

a) if the magnitude of our observed sample statistic $\geq$ its critical value, then we reject H_0. We'd conclude that if H_0 were true, our data and results would be unlikely, therefore H_A is more plausible.
b) if the magnitude of our observed sample statistic < its critical value, then we cannot reject H_0. We'd conclude that if H_0 were true, our results would be quite likely, therefore H_0 is plausible.

SUMMARY

This chapter provided a quick review of some basic statistical concepts, definitions, and equations. The analysis of variance comes next and it builds on these concepts, of a one-sample and a two-sample t-test.

REFERENCES

1. Fox, John (2009), *A Mathematical Primer for Social Statistics*, Thousand Oaks, CA: Sage.
2. Hage, Timothy M. (1995), *Basic Math for Social Scientists: Concepts,* Thousand Oaks, CA: Sage.
3. Iversen, Gudmund R. (1996), *Calculus*, Thousand Oaks, CA: Sage.

CHAPTER 2

ONE-WAY ANALYSIS OF VARIANCE

Questions to guide your learning:
Q_1: Remind me, why are experiments good?
Q_2: What is a one-way analysis of variance (ANOVA) and how is it an extension of a t-test?
Q_3: What are $SS_{between}$, SS_{within}, SS_{total}?
Q_4: How is the F-statistic calculated?
Q_5: How is the F-statistic evaluated; what are the df necessary to find the critical value?

In this chapter, we discuss "one-way" or "one-factor" analysis of variance. We'll see what that means.

Before proceeding, let's deal with a few orders of business.

1. First, while the z-distribution is familiar and easy to understand, from here on, we'll continue with a focus more on the t-statistic than the z, in part because it has degrees of freedom associated with it, which makes it more analogous to future tests we'll use.
2. Second, it's been easy to consider the possibility of unequal sample sizes so far (i.e., $n_1 \neq n_2$), but to keep things simple going forward, we'll assume they are equal ($n_1 \neq n_2 = n$) unless explicitly stated (we'll see different sample sizes in Chapter 10 on "unbalanced designs").
3. Third, the notion introduced at the end of the previous chapter, of pooling sources of error variability to get a better estimate of a standard error, is something that will continue to reappear, along with its required assumption of homogeneity of the variances, in each sample, that we're counting on being comparable in order to aggregate.

In Chapter 1, we began with the consideration of 1 and then 2 samples. Now we'll see what to do if the data represents 3 or more groups. As an example, consider Figure 1 in which 3 populations are depicted with means: $\mu_1 = 0, \mu_2 = 2, \mu_3 = 3$. If we wanted to know whether these means are equal, and the only thing we knew was how to compare 2 groups in a two-sample t-test, then we could compare $\mu_1$to μ_2 and then μ_2 to μ_3 (and by transitivity, we'd learn about μ_1 vs. μ_3). That would be okay in a pinch. However, the two statistical tests would not be independent (we'd be using the sample 2 data more than once). In addition, 2 t-tests are not as powerful (or sensitive in detecting group differences) as one "F-test" in an analysis of variance. This F-test will use all the data from all the samples, so the overall sample size is effectively bigger (and remember, larger samples bring more power). The F-test will test the null hypothesis: $H_0: \mu_1 = \mu_2 = \mu_3$. The alternative hypothesis is merely, $H_A: H_0\ is\ not\ true$. The null says all the means are the same, for however many groups there are (3 or more). The alternative is stated a little

Figure 1
3 Populations

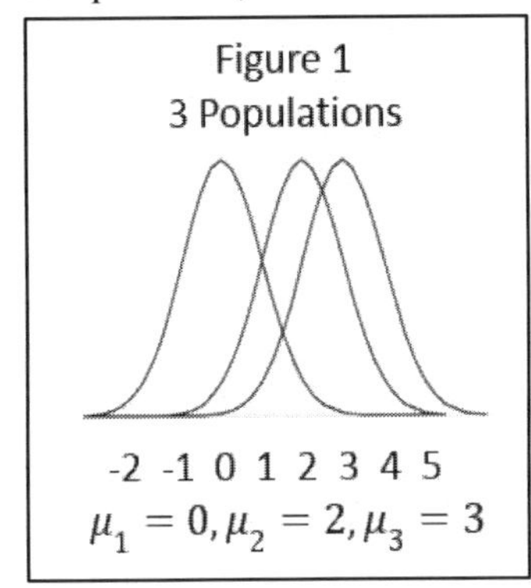

vaguely because there are many ways the null might not be true; e.g., 1 mean might be different from the other 2 (which might be equal), or all 3 means might be different. In the example in Figure 1, it might be that the mean values of 2 and 3 are not significantly different but that both differ from 0. If the within-sample variability is small (i.e., the 3 normal bell curves are narrow), then all three means, 0, 2, 3, might be significantly different. (We'll see how to determine which means are the same or different in Chapter 5 on "contrasts.")

The analysis of variance, or ANOVA for short, can be thought of as an extension or generalization of a t-test to 3 or more groups. (When there are 2 groups, the 2-sample t-test will be the square root of the F-test run on the same 2 groups; i.e., $t^2 = F$, and the qualitative outcome, about rejecting the null or not, is identical. We're getting ahead of ourselves and we'll see this equivalence later.)

ANOVA can be used to test differences between naturally occurring groups. For example, we could compare the GMAT scores in the following populations:

Group 1 = new students in marketing
Group 2 = new students in finance
Group 3 = new students in organizational behavior

The null hypothesis would be, $H_0: \mu_{mkt} = \mu_{fin} = \mu_{ob}$, versus the alternative, H_A: "at least 1 μ is different."

However, ANOVA's greatest strength is when it is used in conjunction with experiments. In the GMAT scores example, we can make descriptive statements about the 3 groups, but we couldn't assess causality. In well-designed experiments, we want to attribute group differences to our "independent variables" or "factors," the things we manipulate. As an example, say we randomly select 90 students, and randomly assign 30 to each of the following conditions:

1) classroom #1 using teaching method consisting of Q&A and discussion
2) classroom #2 using teaching method using lectures
3) classroom #3 using teaching method that is self-paced

Then after the course is over, we'd have all the students in all 3 groups take the same achievement test. We'd posit that $H_0: \mu_1 = \mu_2 = \mu_3$ and wonder whether the alternative hypothesis were true that at least one group mean was different (being better or worse than the others). If one or more groups' means were different, we'd attribute the differences in scores to the teaching method. We can make this causal inference because of the random assignment—prior to being exposed to different teaching methods, presumably the 3 groups of students started out as statistically equal. If the groups begin equal, and at the end of the study they're different, we can fairly cleanly, confidently attribute the differences to the teaching methods' efficacies.

The terminology is this: the teaching method would be referred to as the "predictor," the "independent variable," the "explanatory factor," or the "factor," all terms are used interchangeably. It's the thing we manipulate (or measure) that's used to understand or predict possible differences in the "dependent variable" (or the "response variable" or the "criterion measure"). In the example, the test score is the dependent variable.

Consider another example, at the nexus of marketing and economics. We might manipulate the price of a product such that some consumers see a "low" price, others see a "medium" price and others see a "high" price. The study could be a "lab" study (good for internal validity, the attribution of causality), e.g., we could expose different groups of consumers to different prices in an online experiment in which consumers are shown a picture of the product, a description, and its price, and the picture and description are the same for all study participants, but 1 group sees the screen with the "low" price, and 2 different groups of study participants see the screen with the "medium" or "high" price points. Alternatively, we could run a "field" study (good for external validity, or generalizability), e.g., we'd run a test market across 3 representative markets in the country, wherein the product in all the stores in one market are sold at the low (or medium or high) price point. In these scenarios, the independent variable is "price," which takes on the values low, medium, and high (e.g., $4.99, $10.99, and $15.99). In the lab study, the dependent variable would be a measure of attitude or willingness to buy the product, and in the field study, the dependent variable would be sales. The null hypothesis would be that price didn't matter (so that attitudes or sales would be the same across all 3 groups), and the alternative hypothesis would be that price has an effect.

Experimentation

The "teaching methods" and "price" studies are examples of experiments. Before proceeding to see how the ANOVA model works, let's take a brief tangent to experiments. The ANOVA model is so closely coupled with the analysis of data that arise from experiments, that presumably a reader perusing a book on ANOVA has some basic familiarity with and motivation to run experiments. However, rather than presupposing certain knowledge, let's go ahead and be explicit, offering a reminder of some basic qualities.

Research methods are typically classified into 1 of 3 categories: exploratory (e.g., focus groups, interviews), descriptive (e.g., surveys), and causal (e.g., experiments). Per its name, causal studies are usually designed to study "cause-effect" or "if-then" relationships, e.g., "If we increase price by X amount, how much will sales fall off?," or "If we begin stocking our goods in a certain large discount retailer, how much will sales increase (due to volume) and how much will our brand reputation deteriorate?" If we were to try to find answers to these questions by conducting interviews or surveys, the answers would be less definitive than if we were to run an experiment.

The conditions for establishing a sound if-then conclusion (roughly à la Hume) are usually put forth as:

1) concomitant variation: if X causes Y, then X and Y should be correlated (the reverse is not necessarily true),
2) sequential ordering: X should precede Y,
3) and trickiest of all, elimination of alternative explanations: thus, if we draw a sample and randomly assign each member to 1 condition, we assume that statistically the groups begin the same so that if we observe differences at the end of the study, we may attribute the differences to how we treated the groups, and not to some other extraneous factor.

An experiment can be as simple as a study that involves only 2 groups, e.g., to test the efficacy of some new drug, 1 group of patients would receive the drug, and 1 group would receive the placebo. In business, these simple studies are referred to as "A-B testing"; e.g., 1 group of study participants sees ad A and 1 group sees ad B.

Research studies are characterized by their strengths on internal and external validity. Internal validity is the extent to which we may attribute the study's results to our actions (e.g., sales fell off due to the price increase, not because of seasonal factors). External validity is the extent to which we may generalize our results to the broader population (e.g., studies of a drug in an upscale nursing home might not generalize to patients of different socioeconomic status or age). External validity is enhanced to the extent that we draw a random sample from the population (rather than a convenience sample). Internal validity is enhanced by our randomly assigning study participants to the "A" or "B" conditions. Thus, in the A-B experiments, we draw a random sample, and then randomly assign each person in the sample into 1 of the 2 conditions. (We can flip a coin, or use the function in Excel, =randbetween(1,2). If the coin comes up heads (or Excel says 1), the customer is assigned to condition 1, and if it's tails (or Excel says 2), the customer is assigned to cell 2. If a=5, we'd say =randbetween(1,5).)

The degree to which a researcher values internal or external validity (it is challenging to achieve both in a single study) depends often on the researcher's setting. Academics tend to favor internal validity to test theories, and so might conduct the study "in the lab" to control as much extraneous variation as possible, whereas practitioners tend to favor external validity to test the extensiveness of the applicability, and so might conduct a "field" study to observe and include the likely naturalistic complicating factors.

In an A-B test, we could use a t-test to study whether the group differences are significant. ANOVA is the model that extends the t-test to more than 2 groups (in this chapter) and to studies wherein the groups are defined along more than 1 factor (in Chapters 3 and 7).

The F-test

Let's turn now to the analysis of variance. We'll begin by looking at the F-statistic conceptually. The F-test is a ratio of "between-group" to "within-group" variability; roughly: $F = \frac{variability\ between\ groups}{variability\ within\ groups}$. So F will be large and we'll reject the null ($H_0: \mu_1 = \mu_2 = \cdots = \mu_a$) when the between group variability ($MS_{between}$) is large relative to the variability within groups (MS_{within}). In Figure 2, the means are farther apart so the group differences are clearer than in Figure 3, where the group means are closer together and the group differences are less distinct. Thus, the F-statistic would be larger in Figure 2 than in Figure 3 because there is more "between group" variability in Figure 2 (the numerator of F will be larger).

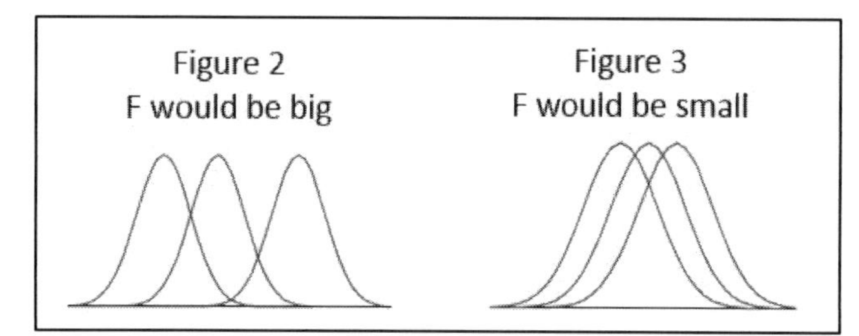

Figure 2
F would be big

Figure 3
F would be small

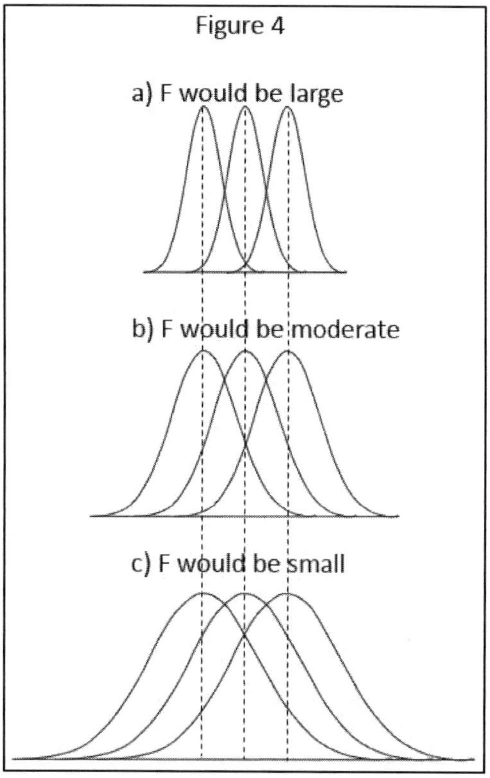

Figure 4

a) F would be large

b) F would be moderate

c) F would be small

The F-test is also affected by the "within group" variability. Figure 4 depicts three scenarios where the mean differences between the 3 populations are held constant (the between-group variability in the numerator of the F-test is the same), and what varies is the within-group variability. The F-statistic would be largest in scenario a, then b, then c. Why? In scenario "a," there is apparently a great deal of consensus in the study participants' perceptions, represented by the fact that within a group, there is little variability in their responses (the denominator in F will be small). In contrast, in scenario "c," there is more heterogeneity or individual differences contributing to greater within group variability (the denominator will be large).

Let's work toward the official equation for the F-test by beginning to understand the ANOVA model. We'll use the notation x_{ij}, where "i" represents what group a person is in (i = 1, 2, ..., "a" for factor A), and "j" represents what person in that group i (j = 1, 2, ..., n, and for a while, we'll assume equal sample sizes across the groups, $n_1 = n_2 = \cdots = n_a = n$, such that the total sample size is N = a×n; later we'll relax that limitation and allow the n's in different groups to vary and be unequal, in Chapter 10).

The one-factor ANOVA model is quite simple: we posit an observation, x_{ij} to be a function of a grand mean, a group effect, and an error term. The grand mean, μ will be estimated as the average over all our data—all observations in all groups. If we knew nothing about the person (j) or the group they're in (i), then our best guess for that person's data point would be μ. In addition, we'll calculate means for each group, i.e., μ_i, and the differences between each group mean and the overall mean will be termed α_i; simply defined as: $\alpha_i = \mu_i - \mu$. (This α_i has nothing to do with the Type I error rate α; stats people just like to make notation similar-looking to be confusing.) The α_i's are the group effects. In our hypothetical guessing game, if we knew which group a person was in, we'd be better off using that group's mean as the best guess for that person's data point. We already know the overall mean μ so we would adjust upward (or downward) by adding (or subtracting) the α_i term for whatever group "i" they're in. Even with that more refined guess, we'll typically be off a bit from the person's real data point, and that error is captured in ϵ_{ij}.

Collecting these terms together, we have the one-factor ANOVA model:

$$x_{ij} = \mu + \alpha_i + \epsilon_{ij} \qquad (1)$$

In the model in equation (1), μ is the "overall effect," α_i is the "incremental group effect" (it is $\mu_i - \mu$) and ϵ_{ij} is the model's "error" (it's how the model captures that we're off a bit in our estimation for person j in group i).

Figure 5 depicts the elements of this ANOVA model. In this illustrative study, the 3rd group has the lowest scores, and the 2nd group has the highest scores. In that 2nd group with the high scores, let's look at some random data point, person j in group 2, hence, x_{2j}. To get to that data point, we start at μ, then adjust a bit using $\alpha_2 = \mu_2 - \mu$, and given that x_{2j} (person j in group 2) is not equal to the 2nd group's mean μ_2, there is a nonzero error term ϵ_{2j}.

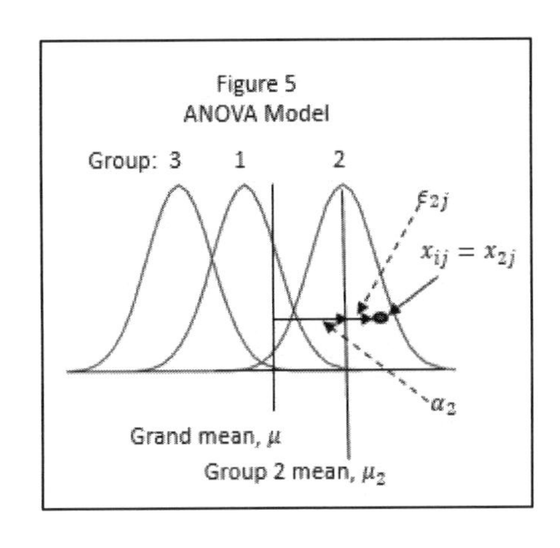

Figure 6 shows all the incremental group effects, the α_i's for all 3 groups. The incremental group effect, $\widehat{\alpha}_i = \bar{x}_i - \bar{x}$ is the model's way of picking up on the "between group variability." If all the groups were the same, then all the $\bar{x}_i$'s would equal each other, all equal to $\bar{x}$ (the grand mean over all groups), so there would be no differences between groups, all α_i's would equal zero, that is, there would be no "between group variance."

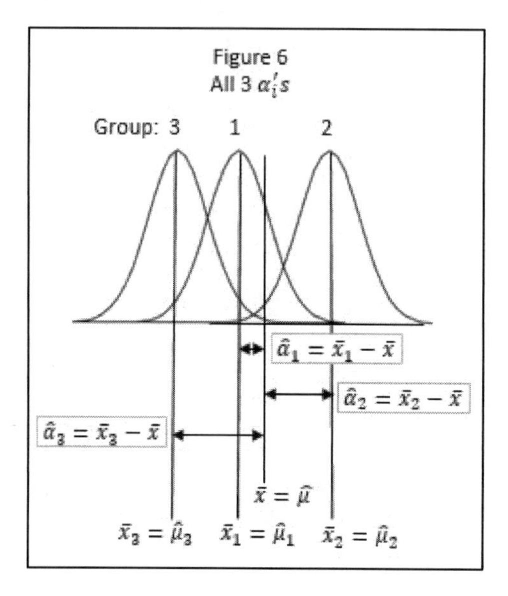

Similarly, the error term is the model's way of reflecting the "within group variability." While all the study participants in group "i" share the same mean, their particular individual data points x_{ij} are not likely to all be identical even within the same group. Each sample respondent's error term is estimated as the difference between his or her data point and his or her group's mean: $\hat{\epsilon}_{ij} = x_{ij} - \bar{x}_i$. This relationship is shown in Figure 7.

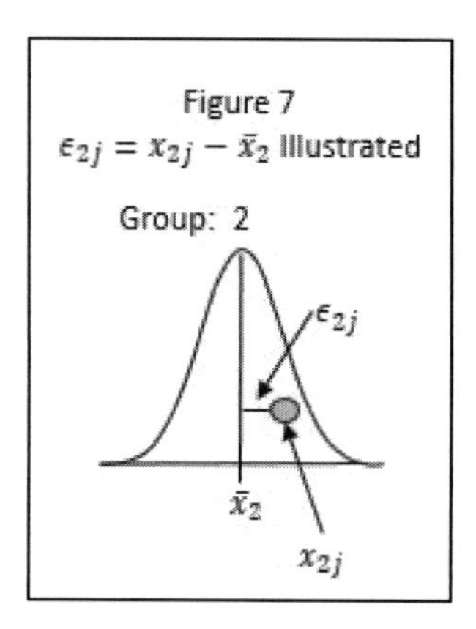

When we say these model elements are beginning to reflect the between-group and within-group variability that we'll need for the F-test, we can see it explicitly if we replace all these estimates into the model statement, from equation (1):

$$x_{ij} = \mu + \alpha_i + \epsilon_{ij}$$

Inserting the estimates to obtain this:

$$x_{ij} = \hat{\mu} + \hat{\alpha}_i + \hat{\epsilon}_{ij} = \bar{x} + (\bar{x}_i - \bar{x}) + (x_{ij} - \bar{x}_i) \tag{2}$$

Next, move the first grand mean to the left hand side:

$$(x_{ij} - \bar{x}) = (\bar{x}_i - \bar{x}) + (x_{ij} - \bar{x}_i) \tag{3}$$

Square both sides of the equation:

$$(x_{ij} - \bar{x})^2 = (\bar{x}_i - \bar{x})^2 + (x_{ij} - \bar{x}_i)^2 + 2(\bar{x}_i - \bar{x})(x_{ij} - \bar{x}_i) \tag{4}$$

Sum everything over all "a" groups and all "n" observations within each group:[3]

$$\sum_{i=1}^{a}\sum_{j=1}^{n}(x_{ij} - \bar{x})^2 = \sum_{i=1}^{a}\sum_{j=1}^{n}(\bar{x}_i - \bar{x})^2 + \sum_{i=1}^{a}\sum_{j=1}^{n}(x_{ij} - \bar{x}_i)^2 \tag{5}$$

Now we're getting somewhere! Those 3 terms are called "sums of squares," because they are differences or deviations (of one sort or another) that have been squared and then summed. These sums of squares, or SS for short, go by specific names:

$$SS_{total} = SS_{between\ groups} + SS_{within\ groups} \tag{6}$$

Each of these sources of variability (total variability, or between- or within-groups) has associated with it a certain "degrees of freedom," or "df." Degrees of freedom are always (in any model) calculated as the number of data points, minus the number of parameters estimated (or the constraints built into the model). So in general:

df = # data point - # parameters estimated (or constrainted) (7)

Specifically, let's see the df for our model. For total df, we begin with "n" observations in each of the "a" groups, and we lose 1 df for estimating the overall grand mean, μ:

$$df_{total} = (a \times n) - 1 \tag{8}$$

For between groups, we begin with "a" groups but we know their means have to mesh with the grand mean. Say $a = 3$. Then we'll estimate $\bar{x}_1$, and we'll estimate $\bar{x}_2$, but because we already have the overall mean $\bar{x}$, we can already determine $\bar{x}_3$, so we only have 3-1=2 df. Thus:

$$df_{between} = a - 1 \tag{9}$$

For within groups, we have n observations, but we know the group's mean, so we have n-1 degrees of freedom, and that's true in each of the "a" groups. Thus:

$$df_{within} = a(n - 1) \tag{10}$$

Just as SS sum up, so do df. That is:

$$SS_{total} = SS_{betw} + SS_{w/in}$$

and

$$df_{total} = df_{betw} + df_{w/in}$$

Next, remember how a simple variance is computed as: $\sigma^2 = \frac{\sum_{i=1}^{n}(x_i - \bar{x})^2}{n-1}$, and we

[3] It's easy to show that the "cross-products" term equals zero: $2\sum_{i=1}^{a}\sum_{j=1}^{n}(\bar{x}_i - \bar{x})(x_{ij} - \bar{x}_i) = 0$.

referred to it as "sum of squares" divided by "degrees of freedom." For the sources of variability we've been considering, we have a "sum of squares" and we have the df to create an analogous form.

We compute the variances, here called "mean squares between," denoted $MS_{between}$, and "mean squares within," denoted MS_{within}, each as their respective SS over df:

$$MS_{between} = \frac{SS_{betw}}{df_{betw}} = \frac{SS_{betw}}{(a-1)}$$

and

$$MS_{within} = \frac{SS_{within}}{df_{within}} = \frac{SS_{within}}{a(n-1)} \tag{11}$$

(12)

Then, the F-test is their ratio, on $(a-1)$ and $a(n-1)$ degrees of freedom:

$$F = \frac{MS_{between}}{MS_{within}} \tag{13}$$

We'd say that the F-value we've calculated is distributed as a theoretical F, $F \sim \mathcal{F}_{(a-1),a(n-1)}$. That simply means that we take our observed F and compare it to critical values in F-tables with $(a-1)$ degrees of freedom associated with the numerator and $a(n-1)$ degrees of freedom associated with the denominator. We can also use Excel to obtain critical values (instead of looking them up in tables in the back of dusty, old statistics books). For a given Type I Error, typically $\alpha = .05$, use Excel's function =F.INV.RT(.05,df1,df2). For example, say we had obtained a sample F-value of 3.50 for 2 and 30 df. To find out if the F-value of 3.50 exceeded its critical value, we'd enter in Excel: =F.INV.RT(.05,2,30), and it would yield the critical value 3.316. Given that our observed F-value had been 3.50, we can see that it exceeded the critical value (associated with a p-value less that 0.05), so the F would be deemed "significant." We'd reject the null, and conclude that the groups differ.

Even more precisely than using a comparison of an F-test to a critical value (from tables or Excel), all the statistical computing packages (SAS, SPSS, etc.) will print a p-value that is the probability of obtaining an F-value as big or bigger than the one we observed, if the null hypothesis is true. The p-values are also easily found in Excel using the =F.DIST.RT(F,df1,df2) function. For our example F = 3.50 with 2 and 30 df, we'd type =F.DIST.RT(3.50,2,30), hit return, and Excel would say: 0.043. That is the p-value. It is less than 0.05, hence the F-test is significant, we reject the null, etc.

How to Do it: An Example

Let's walk through a little example. In this dataset, we have 3 groups, and 4 subjects (study participants, respondents, observations, replications, whatever you want to call them) in each group, for a total sample size of 12.

	group			
subject	1	2	3	
1	4	5	9	
2	5	6	10	
3	6	6	11	
4	5	7	10	
group sums, $\sum_{i=1}^{n} x_{ij}$	20	24	40	grand sum = 84
group means, $\frac{1}{n}\sum_{i=1}^{n} x_{ij}$	$\bar{x}_1 = 5$	$\bar{x}_2 = 6$	$\bar{x}_3 = 10$	grand mean, $\bar{x} = 7$

FYI, the 3 $\hat{\alpha}$'s are: $\hat{\alpha}_1 = 5 - 7 = -2$, $\hat{\alpha}_2 = 6 - 7 = -1$, and $\hat{\alpha}_3 = 10 - 7 = 3$. Note that as promised, the $\hat{\alpha}$'s sum to zero (-2 -1 + 3 = 0).

To continue, the sums of squares are computed:

$$SS_{total} = \sum_{i=1}^{a}\sum_{j=1}^{n}\left(x_{ij} - \bar{x}\right)^2$$
$$= (4\text{-}7)^2 + (5\text{-}7)^2 + (6\text{-}7)^2 + (5\text{-}7)^2$$
$$+ (5\text{-}7)^2 + (6\text{-}7)^2 + (6\text{-}7)^2 + (7\text{-}7)^2$$
$$+ (9\text{-}7)^2 + (10\text{-}7)^2 + (11\text{-}7)^2 + (10\text{-}7)^2$$
$$= 62.00 \qquad (14)$$

$$SS_{between} = \sum_{i=1}^{a}\sum_{j=1}^{n}(\bar{x}_i - \bar{x})^2$$
$$= (5\text{-}7)^2 + (5\text{-}7)^2 + (5\text{-}7)^2 + (5\text{-}7)^2$$
$$+ (6\text{-}7)^2 + (6\text{-}7)^2 + (6\text{-}7)^2 + (6\text{-}7)^2$$
$$+ (10\text{-}7)^2 + (10\text{-}7)^2 + (10\text{-}7)^2 + (10\text{-}7)^2$$
$$= 56.00 \qquad (15)$$

$$SS_{within} = \sum_{i=1}^{a}\sum_{j=1}^{n}\left(x_{ij} - \bar{x}_i\right)^2$$
$$= (4\text{-}5)^2 + (5\text{-}5)^2 + (6\text{-}5)^2 + (5\text{-}5)^2$$
$$+ (5\text{-}6)^2 + (6\text{-}6)^2 + (6\text{-}6)^2 + (7\text{-}6)^2$$
$$+ (9\text{-}10)^2 + (10\text{-}10)^2 + (11\text{-}10)^2 + (10\text{-}10)^2$$
$$= 6.00 \qquad (16)$$

Next, we'll calculate the degrees of freedom:

- $df_{total} = (a \times n) - 1 = (3 \times 4) - 1 = 11$ (17)
  (remember, a×n data points, minus 1 for estimating μ)
- $df_{between} = a - 1 = 3 - 1 = 2$ (18)
  ("a" group means and 1 constraint that their average is μ)
- $df_{within} = a(n - 1) = 3(4 - 1) = 9$ (19)
  (n subjects in each group, minus 1 df for the group mean; done "a" times)

Given the SS and df, we can obtain the mean squares, MS's:

- $MS_{between} = \frac{SS_{between}}{df_{between}} = \frac{56}{2} = 28.000$ (20)
- $MS_{within} = \frac{SS_{within}}{df_{within}} = \frac{6}{9} = 0.667$ (21)

And, drum roll please, the F-statistic is: $F = \frac{MS_{between}}{MS_{within}} = \frac{28.000}{0.667} = 41.98$ on 2 and 9 df. The critical F-value for an $\alpha = 0.05$ significance level is 4.26, so our observed F clearly exceeds that critical value. We reject the null hypothesis, H_0: $\mu_1 = \mu_2 = \mu_3$; i.e., all 3 groups' means are the same; i.e., statistically speaking, 5 = 6 = 10, in favor of the alternative hypothesis (H_A: at least 1 mean is different). So, we'd know that at least 1 group's mean is different from the others (probably the 3rd group is different), and we'll see how to determine the nature of the group differences in Chapter 5 on contrasts and follow-up comparisons.

Note that, as promised:

- $SS_{total} = SS_{between} + SS_{w/in} = 62 = 56 + 6$ (22)
- $df_{total} = df_{between} + df_{w/in} = 11 = 2 + 9$ (23)

The ANOVA is complete. It is standard to collect all this information into a summary table—it's even called "an ANOVA table":

source (of variation)	SS	df	MS	F on 2,9 df
between (factor A)	56	2	28.000	41.98
within (error)	6	9	0.667	
total (corrected for the grand mean)	62	11		
	↑ These will sum.	↑ These will sum.	↑ $MS = \frac{SS}{df}$	↑ $F = \frac{MS_{effect}}{MS_{error}}$

This experimental design had one factor with 3 levels. It was handy to present the data in the 2-way table above, where the groups comprised the columns and the 12 observations, n = 4 in each group formed the rows. However, when we're working with a dataset on the computer, we'll need to string the data out, as vectors in the form below:

obsID	group	x_{ij}
1	1	4
2	1	5
3	1	6
4	1	5
5	2	5
6	2	6
7	2	6
8	2	7
9	3	9
10	3	10
11	3	11
12	3	10

Revisiting the Logic of the F-test

At this point, we've seen how to calculate an F-statistic and we've seen an intuitive approach to understanding it (e.g., Figures 2-4). In this section, we're revisiting the logic and philosophy of the F-test from a slightly more technical angle.

For the simple design we've been working with (1 factor, complete randomization; study participants are randomly assigned to 1 of "a" groups), the model is:

$$x_{ij} = \mu + \alpha_i + \epsilon_{ij}$$

where the terms μ and α_i are structural or fixed model components and the ϵ_{ij}'s are random. The α_i's, or the extent to which the group means μ_i, differ, are the terms of greatest interest.

This model, like any model, carries with it some assumptions on the ϵ_{ij}'s (and therefore in effect on the x_{ij}'s):

a) The ϵ_{ij}'s are mutually independent. That is, we randomly assign study participants to groups and one person's score doesn't affect another's.

b) The ϵ_{ij}'s are normally distributed with mean=0 (e.g., so the errors cancel each other out) in each population or group.
c) We assume the homogeneity of variances: $\sigma_1^2 = \sigma_2^2 = \cdots = \sigma_\epsilon^2$ = a common "error variance" in order to pool the terms into MS_{within} (a.k.a. MS_{error}).

We talk about drawing a sample of size a×n and we collect the x_{ij}'s as sample observations drawn from "a" populations, and the x_{ij} scores vary around the group means because of the ϵ_{ij}'s. In the sample, we compute statistics like the $\bar{x}$'s and the $MS_{between}$, etc. Imagine we did that repeatedly, drawing a very large number of such samples (sort of meta-sampling, or sampling samples), computing statistics like $MS_{between}$, setting it aside, doing it again and again, eventually compiling a distribution of the $MS_{between}$ statistics across the many samples, and the same for the MS_{within} values, etc. Then we'll call the means of those corresponding distributions the "expected values" denoted "E":

$E(MS_{within}) = \sigma_\epsilon^2$ it's an unbiased estimate (UE) of error variance (25)

$E(MS_{between}) = \sigma_\epsilon^2 + \frac{n(\sum \alpha_i)^2}{(a-1)}$ this is not an UE of error variance. It also reflects treatment effects, the α_i's. (26)

And of course, once we have $MS_{between}$ and MS_{within}, we compute an F-statistic: $F = \frac{MS_{between}}{MS_{within}}$. If we were to do that empirically over all those samples (which is what is reflected by the statistical theoretical distributions), we'd expect that MS_{within} essentially converges on σ_ϵ^2 and $MS_{between}$ on $\sigma_\epsilon^2 + \frac{n(\sum \alpha_i)^2}{(a-1)}$, then what's going on in an F-ratio is a comparison of their expected values:

$$\frac{E(MS_{between})}{E(MS_{within})} = \frac{\sigma_\epsilon^2 + \frac{n(\sum \alpha_i)^2}{(a-1)}}{\sigma_\epsilon^2} \tag{27}$$

So, what does that tell us? Imagine $H_0\colon \mu_1 = \mu_2 = \cdots = \mu_a$ were true, that is, there are no group differences. In terms of the model parameters, that H_0 is equivalent to stating $H_0\colon \alpha_1 = \alpha_2 = \cdots \alpha_a = 0$, again saying there are no group differences. When the null is true, then that piece in the right side of the numerator of equation (27) would be zero, $\frac{n(\sum \alpha_i)^2}{(a-1)} = 0$. Stated differently, if the null is true, then all the groups would have the mean: $\mu + \alpha_i = \mu + 0 = \mu$. Thus all α_i's=0, and squaring them doesn't change that; all $(\alpha_i)^2$=0. So if H_0 is true, $\frac{n(\sum \alpha_i)^2}{(a-1)} = \frac{n \sum 0}{(a-1)} = 0$. That's all to say that under the null (when H_0 is true):

$$F = \frac{\hat{\sigma}_{error}^2 + 0}{\hat{\sigma}_{error}^2}.$$

When that holds (when the null is true), the F-test would be a ratio of 2 independent estimates of error variance, so the F-value would be "near" 1.

By comparison, what's going on when F is large? We reject H_0 as not plausible, because…when H_0 is not true, each $(\alpha_i)^2$ will be > or = 0, so that term $\frac{n(\sum \alpha_i)^2}{(a-1)}$ will be >0, so the F-statistic will be: $\frac{\hat{\sigma}^2_{error}+something}{\hat{\sigma}^2_{error}}$, much > 1.

SUMMARY

This chapter introduces the concepts and definitions of analysis of variance (ANOVA). The null hypothesis posits that the group means are identical, for a>2 groups. Sums of squared terms reflect variability in the data, between groups and within, and they're compared in an F-test, which is an extension of a 2-group t-test, for 3 or more groups.

REFERENCES

1. Keppel, Geoffrey and Thomas D. Wickens (2004), *Design and Analysis: A Researcher's Handbook* (4th ed.), Englewood Cliffs, NY: Prentice Hall (or Keppel, Geoffrey (1991), *Design and Analysis: A Researcher's Handbook* (3rd ed.), Englewood Cliffs, NY: Prentice Hall). Great intro, especially for non-quant people; i.e., lots of good verbal explanations (but the notation tends to make quant students crazy).
2. Iversen, Gudmund R. and Helmut Norpoth (1987), *Analysis of Variance*, Sage. Succinct (I love these little green Sage paperback primers!)
3. Hays, William L. (1988), *Statistics* (4th ed.), NY: Holt, Rinehart & Winston.
4. Kirk, Roger (1982), *Experimental design: Procedures for the Behavioral sciences,* Belmont, CA: Brooks/Cole Publishing Co.
5. Snedecor, George W. and William G. Cochran (1980), *Statistical Methods*, (7th ed.), Ames, IA: Iowa State University Press.
6. Scheffé, Henry (1959), *The Analysis of Variance*, NY: Wiley. More "math-stat-y."
7. Also: Iacobucci, Dawn (1994). "Analysis of Experimental Data," in Richard Bagozzi (ed.), *Principles of Marketing Research*, Cambridge, MA: Blackwell, 224-278.

APPENDIX: SAS

To run a one-way ANOVA in SAS, we'd use the following syntax.

```
data oneway; input ID group x; datalines;
 1 1  4
 2 1  5
 3 1  6
 4 1  5
 5 2  5
 6 2  6
 7 2  6
 8 2  7
 9 3  9
10 3 10
```

```
11 3 11
12 3 10
;
proc print data=oneway; proc glm data=oneway; class group;
model x = group / ss3; means group; run;
```

The SAS output follows. Note that it matches the analysis in this chapter.

The SAS System
The GLM Procedure

Dependent Variable: x

Source	DF	Sum of Squares	Mean Square	F Value	Pr > F
Model	2	56.00000000	28.00000000	42.00	<.0001
Error	9	6.00000000	0.66666667		
Corrected Total	11	62.00000000			

Source	DF	Type III SS	Mean Square	F Value	Pr > F
group	2	56.00000000	28.00000000	42.00	<.0001

Level of group	N	X Mean	X Std Dev
1	4	5.0000000	0.81649658
2	4	6.0000000	0.81649658
3	4	10.0000000	0.81649658

HOMEWORK

In homework, always show your work.
This HW covers stats review and one-way ANOVA.

Problem 1. In deriving:

$$\sum_{i=1}^{a}\sum_{j=1}^{n}(x_{ij}-\bar{x})^2 = \sum_{i=1}^{a}\sum_{j=1}^{n}(\bar{x}_i-\bar{x})^2 + \sum_{i=1}^{a}\sum_{j=1}^{n}(x_{ij}-\bar{x}_i)^2$$

$$SS_{total} \quad = \quad SS_{betw} \quad + SS_{w/in}$$

we noted that the cross product term = 0:

$$2\sum_{i=1}^{a}\sum_{j=1}^{n}(\bar{x}_i-\bar{x})(x_{ij}-\bar{x}_i) = 0$$

You don't need to "prove" this analytically, but demonstrate that it is true on the following (arbitrary) data:

X_{ij}'s:	Group 1	Group 2	Group 3
Subject 1	4	5	9
2	5	6	10
3	6	6	11
n = 4	5	7	10

Problem 2. These data were obtained from an experiment that had a=5 treatment conditions, wherein study participants made ratings from 1 to 10 (10 being more favorable):

a_1	a_2	a_3	a_4	a_5
8	3	7	5	8
4	2	6	7	1
3	4	1	5	8
2	2	4	3	7
3	8	1	10	7
1	8	5	9	5
1	2	1	5	3
2	5	3	10	1
1	2	1	9	4
5	4	1	7	6

a) Calculate the treatment means.
b) Calculate the SS_A, $SS_{S(A)}$, SS_T (by hand).

Problem 3. This small data set has a=3 conditions and n=5 respondents per condition. Respondents made ratings on a 1-7 point rating scale (7 = 'strongly agree'):

a1	a2	a3
6	5	4
1	3	4
5	5	5
3	1	6
2	6	7

a) Calculate the mean for each of the groups.
b) Perform an analysis of variance on these data (by hand).
c) What can you conclude from this study?

Problem 4. To get facile with Excel's functions, find the critical F-values for these scenarios:
a) $F(2,30)$ at $\alpha = .05$
b) $F(3,100)$ at $\alpha = .01$
c) $a = 5$, $n = 10$, $\alpha = .05$
d) $a = 3$, $n = 7$, $\alpha = .05$

Hint: use Excel's =F.INV.RT function.

CHAPTER 3

TWO-WAY, THREE-WAY, AND HIGHER-ORDER ANOVA

Questions to guide your learning:
Q_1: What is a two-way ANOVA?
Q_2: What is an interaction?
Q_3: What are the principles to generalize to a higher-way ANOVA?

Chapter 2 introduced the analysis of variance (ANOVA) in its simplest form, a "one-factor ANOVA," that is, an ANOVA for a single factor, a single variable along which the groups may differ. In this chapter, we extend the ANOVA model and application to two factors, then three factors, and then discuss its further generalization to more factors.

Two-way ANOVA overview

In a 2-way (or higher-order) ANOVA, we can study interactions.

Chapter 2 showed how a one-way ANOVA is a simple extension of the comparison of two means. It is appropriate when there is one independent variable (explanatory factor) and we want to compare the means on some continuous dependent variable (response measure) across 2 or more groups. We label the groups i=1, 2, ..., a. For example, an ad agency might propose to run one of several new ad campaigns 1, 2, ..., a, to see which ad performs best, as measured by sales in a market, or consumers' brand attitudes, or whatever criterion serves as the dependent variable. In some applications, such as pharmaceuticals R&D testing, the first groups 1, 2, …, a-1 might be a set of experimental newly proposed treatments and their performance would be compared to the final a^{th} group which consists of the current, best possible treatment, as a control comparison.

There are "n" observations or replications in each of the "a" cells, so the total sample size is N=a×n. In the ad (or pharma) experiment, each of the N experimental units, or subjects, would be randomly assigned to one of the "a" treatment groups so that there are n subjects in each group. The observations (data points) may be denoted x_{ij} for the rating of the ith ad (or group) by the jth subject. The observed means for each of the "a" groups may be computed $(\bar{x}_1, \bar{x}_2, \dots, \bar{x}_a)$ as well as the overall mean $(\bar{x})$ (i.e., the mean over all groups).

Before proceeding to the analyses, let's review the issue of "causality" that we saw in Chapter 2, now as it applies to our current issues. If the experimental units cannot be randomly assigned to the "a" groups, an ANOVA can still be run, but the results cannot be interpreted in a causal sense. For example, if the a=3 groups are comprised of consumers in their 20s, 30s, or 40s, so the independent variable is "the age cohort of the study participant" (and we'd measure some dependent variable like, "how much do you spend per month dining out?"), the consumers cannot be assigned to an age group. If we ran such a study (measuring, not manipulating age) and saw that the groups' means differ on the dependent variable, we could conclude that the age cohorts differ significantly, but we cannot make simple statements of the effect of age per se on the dependent variable, given that there are so many factors confounded with age (health, experience, discretionary income, popular media the cohort grew up with, etc.). Similarly, if the "a" groups in a study consisted of psych majors, econ majors, etc., the experimenter would not have control over assigning students to the groups—in this case, we speak of the students having "self-selected" into one of the groups. Here too, if the group means differ in an ANOVA, we can certainly say that the psych mean is significantly different from the econ mean, and we'd know that something about the psych students was different from the econ students, but we'd need further work to disentangle the precise nature of those differences. For causality, experimentation still reigns supreme; indeed many philosophers of science say it's the only way to test "If-Then" causal statements.

Back to business... If we're running an experiment, we're probably doing so because we expect to see that different experimental treatments yield different results. The differences between the groups on their means is referred to as "between-group differences." Furthermore, even for subjects who receive the same experimental treatment (e.g., see the same ad), there will be some variability (e.g., individual differences). ANOVA compares the "between-group variability" to the "within-group variability" to see if there are group differences (i.e., was the "treatment" effective, e.g., are there differences among the ads), above and beyond the source of variability in the data that is attributable to mere individual heterogeneity.

So far in this chapter, all the logic and terminology true of a one-way ANOVA (per Chapter 2) is true in this chapter as well. What's new in this chapter is that in a two-way ANOVA, we have two independent variables, factor "A" and factor "B" that we manipulate simultaneously, and the total variability will be partitioned:

$$SS_{total} = SS_A + SS_B + SS_{AB} + SS_{error} \quad (1)$$

The hypotheses we would test are three: First, is there a "main effect" for factor A (separate from or independent of whatever effects B has, by itself or with A)? Second is there a "main effect" for factor B (regardless of A)? And third, is there an A×B interaction? The main effect questions (for A or B) are just like those tested in Chapter 2 in which there are "a" (or "b") groups, and the null hypothesis states that all groups that saw different levels of factor A (or B) are nevertheless the same and we'll test whether to reject that null. Specifically, those main effects test $H_0: \mu_1 = \mu_2 = \cdots = \mu_a$ for factor A, and $H_0: \mu_1 = \mu_2 = \cdots = \mu_b$ for factor B.

The interaction hypothesis is the one that's new, and it's important and fun. This term reflects whether it is the case that, separate from or above and beyond the individual main effects of factors A and B on the dependent variable, do the factors somehow work together to have some joint impact on the dependent variable that cannot be explained by just the sum of their individual parts (the main effects of A and B). All of these effects, the main effects for A and B and their interaction, are statistically independent, so when we run the ANOVA, we might see that the main effect for A is significant or not, same for B, and the same for the interaction term. So, an interaction effect takes into account the effects of factors A and B (statistically partialling them out) and asks, regardless of the effects of factors A and B by themselves, do they do anything together?

For example, maybe there is a main effect of factor A such that, in general, the means for a=1 and a=2 are low but the mean for a=3 is higher. An interaction effect might show that when factor B takes on value b=1, then that pattern in factor A holds but when b=2, then the mean for a=3 is really really high, much more exaggerated than the main effect of A would have suggested. So, the interaction term captures whether there is some particular combination(s) of factors A and B for which the dependent variable is especially high or particularly low (e.g., an ad or medicine is super effective or completely useless).

Let's see how it's all done. We'll spend time on terminology and notation, and quite a bit of time nailing down the notion of an interaction. Then we'll see the model, how to estimate the pieces, and test the hypotheses.

Two-way ANOVA with replications (equal n's)

To be precise, we are now considering a "two-way ANOVA with replications." The "replications" part of that title simply means that there are "n" observations in each cell of the two-way design, where n>1. We'll see shortly what happens if n = 1, but for the most part, we don't have to worry about that as an issue. Usually n is much larger than 1, like, 10 or 30, or even 100 in each cell, and then our major concern is to strive to have the n's be equal to each other across all the cells of the design. When they are unequal, that's an unbalanced design and that causes enough problems that we'll look at that scenario in some detail in Chapter 10 (luckily the solution there is easy).

The full name of the experimental design that is most commonly encountered (probably 99% of the time) is a "completely randomized, 2-factor factorial." By "completely randomized," we mean that we have randomly assigned each study participant to 1 (and only 1) of the experimental conditions. The "2 factor" means that those experimental conditions are defined by two independent variables, A and B. "Factorial" means that the two independent variables are completely crossed, that is, the study includes all combinations of all the levels of factor A with all the levels of factor B. With factor A having "a" levels and B having "b," there will be a×b cells in the design.

Full Name of Design

"completely randomized	2-factor	factorial"
randomly assign each subject to 1 of our experimental conditions	experimental conditions are defined by 2 factors or independent variables	factors are completely crossed; all levels of both variables

For example, imagine a psych experiment in which the experimenter is interested in the constructs of "cognitive resources" and "task difficulty" on a memory recall task. Factor A might be operationalized as the amount of time given to a study participant to read and try to memorize a list of words (e.g., 2 vs. 5 minutes). Factor B might be operationalized as the level of challenge of those lists (e.g., easy words, complicated words, very complicated words). A subject would be randomly assigned to a = 1 or a = 2, and b = 1, 2, or 3, e.g., a subject assigned to the cell (1,3) would be given 2 minutes to study really hard words. At the end of the word exposure time period, the experimenter would ask the study participants in every condition (every cell) to recall all the words they can from their lists. We could map the design like this:

		B = task difficulty		
		easy b = 1	hard b = 2	very hard b = 3
A = cognitive resources, e.g., time to study a list of words	a = 1			
	a = 2			

Or for a marketing example, consider the following. We might create 6 ads, 3 of which look like our currently running ad, and 3 of which have more of an upscale, luxurious appeal. Each of the ads then mentions a price point that is low or medium or high:

A = ad	B = price: low	medium	high
standard ad			
luxury appeal			

In this 2-factor experimental design, we'd have 3 research questions or 3 ANOVA hypotheses:

- Is one ad more effective than another?
- Is one pricing level more effective than the others?
- Do the ads and prices "interact"?; is there any particular combination of the ads and prices that is especially effective, or detrimental?

Each of these 6 cells is a treatment or experimental condition. The experimenter creates all 6 conditions, but any given subject is run in only 1 of the 6 conditions. All levels of factor A are crossed with all levels of factor B—so there are a×b cells.

There are n subjects per cell, so the total sample size is N=n×a×b. Note that the Central Limit Theorem (CLT) kicks in around n=30, that is, even if a population distribution isn't particularly normal-looking, a sample of size 30 or more will start making the statistics behave more like a normal distribution. The rule of thumb however would not be to obtain a sample of size 30 total and spread them over the experimental design (that would only yield 7 or 8 subjects in each cell in a 2×2 factorial). Rather, we'd aim for at least 30 observations per cell.

If the full name of that experimental design was a "completely randomized, 2-factor factorial," the short name of the design is a 2×3 (read "2 by 3") factorial. The "2" refers to two levels of factor A, that is, a=2, and the "3" refers to the number of levels of the second factor, b=3.

The design and terminology is easily extended, say to three factors, A, B, and C, with design size a×b×c. For example, a 2×2×2 (or 2^3) factorial would look like this:

	c = 1: b = 1	b = 2	c = 2: b = 1	b = 2
a = 1				
a = 2				

If a factorial design is not fully crossed, then it's not a true factorial, and the analyses are more complicated. For example, if there is a missing cell (a condition with no observations in the cell marked with the "*" below), that design is not fully crossed. A missing cell is different from and more severe than a few missing data points in each cell. If all cell sample sizes, $n_{ij} > 0$, and all $n_{ij} = n$, that's ideal, that's what we shoot for. If all $n_{ij} > 0$ but the n_{ij}'s vary, that's an unbalanced design, but still quite do-able and common (it's discussed in Chapter 10). But if one or more $n_{ij} = 0$, the design is not a factorial and it will be trickier to analyze (also discussed in Chapter 10).

	b = 1	b = 2	b = 3
a = 1			
a = 2		*	

Why Run a Two-Factor (or More) Design?

There are several advantages of factorials over a series of single factor experiments. The first is economy. Hypotheses can be tested with comparable power using fewer subjects.

For example, if we were interested in the effects of Factor A and Factor B on some dependent variable, we could run 2 separate one-factor designs. Say we wanted n=30 subjects in each cell. Executing both studies would require 90 + 90 = 180 total subjects:

Factor A		
	a = 1	30
	a = 2	30
	a = 3	30
		90

Factor B			
b = 1	b = 2	b = 3	
30	30	30	/ 90

However, if we ran a single study that was a two-factor design, even if we inserted only n=10 subjects in each cell, the marginal sample sizes would all be 30, so when we test the A and B main effects, we'd have just as much power as the previous 2 studies had. Note that the total sample size is a mere 90:

	b = 1	b = 2	b = 3	
a = 1	10	10	10	30
a = 2	10	10	10	30
a = 3	10	10	10	30
	30	30	30	90

However, there's no question that the biggest advantage of running a two-factor design is that it allows for the estimation and study of the interaction of the two factors. In addition to studying the main effects of factor A and B independent of the other, we can also examine the effect of their "interaction"—the joint influence of factors A and B together on the dependent variable.

The interaction, how A and B work together in their impact on the dependent variable, is where most social scientists believe the interesting theoretical and empirical action is. Yes, factor A might have an effect, and yes B might have an effect, but it's cool when A and B work together in ways that cannot be predicted by their individual main effects. Interactions are also referred to as moderators, as in, the effect of one factor A (or B) is moderated by the level of the other factor B (or A). They're also referred to as boundary conditions, as in, the effect of A on the dependent variable acts like such and such up until the point that factor B takes on another value, etc. The interactions are what make most theories interesting. A theoretical conclusion based on a main effect would be of the order, "factor A has an effect on the dependent variable." That's important to know, certainly. But an interaction is what allows us to say, "It depends!" What's the effect of factor A (or B) on the dependent variable? Well, it depends on the level of factor B (or A)." Cool, right? Now, more details.

Envisioning Interactions

To understand the workings of an interaction, think first about the raw data in a simple 2×2 design, with n subjects in each cell, x_{ijk}, where i represents factor A, j represents B, and k represents which subject's data in the i,j cell:

	B_1	B_2
A_1	n x_{11k}'s	n x_{12k}'s
A_2	n x_{21k}'s	n x_{22k}'s

That is:

	B_1	B_2
A_1	x_{111} x_{112} … x_{11n}	x_{121} x_{122} … x_{12n}
A_2	x_{211} x_{212} x_{21n}	x_{221} x_{222} x_{22n}

Next, take the mean over all n subjects in each cell to get 1 mean in each cell:

	B_1	B_2
A_1	$\bar{x}_{11}$	$\bar{x}_{12}$
A_2	$\bar{x}_{21}$	$\bar{x}_{22}$

The most vivid way to see an interaction is to plot those 4 means.

- In Figure 1, we see no interaction. The effect of factor B is the same for both levels of factor A. Specifically, as we go from b_1 to b_2 the effect is the same (statistically) for a_1 as a_2. No interaction also simultaneously means that the effect of factor A is the same for both levels of B; the a_1 means are a little higher than the a_2 means, regardless of, or independent of, the level of factor B.
- In Figure 2, we see an interaction. The effect of factor B is different at different levels of A; or, the effect of factor A is different at different levels of B (probably due to something going on in the a_1b_2 cell).

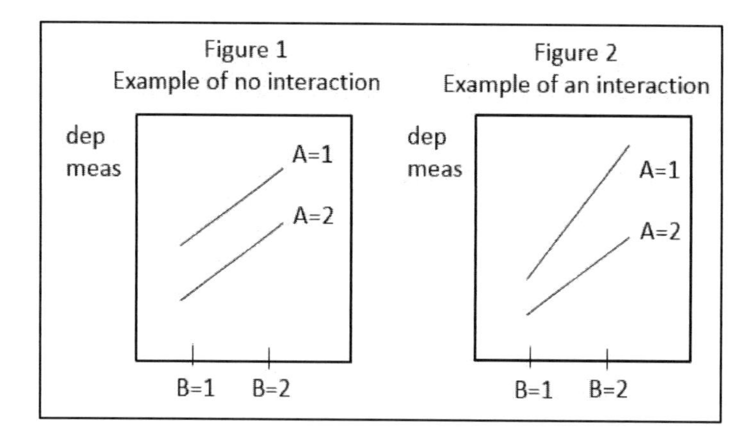

Interaction plots can have factor A as the horizontal axis and factor B forming the profiles, or the reverse. Use whichever makes most sense to communicate your interpretations. Here is an example in which the means are $\bar{x}_{a1b1} = 3.2$, $\bar{x}_{a1b2} = 6.0$, $\bar{x}_{a2b1} = 1.1$, $\bar{x}_{a2b2} = 1.5$, and they are plotted both ways:

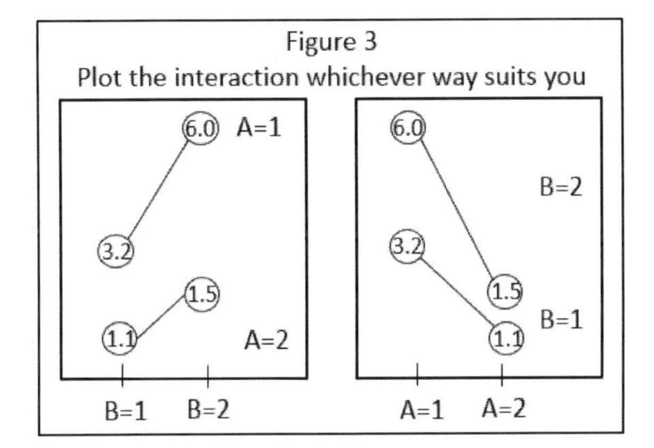

In general, a plot of means will indicate that there is no interaction when the lines are parallel (or roughly, statistically speaking, parallel), such as any of these:

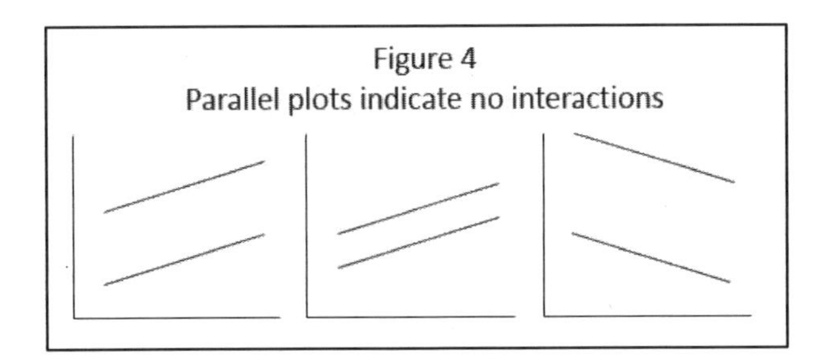

When there exists an interaction in the data, the lines will appear to diverge, converge, or cross-over:

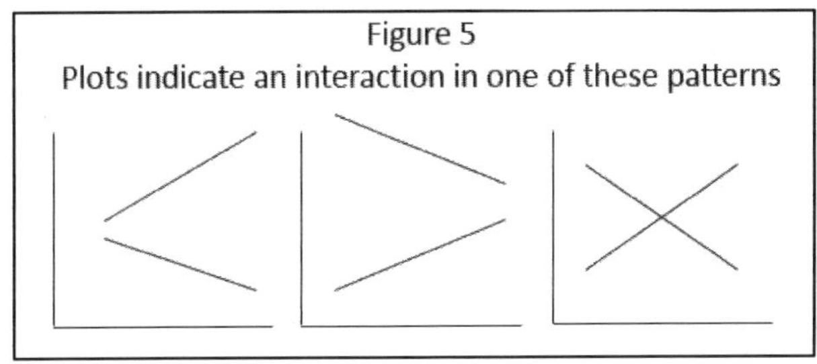

Of course, while interaction plots are helpful in studying the nature of the interaction, the effect must be tested statistically. For example, there may be an interaction that is statistically significant but looks parallel, and conversely, there may be a non-significant interaction effect that corresponds to a plot like one in Figure 5, due to the data being noisy.

We'll look at the F-test for the interaction, and if it's significant, we'll plot the cell means to see the characteristics of the interaction (at that point, we're still not done yet—we have further tests to examine and test the so-called "simple effects" as discussed in Chapter 5).

Some researchers seem to think that one of the forms of interactions depicted in Figure 5 is superior to another, but that's simply not true. Think about lines in geometry—if they're parallel, they won't ever diverge, converge, or cross-over, whereas if the lines are not parallel, they'll cross-over somewhere, and they'll be convergent to that point's left, and divergent to the right.

For example, say we're running a study in which the experimental factor is something like how many minutes a consumer is given to read information on a website to learn about a product: 1 minute, 2 minutes, 3, etc. If the population effect is like that in Figure 6, but we ran a study varying only 1, 2, and 3 minutes, we'd get the convergent pattern in Figure 7, and if we ran a studying varying only 5, 6, 7 minutes, we'd get the divergent pattern. If we ran levels 1, 4, 7 or 2, 3, 6, etc., we would see the cross-over pattern in our data. Very often we'll choose the levels of each independent variable to cover a range as wide as we expect changes, or wide enough to cover the range of effects of interest.

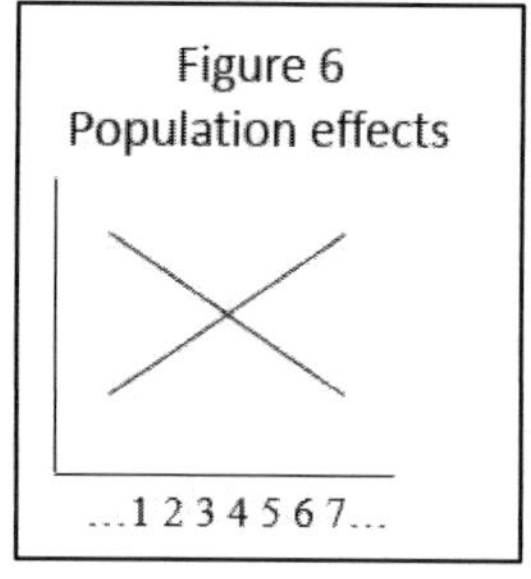

Figure 6
Population effects

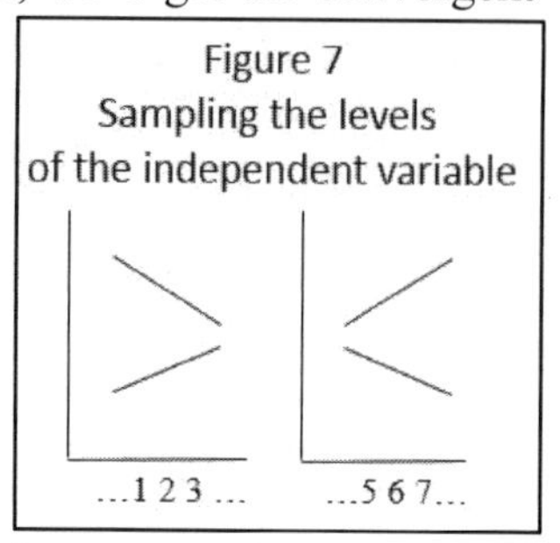

Figure 7
Sampling the levels of the independent variable

There is another observation that makes it clear that the pattern (cross-over or convergent or divergent) is unimportant. It is very often the case that if we plot factor A on the horizontal axis, with profile lines depicting different levels of factor B and we get, say, a cross-over appearance, just by plotting the interaction the other way, with factor B on the horizontal axis and profile lines for different levels of factor A, we might get a convergence appearance, for example, even though it's obviously the same data and the same interaction.

We'll see how to officially test for an interaction in just a moment. Before turning to the technique, however, there are several issues about interactions that are worth mentioning. First, the interaction effects are often those of greatest interest in a study. Later in the book we will see some other types of experimental designs that do not allow for the estimation of certain interactions. If those interactions are of interest, clearly we'd need to find a more suitable design.

Second, particularly when studying human behavior, where phenomena are complex, we don't expect to completely model behavior by studying only a few factors. And while, say, a 6-factor interaction term sounds intimidating, its interpretation is logically no more complicated than the interpretation of a 2-factor interaction; i.e., the relationship among the 5 factors on the dependent variable depends on the level of the 6th factor.

Third, it is rare, but occasionally a paper might discuss pooling higher-order interactions into the error term, presumably because the researchers find those higher-order interactions intimidating (of course, they'll say the terms are theoretically uninteresting). We can no longer study the interactions that have been pooled, but it's not inappropriate statistically. In fact, adding effects like the higher-order interactions to the error term increases the error SS and MS, so the denominator in an F-test increases, making it more difficult to reject a H_0. So statistically at least the pooling is conservative in that it works against finding significant results. Still, it's cleaner to simply let the computer estimate all the pieces distinctly (i.e., don't pool), that way every effect is partialled out or statistically controlled for all the other effects. If something isn't significant, we don't have to discuss it, but we also don't need to bury it into the error term.

Plots are also helpful in reminding us that any statements interpreting "lower-order" effects (simpler effects, like main effects) must be qualified by the presence of any significant higher-order interactions. Specifically, in a 2-way design, each main effect is qualified by a 2-way interaction (if it is present, i.e., significant).

For example, say we have 30 subjects per cell, with the following cell means and marginal means (the means in the margins are the main effect means):

	b_1	b_2	
a_1	3	7	$\bar{x}_{a1} = 5$
a_2	11	7	$\bar{x}_{a2} = 9$
	$\bar{x}_{b1} = 7$	$\bar{x}_{b2} = 7$	

We'd plot the marginal means to examine the main effects.

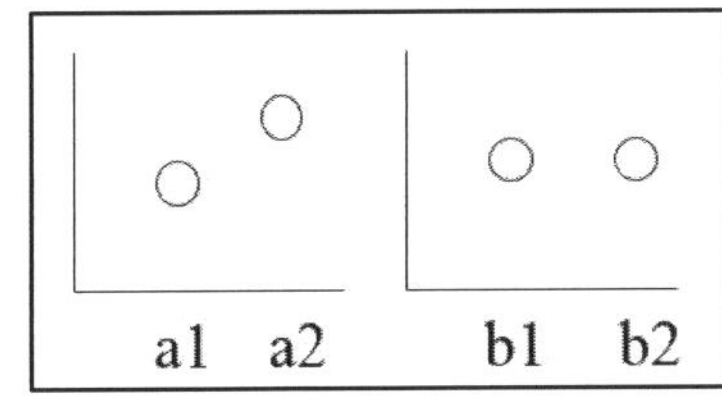

From the table or plots, the main effects suggest that there may be an effect of factor A (we still must test whether it is significant) in that a=2 results in higher scores than a=1. It also appears that there is no main effect for B. However! Let's next plot the interaction cell means to look at the nature of the interaction:

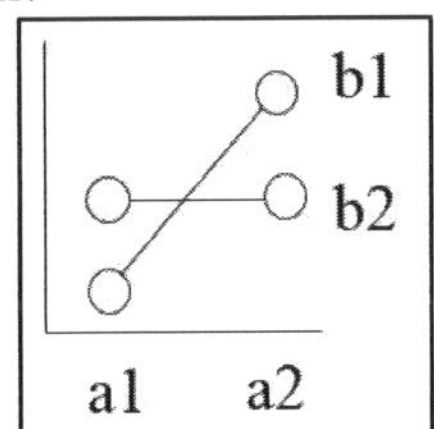

Here we see what certainly looks like an interaction (which, again, we must test statistically). In this interaction plot, we see that it a_2 is not always $>a_1$ (that's true only for b_1), and b_1 is not $= b_2$ if we separate out a_1 and a_2. The best performance (i.e., the highest score on the dependent measure) occurs in the (a_2,b_1) cell. The lowest scores occur at the combination of (a_1, b_1).

In terms of interpreting interactions, consider the following table and the accompanying figure. Both show all possible outcomes of a 2×2 ANOVA, the most commonly employed experimental design (cf., Keppel and Wickens, 2004, p.205; Keppel 1991, p.199). In the chart, 8 scenarios are depicted in the columns. For example, scenario "a" is a dataset in which no effects are significant, whereas scenario (column) "f" indicates a dataset in which there is a main effect for factor A and an A×B interaction. Figure 8 then plots what those patterns might look like. (It takes some time to eyeball an interaction plot and discern which effects seem to be present in the data, so this is good practice.)

Is the effect significant in Figure a through h? Yes or no:

	a	b	c	d	e	f	g	h
A	n	y	n	n	y	y	n	y
B	n	n	y	n	y	n	y	y
A×B	n	n	n	y	n	y	y	y

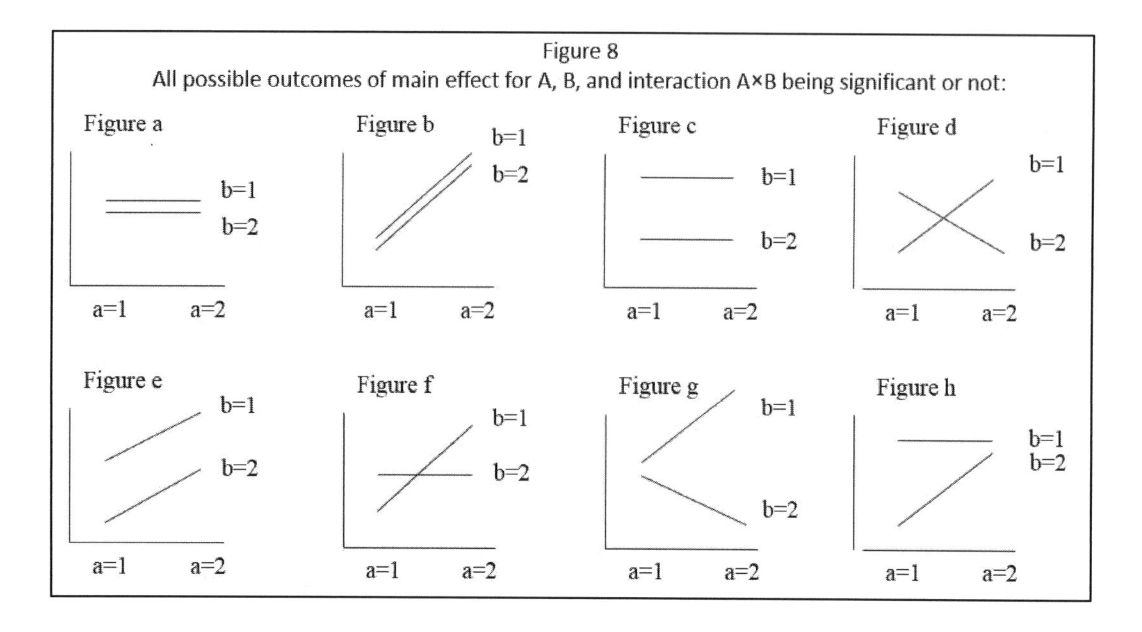

The Two-Way ANOVA Model

So, with that preview of interactions by way of motivation, let's get into the two-factor ANOVA model. The model states:

$$x_{ijk} = \mu + \alpha_i + \beta_j + (\alpha\beta)_{ij} + \epsilon_{ijk} \tag{2}$$

The data point is x_{ijk}, representing:

i = 1…a for factor A

j = 1…b for factor B

k = 1…n (all $n_{ij} = n$; i.e., equal cell sizes, equal #subjects in each A×B cell).

The terms in the model reflect the following:

μ is the overall grand mean

$\alpha_i = \mu_i - \mu$, the main effect for group "i" of factor A

$\beta_j = \mu_j - \mu$, the effect at level b_j of factor B

$$(\alpha\beta)_{ij} = \mu_{ij} - (\mu_i - \mu) - (\mu_j - \mu) - \mu \tag{3}$$

$= \mu_{ij} - \mu_i + \mu - \mu_j + \mu - \mu$

$= \mu_{ij} - \mu_i - \mu_j + \mu$, the interaction effect of factors A and B in cell $a_i b_j$

$\epsilon_{ijk} = x_{ijk} - \mu_{ij}$, the within cell deviation (a person's score as it varies from the mean of the cell (i,j) that person k is in).

These terms are mostly familiar. There is an overall grand mean like in the one-way ANOVA model, and the two main effects, for factors A and B, are each like the one-factor effect in the one-way ANOVA, and the error term is just like that in a one-way ANOVA, but with 1 more subscript to capture the added factor (3 subscripts to capture the 2 factors and the individual within the cell).

The new term in the model is the interaction term. There are two neat heuristics about these interaction parameters. In the first line of (3), the interaction term $(\alpha\beta)_{ij}$ is defined as the effect that's left in the data after subtracting off all the lower-order effects. Specifically, for a 2-way interaction, the lower-order effects are the A and B main effects, and the grand mean. So that line begins in the (i,j) cell with mean μ_{ij}, then we subtract off the main effects for factors A and B, that is, α_i and β_j, or $(\mu_i - \mu)$ and $(\mu_j - \mu)$. Lastly the line also subtracts off the grand mean (the simplest, most aggregate term), μ.

The next heuristic shows up in the last line of (3) where the terms begin with the cell mean (a 2-factor effect), and the next level (the main effects) are subtracted off, and then the next level (the grand mean) is added back, etc. It's minus, plus, minus, plus until we hit μ. That might not seem impressive yet because the model is still fairly simple and straightforward, but when the number of factors increases, the heuristics will be handy.

So, given the two-factor ANOVA model, what are the hypotheses we're trying to test? There are three questions:

1) Is there a main effect for factor A?
H_0: $\alpha_1 = \alpha_2 = \cdots \alpha_a = 0$ (or, $\alpha_i = 0$ for all i = 1...a)
vs. H_A: at least 1 $\alpha_i \neq 0$

2) Is there a main effect for factor B?
H_0: $\beta_1 = \beta_2 = \cdots \beta_b = 0$ (or, $\beta_j = 0$ for all j = 1...b)
vs. H_A: at least 1 $\beta_j \neq 0$

3) Is there an interaction between factors A and B?
H_0: $(\alpha\beta)_{ij} = 0$ for all i and j
vs. H_A: at least 1 $(\alpha\beta)_{ij} \neq 0$

Recall the "ad by price" experiment at the start of this chapter, for which those hypotheses translate into the substantive questions:

1) Is there an effect of ad (is one ad better or worse than the other)?
2) Is there an effect of price (is one price point better or worse than the others)?
3) Is there a joint effect of ad and price such that some combination is particularly good or bad (yields especially high or low scores on the dependent variable)?

To test these hypotheses, we'll begin with the raw data: x_{ijk} and calculate the means. First, the means in each cell will reflect possible interaction effects:

$$\bar{x}_{ij} = \frac{1}{n}\sum_{k} x_{ijk}$$

The trick with any model and notation is to watch the subscripts. The mean $\bar{x}_{ij}$ has subscripts "i" and "j" so we'll have a mean for every factor A (i) and B (j) combination. It's missing the "k" subscript because we're averaging over k, the subjects within the "i,j" cell.

Next, we'll get the means for each level of factors A and B and they'll reflect the possible main effects:

$$\bar{x}_i = \frac{1}{bn}\sum_{j=1}^{b}\sum_{k=1}^{n} x_{ijk}$$

Those means have only the subscript "i" so there will be one for every level of factor A (i=1,…,a), and we've averaged over all the subscripts that are missing, all levels of factor B (j) and all subjects within all cells (k). These next means are for factor B (j=1,…b) (and similarly, we're summing up over factor A because for the factor B main effect, we temporarily ignore factor A.)

$$\bar{x}_j = \frac{1}{an}\sum_{i=1}^{a}\sum_{k=1}^{n} x_{ijk}$$

Finally, for the estimate of the grand mean, we compute the following. This mean has no subscripts which indicates that it doesn't depend on any level of factor A ("i") or factor B ("j"). We've averaged over all the data points: all levels of A, all levels of B, and all the observations within each AB cell.

$$\bar{x} = \frac{1}{abn}\sum_{i=1}^{a}\sum_{j=1}^{b}\sum_{k=1}^{n} x_{ijk}$$

So now we have all the means. We can look at them and we can plot them. But we still need to test the 3 hypotheses to see whether the means are significantly different or not.

Creating the ANOVA Table

To generate the structure and entries in an ANOVA table, we proceed through the following steps. The 1st column in the ANOVA table is called "source" or "sources of variability," and that's where we list all the factors. We manipulate factors A and B, so they're on the list. We also know there will be an error term. In a one-way ANOVA, it was easy to refer to $SS_{between}$ and SS_{within}, so the "within" was the error term. We will see different kinds of experimental designs however where things aren't so simple, so to allow for easier generalizability, at this point, we're going to refer to the error term more precisely as "S(AB)" which stands for "subjects nested in one AB cell." That is, while factors A and B are completely crossed (all combinations exist in the experimental design), each subject is in one and only one cell in the A×B design. So far our "source" list consists of: A, B, S(AB).

Next, we cross all possible sources from that list. Doing so creates new terms: A×B, A×S(AB), B×S(AB), A×B×S(AB). However, it's logically impossible for a subject to be nested in a factor S(A) or here S(AB), as well as crossed with that factor, so we strike out all terms in which there is a factor, here A or B, that appears more than once, both inside of and outside of parentheses. So these terms make no sense: ~~A×S(AB)~~, ~~B×S(AB)~~, ~~A×B×S(AB)~~ which leaves only the interaction term that is familiar to us: A×B, Now the "source" list consists of: A, B, A×B, S(AB). And of course the last line in the "source" column will be that for "total" (total variability).[4]

Moving next to the degrees of freedom column, we compute df using the following rules:

1) For a main effect like A (or B), the df are the number of levels of that factor minus 1, so $df_A = a-1$, $df_B = b-1$.

[4] $SS_{between}$ for a one-way ANOVA now gets explicated into SS_A, SS_B, and SS_{AB}, each of which contrast means between groups, but the groups are defined in different ways, first by A, then B, then by the interaction cells.

2) For an interaction, the df are the product of the df of the factors that are in the interaction term, so $df_{A\times B}$ = (a-1)(b-1). (Very handy that $df_{A\times B} = df_A \times df_B$.)
3) For the error term we're calling S(AB), we use the levels of the factor(s) inside the parentheses, and we subtract 1 from any factors outside of the parentheses. In S(AB) we have factors A and B in parentheses, so we'll use "a" and "b." There is "S" outside the parentheses denoting the subjects in each cell, and we know there are "n" of them, so when we subtract 1, we have n-1. Altogether then, the error df for S(AB) are $df_{S(AB)} = ab(n-1)$.
4) Total df are always "number of all observations in the data set minus 1," so we have df total = abn – 1.

Next, we estimate the Sums of Squares (SS) terms, defined as follows. Note that for each SS, we sum over everything—all levels of factor A, all levels of B, and all subjects in each A,B cell. Also note that the term in the parentheses in each SS below correspond to how we defined the parameters in the model statement (2) and (3), but of course we insert sample means here, $\bar{x}$'s instead of μ's.

source	df	SS definitional formula
A	a-1	$SS_A = \sum_{i=1}^{a}\sum_{j=1}^{b}\sum_{k=1}^{n}(\bar{x}_i - \bar{x})^2$
B	b-1	$SS_B = \sum_{i=1}^{a}\sum_{j=1}^{b}\sum_{k=1}^{n}(\bar{x}_j - \bar{x})^2$
A×B	(a-1)(b-1)	$SS_{AB} = \sum_{i=1}^{a}\sum_{j=1}^{b}\sum_{k=1}^{n}(\bar{x}_{ij} - \bar{x}_i - \bar{x}_j + \bar{x})^2$
S(AB) = error	ab(n-1)	$SS_{S(AB)} = \sum_{i=1}^{a}\sum_{j=1}^{b}\sum_{k=1}^{n}(x_{ijk} - \bar{x}_{ij})^2$
total	abn-1	$SS_{total} = \sum_{i=1}^{a}\sum_{j=1}^{b}\sum_{k=1}^{n}(x_{ijk} - \bar{x})^2$

Next, let's proceed to the MS and F-tests to complete the ANOVA table. When learning something new, it's comforting to see that it is familiar to something we already know. So as a reminder, the 1-way ANOVA table from Chapter 2 looked like this:

source	SS	df	MS	F
A (between groups)	SS_A	a-1	$MS_A = \frac{SS_A}{a-1}$	$\frac{MS_A}{MS_{S(A)}}$
S(A) (w/in group; error)	$SS_{S(A)}$	a(n-1)	$MS_{S(A)} = \frac{SS_{S(A)}}{a(n-1)}$	
total (corrected for grand mean)		an-1		

Now for a 2-factor ANOVA, the ANOVA table looks like this. It's all the same logic, there are simply more rows to the table because there are more effects to test:

Source	SS	df	MS	F
A (main effect)	SS_A	a-1	$MS_A = \frac{SS_A}{a-1}$	$\frac{MS_A}{MS_{S(AB)}}$
B (main effect)	SS_B	b-1	$MS_B = \frac{SS_B}{b-1}$	$\frac{MS_B}{MS_{S(AB)}}$
A×B (interaction)	SS_{AB}	(a-1)(b-1)	$MS_{AB} = \frac{SS_{AB}}{(a-1)(b-1)}$	$\frac{MS_{AB}}{MS_{S(AB)}}$
S(AB) (w/in group; error)	$SS_{S(AB)}$	ab(n-1)	$MS_{S(AB)} = \frac{SS_{S(AB)}}{ab(n-1)}$	
total (corrected for grand mean)	SS_{total}	abn-1		

We know the "source" column because we developed it. We know the "df" column as well. We know degrees of freedom will sum up, and SS's will sum up, and we know that mean squares are equal to the effect's SS divided by its respective df.

The last column in the ANOVA table shows that all 3 F-tests take on the form of "MS_{effect}" where the "effect" is the main effect for factor A or B, or the interaction A×B, whichever effect we're testing. The error term for all 3 F-tests is $MS_{S(AB)}$. To verify that those F-tests are using the proper error term, let's again examine the expected mean squares. If we were to take a sample and compute an MS_A or MS_{AB} etc. many times over many samples, the expected mean squares for the error variance is an unbiased estimate (UE) of the population error variance:

$$E\left(MS_{S(AB)}\right) = \sigma_\epsilon^2$$

The expected mean squares for the effects we're interested in testing include the error variance along with a variance that reflects the appropriate treatment effects:

$$E(MS_A) = \sigma_\epsilon^2 + \frac{bn\sum(\alpha_i)^2}{a-1}$$

$$E(MS_B) = \sigma_\epsilon^2 + \frac{an\sum(\beta_j)^2}{b-1}$$

$$E(MS_{AB}) = \sigma_\epsilon^2 + \frac{n\sum(\alpha\beta_{ij})^2}{(a-1)(b-1)} \quad (4)$$

Once we know all the EMS, we can form the F-ratios, as they appear in the final column of the ANOVA table:

$$\frac{MS_A}{MS_{S(AB)}}, \frac{MS_B}{MS_{S(AB)}}, \frac{MS_{AB}}{MS_{S(AB)}} \quad (5)$$

Those F-tests compare the expected values of the effect (for A, B, and the interaction) to the error variance. For example, the F-test for the main effect of factor A compares these 2 EMS:

$$\frac{E(MS_A)}{E\left(MS_{S(A)}\right)} = \frac{\sigma_\epsilon^2 + \frac{n(\sum \alpha_i)^2}{(a-1)}}{\sigma_\epsilon^2}$$

If there is no main effect for factor A, the right-hand term in the numerator will be zero (or very small) so that the F-test will be approximately 1. To the extent that there are group differences, then the α_i's will be nonzero, their squares $(\alpha_i)^2$ will be >0, and the F-statistic will be large.

Ad and Price Example

Let's return to the 2×3 example of 2 ads (standard and luxury appeal) and 3 price points (low, medium, high). Recall that we will want to test 3 hypotheses:

1) Is one ad more effective than another?
2) Is one pricing level more effective than the others?
3) Do the ads and prices "interact"?; is there any particular combination of the ads and prices that is especially effective?

Say the data resulted in means that looked like this:

A = ad		B = price			
		low	medium	high	
	standard ad	21.4	10.6	10.4	14.13
	luxury appeal	21.0	11.2	9.6	13.93
		21.2	10.9	10.0	

It's fine to provide that information in table form. It's also helpful to plot the means. The plots of the main effect means for factors A and B follow (the means in the margin of the table of means), and at the right we see the interaction (the cell means).

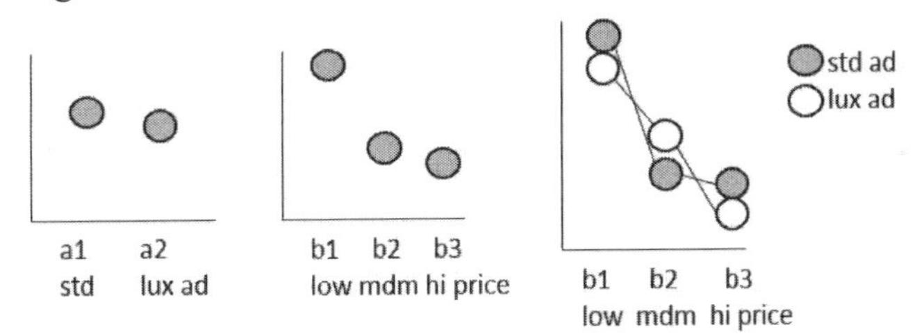

These means, whether in table or plot form, look like there is no effect for ad, or the effect is small. There may be an effect for price, where the low price yields higher scores (willingness to buy, or actual sales) and the results for medium and high prices look smaller and similar. There might be an interaction given that the plots are not parallel in the final figure. To officially test these hypotheses, we run the ANOVA. (The raw data were not provided, but the ANOVA results follow.)

$$SS_{total} = SS_{ad} + SS_{price} + SS_{ad\times price} + SS_{error} \tag{6}$$

$$SS_{total} = \sum_{i=1}^{a}\sum_{j=1}^{b}\sum_{k=1}^{n}\left(x_{ijk} - \bar{x}\right)^2$$

$$SS_{ad} = \sum_{i=1}^{a}\sum_{j=1}^{b}\sum_{k=1}^{n}(\bar{x}_i - \bar{x})^2$$

$$SS_{price} = \sum_{i=1}^{a}\sum_{j=1}^{b}\sum_{k=1}^{n}(\bar{x}_j - \bar{x})^2$$

$$SS_{ad\times price} = \sum_{i=1}^{a}\sum_{j=1}^{b}\sum_{k=1}^{n}(\bar{x}_{ij} - \bar{x}_i - \bar{x}_j + \bar{x})^2$$

$$SS_{within} = \sum_{i=1}^{a}\sum_{j=1}^{b}\sum_{k=1}^{n}(x_{ijk} - \bar{x}_{ij})^2$$

Source of variation	df	SS = Sum of Squares	MS = Mean Squares	F-test	p-value
Advertisement	1	0.300	0.300	0.04	0.839
Price	2	774.467	387.233	54.80	<.0001
Ad × Price	2	2.600	1.300	0.18	0.833
Error	24	169.600	7.067		
Total	29	946.967			

We could compare our calculated F-test to critical values, but SAS, SPSS, or any statistical computing package will provide p-values that indicate whether the effect is significant. We reject the null H_0 if $p<0.05$. Hence, there appears to be no ad effect, as we thought. There is an effect for price, so somehow the 3 price means differ. There is no interaction, even though the plot looked like the lines were not parallel. Apparently, the cross-over is too minor; if we put confidence intervals around all 6 of the means in that interaction plot, the results would be consistent with parallel lines. See? Statistics can help where our eyeballing of the data might mislead us!

Notice that the df sum up, as do the SS. Each MS = SS/df, and each F-test is of the form: $F = MS_{effect}/MS_{error}$.

We look at that ANOVA table and conclude: great news! Something (price) is significant! Woohoo, time to celebrate!

So, after the statistical conclusion comes the substantive interpretation: which ad should we choose for roll-out? Which price level should we choose?

The non-significant result for the ad factor implies that statistically speaking, either ad will perform equally well. We might think we should choose the standard ad because its mean is slightly higher, but the ANOVA tells us that it's not significantly higher. So it doesn't matter. The choice between the ads will have to be made on extra-statistical grounds. For example, we could choose the standard ad if we don't want to rock the boat (it's the one that is currently running), or we could choose the luxury ad if it was the boss's favorite, etc.

For price though, it's significant, so that indicates that the choice of price point matters. Note that because the interaction is not significant, we don't have to answer this question with the more complicated, "well, it depends on which ad we selected." The price means look like the low price mean is much higher than the medium and high price points, and it's not clear whether the medium price mean is greater than the high price mean or that the latter 2 are flat (not significantly different).

Given that ANOVA is a way of partitioning variance, think of the data's variability in the form of Venn diagram. The total SS can be apportioned into that which is attributable to the ad factor (not much variance, since we know it's not significant), the price factor (most of the variance is explained by this factor), the interaction SS (which, like ad, is not significant so it's not much of the total SS either), and the rest is error.

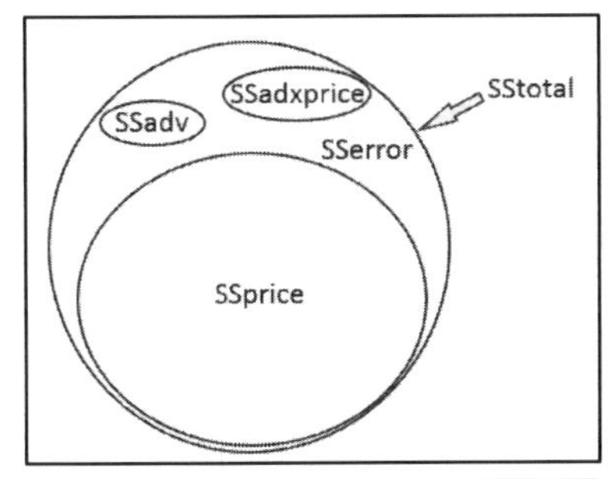

So at this point, in trying to understand the main effect for price, what we'd like to do is break down those price SS further to find out more precisely how the price null hypothesis is wrong. The null hypothesis is H_0: $\mu_{pricelow} = \mu_{pricemdm} = \mu_{pricehigh}$ and given the F-test was so large, we'll reject that null, but we don't know how it's wrong. Is it simply that the low price mean is higher than the other two and the other two are the same, or are all 3 means significantly different?

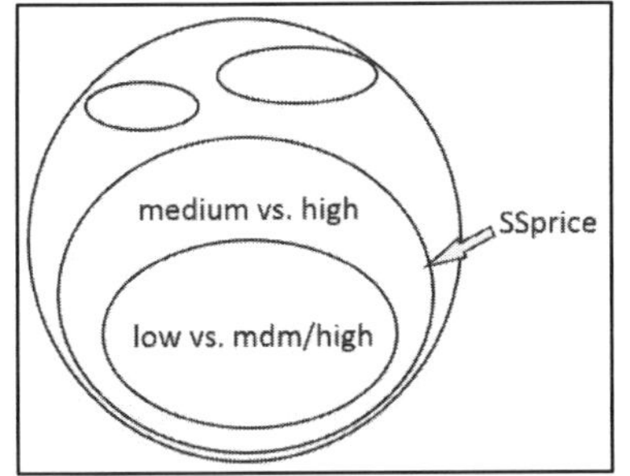

We'll learn how to test these comparisons in Chapter 5 on "contrasts," but basically it will refine our statistical testing. The ANOVA table is essentially expanded, to use the 2 df for price to test 2 portions of the price SS.

Source of variation	df	SS = Sum of Squares	MS = Mean Squares	F-test	p-value
Advertisement	1	0.300	0.300	0.04	0.839
Price	2	774.467	387.233	54.80	<.0001
→*Med vs. Hi*	*1*	*4.050*	*4.050*	*.57*	*0.456*
→*Low vs. M/H*	*1*	*770.417*	*770.417*	*109.02*	*.0001*
Ad × Price	2	2.600	1.300	0.18	0.833
Error	24	169.600	7.067		
Total	29	946.967			

Note the two new rows for the price contrasts. They indicate that the medium and high price means are not significantly different, so in the next comparison, we just combine them and contrast them with the low price mean, and that contrast is indeed significant. So, to return to the question of what price point to choose—we could price low because that mean is the highest. For other reasons, we could of course price high to obtain better margins (even if we know that the number of sales will fall off, significantly). We would not ever price medium though—it's true that sales fall off compared to the low mean, but if we price a little higher (for better margins and profitability), the means don't drop off anymore, at least not significantly. So price low for volume or high for margins, but a compromise medium price would be suboptimal.

Thus far in this chapter, we have looked at the two-factor ANOVA. ANOVA can easily be extended by adding experimental factors, to a 3-factor design, testing ad and price, and something new like package design, etc. In theory we can go to 4, 5, 10 factors and ANOVA will work with the same logic. We'll see a 3-factor ANOVA in a moment and we'll see how to generalize beyond that. However, we have a little tangent to deal with first.

Two-Way ANOVA without Replications (you'd never do this if you had a choice!)

The 2-way ANOVAs we've been working with have had all cell sizes, $n_{ij} > 1$. It is important to see what happens if there is only 1 subject in each cell, n=1. Usually social sciences don't have this problem. Experiments without multiple replications per cell can be found in engineering applications (e.g., there are only 6 machines to test in a 3×2 design) or medical testing (e.g., they've only made 6 artificial heart machines, or whatever). Still, it's good to see why replications are important, and what happens if each n=1.

So, this first table shows a tiny dataset where each n=2. Even that small n is better than an n=1. In the notation, the 1st subscript is for factor A, the 2nd subscript is for factor B, and the 3rd subscript denotes which subject in the (i,j) cell.

	b1	b2
a1	x_{111} x_{112}	x_{121} x_{122}
a2	x_{211} x_{212}	x_{221} x_{222}
a3	x_{311} x_{312}	x_{321} x_{322}

But what happens if n=1; we only have 1 subject per cell?

	b1	b2
a1	x_{111}	x_{121}
a2	x_{211}	x_{221}
a3	x_{311}	x_{321}

such as:

4	7
5	6
5	5

Note that those aren't cell means or sums. They're single data points. Here's the problem. Look at the df for this 3×2 example:

source of variability	df in general	df for this 3×2 example
A	(a-1)	2
B	(b-1)	1
A×B	(a-1)(b-1)	2
S(AB)	ab(n-1)	0 ←
total	abn-1	5

Zero df. That doesn't sound good. Why not? Because we cannot estimate a term associated with no df. In this case, it's the error term that would have df=0, and obviously we need an error term to form F-ratios. If this really was our problem, the solution would be to use the interaction MS as the error term, essentially confounding the notion of an "interaction effect" and "error." The ANOVA table for a 2-way design without replications would look like this:

source	df	SS	MS	F-test
A	(a-1)	SS_A	$SS_A/(a-1)$	$MS_A/\ MS_{\text{"error"}}$
B	(b-1)	SS_B	$SS_B/(b-1)$	$MS_B/MS_{\text{"error"}}$
error	(a-1)(b-1)	SS_{AB} (really)	$SS_{AB}/[(a-1)(b-1)]$	
total	ab-1	SS_{total}		

So in this scenario, we cannot study the interaction without replications because we'll have no F-test for the interaction, and because we'll be using that source of variability to serve as an error term for the main effects, even though the interaction SS term isn't really quite the error SS.

That's it on considering what happens if we do not have replications. Throughout the book, we'll continue to assume each $n_{ij} > 1$. Usually the cell sizes are much greater, and this issue is not a problem.

We've seen that ANOVA is an extension of the t-test, both in application to one-factor but more than 2 groups, and also for designs with more than 1 factor. We will randomly sample (to be in the position to generalize to the population per external validity) and randomly assign subjects to conditions (to be in the position to confidently make statements of causality per internal validity). Let's see the logical and statistical extensions to 3 or more factors.

Three-Way ANOVA (we'll always assume WITH replications)

In a 3-factor ANOVA, we'll work with factors A, B, and C, with levels a, b, and c. A simple example is a 2×2×2 (or 2^3) factorial:

	c1		c2	
	b1	b2	b1	b2
a1				
a2				

With 3 factors, now we'll want to study:

- the main effects for factors A, B, and C
- the 2-factor interactions: A×B, A×C, B×C
- and also a 3-factor interaction A×B×C (as long as n>1)

A 3-way interaction is interpreted as: the effects of 2 factors on the dependent variable depend on the levels of the 3rd factor. Any of these might hold:

- the relationship between A and B on the dependent variable is different for different levels of C
- the effect that A and C have is different for different levels of B
- the effect that B and C have is different for different levels of A.
- the effect of A is different for different levels of the B×C interaction
- the effect of B is different for different levels of the A×C interaction
- the effect of C is different for different levels of the A×B interaction

Here are two figures to illustrate various forms of a 3-way interaction. In this first example, there is an A×B×C interaction because there is no A×B interaction for c=1, but there is an A×B interaction for c=2:

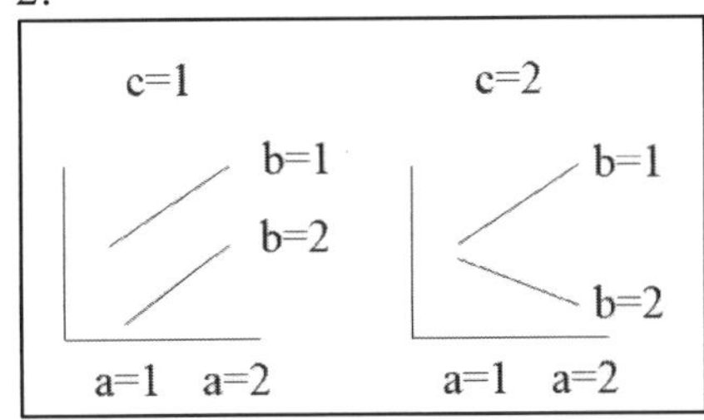

In this second example, there is an A×B×C interaction because the A×B interaction for c=1 looks different from the form of the A×B interaction for c=2:

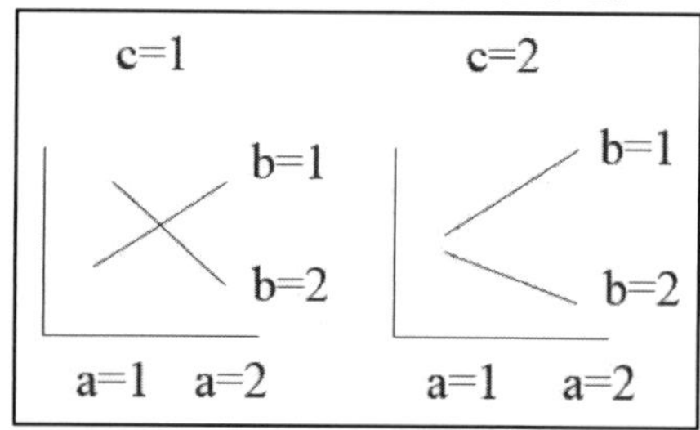

The 3-factor ANOVA model must reflect all these new effects—the 3 main effects, the three 2-way interactions, and the one 3-way interaction:

$$x_{ijk\ell} = \mu + \alpha_i + \beta_j + \gamma_k + (\alpha\beta)_{ij} + (\alpha\gamma)_{ik} + (\beta\gamma)_{jk} + (\alpha\beta\gamma)_{ijk} + \epsilon_{ijk\ell} \quad (7)$$

The new term, the 3-way interaction, is defined as:

$$\begin{aligned}(\alpha\beta\gamma)_{ijk} &= \mu_{ijk} - (\alpha\beta)_{ij} - (\alpha\gamma)_{ik} - (\beta\gamma)_{jk} - \alpha_i - \beta_j - \gamma_k - \mu \\ &= \mu_{ijk} - (\mu_{ij} - \mu_i - \mu_j + \mu) - (\mu_{ik} - \mu_i - \mu_k + \mu) - (\mu_{jk} - \mu_j - \mu_k + \mu) \\ &\quad -(\mu_i - \mu) - (\mu_j - \mu) - (\mu_k - \mu) - \mu \\ &= \mu_{ijk} - \mu_{ij} - \mu_{ik} - \mu_{jk} + \mu_i + \mu_j + \mu_k - \mu \end{aligned} \quad (8)$$

Recall the heuristics we saw with the 2-way interaction? They hold here also (good thing, otherwise they wouldn't be very good heuristics!). First, the line that defines the $(\alpha\beta\gamma)_{ijk}$ term at the start of equation (8) begins with the cell mean μ_{ijk}, now defined by 3 factors, A, B, C, with subscripts i, j, and k (where we've averaged over the subject subscript ℓ). We subtract off all "lower order" effects. So we subtract off the 2-ways (i.e., $(\alpha\beta)_{ij}, (\alpha\gamma)_{ik}, (\beta\gamma)_{jk}$) and all the main effects ($\alpha_i,\ \beta_j,\ \gamma_k$) and of course the grand mean μ. When the pluses and minuses cancel, the last line has that pattern where we begin with the cell mean, subtract off the next level (the 2-ways), add back the next level (the main effects), subtract off the next level (the grand mean). Well, it amuses me anyway.

Next, let's look at the ANOVA notation and definitional formulae, all generalizing from one-way and two-way ANOVAs. We begin with the data point $x_{ijk\ell}$ for i=1...a for factor A; j=1...b for factor B; k=1...c for factor C; and ℓ=1...n for the n subjects within each cell, and a total sample size of N=abcn. The means are:

For A×B×C:	$\bar{x}_{ijk} = \frac{1}{n}\sum_\ell x_{ijk\ell}$
For A×B:	$\bar{x}_{ij} = \frac{1}{cn}\sum_k \sum_\ell x_{ijk\ell}$
For A×C:	$\bar{x}_{ik} = \frac{1}{bn}\sum_j \sum_\ell x_{ijk\ell}$
For B×C:	$\bar{x}_{jk} = \frac{1}{an}\sum_i \sum_\ell x_{ijk\ell}$
For the main effect of A:	$\bar{x}_i = \frac{1}{bcn}\sum_j^b \sum_k^c \sum_\ell x_{ijk\ell}$
For the main effect of B:	$\bar{x}_j = \frac{1}{acn}\sum_i^a \sum_k^c \sum_\ell x_{ijk\ell}$
For the main effect of C:	$\bar{x}_k = \frac{1}{abn}\sum_i^a \sum_j^b \sum_\ell x_{ijk\ell}$
Grand mean:	$\bar{x} = \frac{1}{abcn}\sum_i^a \sum_j^b \sum_k^c \sum_\ell x_{ijk\ell}$ (9)

The last piece is the ANOVA table. This table includes an extra column to specify the null hypothesis being test by the F-test:

source	df	SS	MS	F-test	H_0 tested
A	(a-1)	$\sum_{i=1}^{a}\sum_{j=1}^{b}\sum_{k=1}^{c}\sum_{\ell=1}^{n}(\bar{x}_i - \bar{x})^2$	SS/df	$\frac{MS_A}{MS_{S/ABC}}$	all $\alpha_i = 0$
B	(b-1)	$\sum\sum\sum\sum(\bar{x}_j - \bar{x})^2$			all $\beta_j = 0$
C	(c-1)	$\sum\sum\sum\sum(\bar{x}_k - \bar{x})^2$			all $\gamma_k = 0$
AB	(a-1)(b-1)	$\sum\sum\sum\sum(\bar{x}_{ij} - \bar{x}_i - \bar{x}_j + \bar{x})^2$			all $\alpha\beta_{ij} = 0$
AC	(a-1)(c-1)	$\sum\sum\sum\sum(\bar{x}_{ik} - \bar{x}_i - \bar{x}_k + \bar{x})^2$			all $\alpha\gamma_{ik} = 0$
BC	(b-1)(c-1)	$\sum\sum\sum\sum(\bar{x}_{jk} - \bar{x}_j - \bar{x}_k + \bar{x})^2$			all $\beta\gamma_{jk} = 0$
ABC	(a-1)(b-1)(c-1)	$\sum\sum\sum\sum(\bar{x}_{ijk} - \bar{x}_{ij} - \bar{x}_{ik} - \bar{x}_{jk} + \bar{x}_i + \bar{x}_j + \bar{x}_k - \bar{x})^2$			all $\alpha\beta\gamma_{ijk} = 0$
S(ABC)	abc(n-1)	$\sum\sum\sum\sum(x_{ijk\ell} - \bar{x}_{ijk})^2$			
Total	abcn-1	$\sum\sum\sum\sum(x_{ijk\ell} - \bar{x})^2$			

All the ANOVA logic, the model statements, the notions of interactions and their interpretations, the computation of means and SS and MS and F-tests, all of it, generalizes to 4 and more factors, which would allow testing for even higher-order interactive effects. (In the homework at the end of the chapter, you can try your hand at the 4-way ANOVA model!)

SUMMARY

This chapter extends the principles and statistics of a one-factor ANOVA to a two-way ANOVA, and to a three-way ANOVA. Interactions were defined and illustrated; they are an important statistical tool for advancing theories, specifying how more than one factor might work in conjunction together in their joint effect on the dependent variable, above and beyond their individual effects.

Two-way ANOVAs abound in the journals. Three-way designs and ANOVAs are rarer. Most factorials in published journal articles are merely 2×2's, but this chapter empowers so much more!

The appendix to this chapter presents several recent real world experiments. These examples are light on details but rich as apps.

REFERENCES

1. Keppel, Geoffrey (1991), *Design and Analysis: A Researcher's Handbook*, 3rd ed., Pearson.
2. Keppel, Geoffrey and Thomas D. Wickens (2004), *Design and Analysis: A Researcher's Handbook*, 4th ed., Pearson.

APPENDIX

In academia, experiments are an essential tool of the trade, being probably the cleanest and sometimes even the simplest way to make progress in theory development and testing. By comparison, in the real world, experiments come and go in their popularity—in today's environment of marketing analytics, they are currently popular. In business settings, the research questions often ultimately seek answers to questions about ROI, "How much payoff did we see for the money spent against [some project, such as a new website, or the introduction of a new product, or the testing of a new ad, etc.]." Here are 5 recent applications and examples of real world experiments conducted by real companies.

First, a company was trying to induce more consumers to download and pay for more of its music library. The company ran a 3-factor factorial. Factor A was a series of different referral programs (1: control group "tell your friends!" 2: tell 3 friends and get 50% off a download, 3: tell 1 friend and get a free download). Factor B was the price deal (5 downloads for $0.99, or 2 for $0.99, or 1 for $0.99). Factor C was whether the study participant was recruited into the study via a group at Facebook (yes or no; if no, their email had been obtained via a purchased email list). The dependent variable was the percent of contacts in each market reach (each cell) who bought the deal. The cell n's varied from just under 100 to 250. The study found significant effects for A, B, C, and A×C. Specifically, in factor A, "tell 1 friend and get a free download" fared best, in factor B, the "5 for $0.99" pulled best, no surprise, and for factor C, the Facebook sample was most likely to buy. For the A×C interaction, the Facebook sample was least responsive to the control appeal, more so to the "tell 3 friends" and most to the "tell 1 friend." For the non-Facebook sample, the "tell 3 friends" was no better than the control, but the "tell 1 friend" did well.

Second, a software company ran a 4×2×2 factorial. There were 4 levels of price × 2 levels of promotion × 2 levels of ad message. The promotion factor manipulated a "30 day trial" vs. "receive free gift software." The ad message showed study participants either a speed-focused message, "<Our software> helps you manage customer relationships in just minutes a day" or a power-focused message, "<Our software> helps you handle a virtually infinite number of customer files." The dependent variable was the % of those who were contacted who said yes to the offer in each cell. The power message was the only effect that popped.

Third, the crayon people, Crayola, wanted to drive more traffic to their website and hopefully induce more purchases there. They ran a 2×3×2×3 factorial of email characteristics (sent to parents and teachers). Factor A was whether the email subject line said: "Crayola.com" or "Help us help you." Factor B varied the salutation in the email: "Hi [user name] ☺," "Greetings!" or "[user name]." Factor C asked for one of two calls to action: either "We'd like to get your thoughts about arts and art materials," or "Because you understand the arts, we invite you to help build crayola.com." And finally, Factor D offered three promotional appeals: a monthly drawing for $100 Crayola products, a monthly drawing for an Amazon gift certificate, or nothing. For factor A, "Crayola.com" was the best, for factor B, "greetings!" was the worst, for factor C, "build" was better, and for factor D, $100 for Crayola products was best. (They only tested main effects; what a waste.)

Fourth, currently an area ripe for extensive use of experiments is in resource allocation. One company was studying its deployment of its salesforce, e.g., "How *does* this salesperson spend his/her time, allocated across which clients or brand accounts?" and "How *should* the salesperson spend his/her time?" The current salesforce size was comprised of 450 sales reps, allocated across 7 drugs (Anaprox, Naprosyn, Norinyl, etc.), and 9 physician specialties (family practice, internal medicine, ob/gyn, etc.). Results were held confidential, reported only to the company and FDA.

Fifth, another resource allocation question arises in "integrated marketing communications." The question here takes the form, "How should we spend our advertising budget across the many possible media?" In this study, the company examined how much they had spent on ads on the radio, in magazines, at events, online, etc., last year versus this year. They tracked consumers' awareness of the brand and the source of their brand knowledge. In industry, the differences from last year to this year are referred to as upticks.

More examples can be found at quirks.com and in the white papers at ama.com.

HOMEWORK

This HW covers 2-way and 3-way ANOVA, and gets you into SAS.

Problem 1. Define a two-way interaction.

Problem 2. Consider the following 5 datasets. The numbers below are the cell means for a two-factor factorial experiment. Plot the data, and assume that the MS_{error} term is tiny, so that numbers that appear different are indeed significantly different. Note which effects are present in the data and which are not.

a) data set (a):

	a1	a2
b1	1	4
b2	2	5

b) data set (b):

	a1	a2	a3
b1	3	6	4
b2	2	5	3

c) data set (c):

	a1	a2	a3	a4
b1	2	5	3	4
b2	1	2	1	3

d) data set (d):

	a1	a2
b1	50	90
b2	70	30

e) data set (e):

	a1	a2
b1	1	2
b2	3	6

Problem 3. These observations come from a 3×3 factorial, with, coincidentally, n=3 observations per cell. Conduct an ANOVA (in SAS) for these data and describe the results—what's significant, what's not, what are the means, what's going on. (Turn in a hardcopy of your input and outputs.):

a1			a2			a3		
b1	b2	b3	b1	b2	b3	b1	b2	b3
2	5	3	6	7	9	2	4	8
2	4	3	5	6	8	3	3	9
3	3	2	4	8	9	4	3	9

Hint: reformat the data from that table into a structure like this for the computer:

```
ai bj k xijk
1 1 1 2
1 1 2 2
1 1 3 3
1 2 1 5
1 2 2 4
1 2 3 3
1 3 1 3
1 3 2 3
1 3 3 2
2 1 1 6
2 1 2 5
2 1 3 4
2 2 1 7
2 2 2 6
2 2 3 8
2 3 1 9
2 3 2 8
2 3 3 9
3 1 1 2
3 1 2 3
3 1 3 4
3 2 1 4
3 2 2 3
3 2 3 3
3 3 1 8
3 3 2 9
3 3 3 9
```

Problem 4. In the table below, there are 5 datasets, one per row. Each row contains the cell means from a 3-way 2×2×2 factorial experiment. Plot the 3-way and note which effects—main effects, 2-way interactions, and 3-way interactions are present. Also note when the main effects can be interpreted unambiguously.

	a1 b1 c1	a1 b1 c2	a1 b2 c1	a1 b2 c2	a2 b1 c1	a2 b1 c2	a2 b2 c1	a2 b2 c2
e.g., 1	4	3	3	2	6	5	5	4
2	5	5	8	8	8	8	5	5
3	3	4	4	5	3	4	2	3
4	2	4	1	3	1	4	2	5
5	2	1	3	4	4	5	3	2

Problem 5. The following data arise from a 3×2×2 factorial (this dataset is based loosely on a subset of data in Keppel and Wickens, p.484, #3). In this setting, imagine the data come from a new products testing study, in which 3 ads were tested (factor A), 2 package designs (B), and 2 product flavors (C). Analyze these data in SAS (don't forget to reformat the data—string them out like in Problem #3 above, and turn in a hardcopy of your input and outputs).

	a1		a2		a3	
	b1	b2	b1	b2	b1	b2
c1	7	2	10	4	13	9
	4	4	7	6	10	8
	5	3	6	3	13	9
	6	3	8	5	8	10
	7	7	6	5	12	4
c2	5	6	5	8	13	6
	5	5	5	7	11	4
	6	5	6	6	12	5

Problem 6. Use what you know of ANOVA and extend it to a 4-factor design. Name the factors A, B, C, and D with a, b, c, & d levels, respectively, and main effect parameters, α_i, β_j, γ_k, δ_ℓ. Write the model statement and an ANOVA table, including the following: source, df, expanded df, computational formula for the SS terms, MS's, F's, hypothesis tested by each test. Include all sums and means. Finally, write the definitional formula for SS_{ABCD}.

Problem 7.

The following ANOVA table lists the results from a two-factor experiment. Factor A was whether shelf price was raised or not, and factor B was whether the item was contained in a red or blue package. Consumers rated their intentions to buy the product (on a 7-point scale where 7="would definitely buy" and 1="would definitely not buy") as a result of having seen the product in a virtual grocery store aisle displayed online. Complete the ANOVA table, indicate which of the null hypotheses (i, ii, iii) you would reject, and interpret the findings in the ANOVA table using those means. The critical value of the F-statistic for all the tests is 4.35 (α=.05).

Source	df	SS	MS	F
A	1		65.10	
B	1			
A×B	1		13.50	
error		60.00		
total	23	148.05		

Hypotheses:

(i) H_0: H0: no effect for A (shelf price)
(ii) H_0: no effect for B (package color)
(iii) H_0: no interaction between A & B

Main effect means:

Factor A: price increase mean = 5.30, no price increase mean=6.50
Factor B: red package mean = 2.30, blue package mean = 3.70

Interaction cell means:

	Factor A (price)	
Factor B	increase	no increase
red	3.45	4.50
blue	2.70	3.00

CHAPTER 4

OMEGA-SQUARED AND EFFECT SIZES

Questions to guide your learning:
Q_1: What is an effect size?
Q_2: How is its information complementary to a significance test?
Q_3: What is omega squared?

In this chapter, we consider the notion of "effect sizes." Most statistics like t-tests, z-tests, F-tests, etc. are used to test a null hypothesis; if the test is significant, we reject the null. Hence these indices are called "test statistics." If the null hypothesis is true, the question is how likely is the result we obtained in our data. If the likelihood (probability, p-value) is small ($p<.05$), then we'll say that it seems implausible that the null holds, so let's reject the null, and at least for the moment (until new studies are run, new data collected, etc.), we'll conclude that the alternative hypothesis is more plausible.

The thing is…whether a test statistic is big (significant) or not depends in part on the effect we're studying but also frequently (for many test statistics) on the sample size of the dataset. Recall from Chapter 1, where we took the equation for a simple t-test, $t = \frac{(\bar{x}-\mu)}{s_{\bar{x}}}$, and rearranged the equation a bit so it took on the form $t = \frac{\sqrt{n}(\bar{x}-\mu)}{s_x}$, which makes it clear that the size of t is determined by how far the sample mean $\bar{x}$ is from the population (hypothesized) mean μ and also by n (and also by the standard deviation). Even if $\bar{x}$ is very close to μ, the t-value will be significant if we hike up our sample size n.

Obviously every study we run should have a large enough n that intuitively the findings seem reasonable; if we conducted a political poll of only 50 voters and claimed we could predict an election, that doesn't seem likely; if we ran a 2×2 study with 3 or 4 subjects per cell, it's not certain that the results would be convincing to reviewers or to ourselves. (A standard of "intuitively reasonable" is a little squishy—that's a subjective "face validity" assessment. We will be more precise in discussing power tests later in the chapter.) Yet resources are always limited, and we wouldn't want to set some arbitrary standard like "one must have n=100 observations in each cell" because the study would be more costly and it would take longer to run than a study with 30 or even 10 observations per cell that might suffice (not to mention that sometimes 100 might not be sufficient). (Online or lab studies are conducted so quickly that it might seem unlikely that threats to validity based on time could arise, and yet they certainly could; e.g., a study using celebrities as stimuli could well be interrupted by events of poor celebrity behavior in real life.)

The point is, if n is big enough, the t (or z or F) will be large, the result will be significant, we will reject the null, all of that. Yet sometimes we look at the data (our own or journal articles or manuscripts to review) and we seem to have a sense of, "Yes, the effect is 'significant' but it seems so small as to be negligible." In such cases, we might well wonder whether the significance is attributable primarily to a large sample size.

> Omega-squared is an "effect size" index. Effect size information can complement significance tests.

Sometimes the question is even posed as, "is the effect significant but not important?," a question that is unfortunate given the value-laden connotation of "importance."

To get a sense of how to tease apart a test of significance with an assessment of an effect size, let's step out of ANOVA for a moment and look at the simple correlation coefficient (the issue of effect sizes is definitely not endemic just to t-tests or ANOVA). For a correlation coefficient, if we wish to test the typical null hypothesis about the population correlation, H_0: $\rho = 0$, we use the test-statistic $t = r/\left\{\sqrt{[(1-r^2)/(N-2)]}\right\}$ (with df=N-2). Note that the t-statistic will be large (positive or negative) for large r (positive or negative) and/or for large sample size, N. Aside from whether the t-test is large (in magnitude) and significant, we can look at the correlation itself to judge whether it's large or small—the first being the assessment of significance and the second being an estimate of the effect size. The t-test can be significant for r = 0.9 and we'd say, yes, huge effect, strong correlation. A t-test could be significant even if r = 0.2, and there we'd presumably conclude that the effect is weaker. (We return to the issue of evaluating effect sizes at the end of the chapter.)

This chapter is about how to complement the information in a significance test with an "effect size." In the simple t-test, that effect size is essentially the difference: $(\bar{x} - \mu)$ (or $\frac{(\bar{x}-\mu)}{s_x}$). It has equal impact with "n" in whether the t-test will be significant or not (once more, $t = \frac{\sqrt{n}(\bar{x}-\mu)}{s_x}$); even for a modest sample size, if the effect or difference is large enough, the t-test will be large and significant. And vice versa, even for a modest effect size, if the sample size is large enough, the t-test will be large and significant.

Before proceeding, let's first switch to F-tests, given that this book is on ANOVA (the t-test equation was simply easier to take apart). In addition, there are several kinds of indices that have been proposed as effect sizes. In this chapter, we'll focus primarily on the omega-squared, ω^2, given its frequency of use within the ANOVA setting, though we'll consider alternatives at the end of the chapter.

Relationship between Significance Tests and Effect Sizes

A test of significance is often correlated with an effect size, yet they are conceptually orthogonal. That is, a test-statistic might be large (significant) or small, and an effect size might be large or small. If we consider those binary states (just to keep things simple), we can see that there are four possibilities:

1. We might have a non-significant F-test and a small effect size (a small ω^2). That makes sense. Nothing's apparently going on in the data.
2. We might have a significant F-test and a large ω^2. That makes sense too; something is really happening in the data.

Scenarios 1 and 2 are quite common, and they reflect the situations where it seems like a test-statistic and effect size index are correlated—they're both big or they're both small. But there are two more possibilities and at first they might seem a bit counter-intuitive:

3. We might have a non-significant F-test but a large ω^2.
4. We might have a significant F-test and a small ω^2.

If scenarios 3 or 4 occur, it doesn't necessarily signal an error in the analysis. In scenario 3, it could be that we have a decent-sized treatment effect, but a small sample size. Perhaps the study was somewhat exploratory, so we had been tentative in allocating resources to it. Then the effect size would be estimated as large but there would be insufficient power for the test-statistic to pop up as significant. In scenario 4, it could be the opposite, that the treatment effect in the population is rather small, and the F-test was significant only because, or primarily because, the sample size was large.

By the way, there is nothing necessarily wrong with a small effect size. Most of the phenomena scientists study are complex, and the particular part we're examining and estimating in any particular study is likely to be only a small part of the bigger picture. As stated elsewhere (Iacobucci et al., 2015):

> "… with respect to issues about effect sizes in general, it is important to note that it is typical in the maturation of any area within a discipline to see large effect sizes characteristic of main effects being established early on, and as researchers progress toward more refined studies to examine subtle interactions, by definition, such conditional effects will be smaller (Chow, 1988; O'Grady, 1982). Popper (2002) refers to the calculus of probability, that in studying factors A and B, p(A) and p(B) will equal or exceed p(A&B) (cf., rules of conjunctive probabilities of Tversky and Kahneman, 1983), and says that with the growth of scientific knowledge, scholars want to work with theories of increasing content, thereby implying decreasing probabilities or effect sizes. Hence ironically, relatively small effect sizes can be indicative of mature theories and mature literatures."

Omega-Squared, ω^2

Let's define the effect size index of omega-squared, ω^2. The primary motivation in calculating an effect size is to have a summary statistic that reflects the size of the effect and that is independent of sample size, unlike the F-test. We want ω^2 to reflect the magnitude of treatment effects or group differences, which is a different question from that of statistical significance.

We'll begin with the ω^2 for a one-way ANOVA, and then we'll extend it. In a one-way ANOVA, when we talk about "effects" or "treatment differences," we're talking about whether $\bar{x}_1 = \bar{x}_2 = \cdots = \bar{x}_a$. Recall from the ANOVA model that the effect for group "i" was α_i and in our data, we'd estimate that as: $\hat{\alpha}_i = \hat{\mu}_i - \hat{\mu} = \bar{x}_i - \bar{x}$ for each group. Over all the "a" groups, we'd have a collection of a α_i's, that is, a distribution of α_i's. Any distribution has a mean and variance. The mean of the distribution of α_i's would be zero (the α_i's are unbiased estimates because the group means are, and they're constrained to sum to zero in any given dataset). The variance would be: $\frac{\sum(\alpha_i-\bar{\alpha})^2}{a}$, which is just a standard definition of a variance, and because the mean of the α_i's is zero ($\bar{\alpha} = 0$), that equation simplifies a little to: $\frac{\sum(\alpha_i)^2}{a}$.

Okay, so we have a variance on the α_i's. Now what…? Well, it's difficult to evaluate any stand-alone variance σ^2 in absolute terms as large or small, so we'll compare this source of variability $\frac{\sum(\alpha_i)^2}{a}$ that reflects between groups variability to another source of variability, yep, error variance, σ_ϵ^2 that reflects within group variability.

The definitional formula for ω^2 for a one-factor ANOVA (or for the main effect of factor A in a two-way or higher-way ANOVA) is defined as the relative size of the variance attributable to how the groups differ (the "between" group variability) to that plus the individual differences within the groups (so the "between" plus "within" or "error" variability, which, for a one-way ANOVA is the "total" variability in the denominator, see Hays, 1994, p.499; introduced by Hays, 1963, pp.325, 382):

$$\omega_A^2 = \frac{\frac{\sum(\alpha_i)^2}{a}}{\frac{\sum(\alpha_i)^2}{a} + \sigma_\epsilon^2} \tag{1}$$

The effect size ω_A^2 ranges from 0 to 1. If there are no effects, that is, all the groups' means are the same, then all the α_i's = 0, and:

$$\omega_A^2 = \frac{0}{0 + \sigma_\epsilon^2} = 0 \tag{2}$$

If there are group differences, then there is an effect:

$$\omega_A^2 = \frac{effect}{effect + \sigma_\epsilon^2} > 0 \tag{3}$$

Equation (3) says $\omega^2 > 0$ but it's rare to see an ω^2 that is close to 1, its upper-bound. It's clear in equation (3) that whatever the effect size is, ω^2 approaches 1 as σ_ϵ^2 decreases. But σ_ϵ^2 is the term for within-group variability (or individual differences or study participant heterogeneity or just plain "noise") and $\sigma_\epsilon^2 \neq 0$ in behavioral data (i.e., on human subjects).

The fact that ω^2 has a range of 0 to 1 makes it seem like an r^2 in regression applications. In regression, r^2, or the "coefficient of determination" tells us the proportion of variance that is accounted for (VAF) in the dependent variable by the regression model, that is to say, by the predictor variables. That interpretation is fine for the ω^2 as well—the proportion of variability in the dependent variable that is accounted for by the differences in the group means. We'll address the topic of magnitudes of effect sizes at the end of the chapter.

Other Omega-Squared Equations

Equation (1) is known as the definition formula for ω^2, because its form expresses how the effect size is defined—as a relative comparison of between group variability to the total variability. The equation allows users to see the pieces that enter into the logic of the index. However, there is a computational formula that is a little easier to use when calculating the index because all of the pieces are produced by the computer in the ANOVA table (e.g., Hays, 1994, p.499; Hays, 1963, pp.327, 382):

$$\omega_A^2 = \frac{SS_A - (a-1)MS_{S(A)}}{SS_T + MS_{S(A)}} \tag{4}$$

It is not immediately obvious how the computational formula in equation (4) derives from the definitional formula in equation (1); they are rather different-looking equations. To see the intermediate steps from equation (1) to equation (4), please see Appendix A to this chapter.

Another equation for ω^2 is to write it as a function of the F-test. Doing so might first seem counter-intuitive to the goal of having an effect size that is not impacted by sample size when we know that the F-test is so impacted, but as you'll see in equation (5), there is an F in the numerator and denominator, so any effect of sample size is sort of canceled out (see Appendix B to this chapter for more information):

$$\omega^2 = \frac{F-1}{F - 1 + \left(\frac{an}{a-1}\right)} \tag{5}$$

Equation (5) allows us to view other properties of the ω^2. Recall that the logical range of ω^2 is 0 to 1, and we have seen that it's rare to see an ω^2 close to 1. Let's look at the lower-bound now. In theory, ω^2 cannot be negative, but if F is very small, like <1, for instance in a dataset with lots of error variance, then ω^2 can dip below 0 in practice.

Next, let's consider a general form for the omega-squared equation. That is, we've been looking at ω^2 for a one-factor ANOVA, and let's extend it. Let's say we're looking at factor A or B or the interaction in a two-way ANOVA, and we'll call whatever our focus is the "effect." Then, in general, for some "effect," the equation for ω^2 for that effect is:[5]

$$\omega_{effect}^2 = \frac{SS_{effect} - df_{effect}(MS_{error})}{SS_T + MS_{error}} \tag{6}$$

where the F-test for that "effect" had been $F = \frac{\left(\frac{SS_{effect}}{df_{effect}}\right)}{MS_{error}}$.

For example, in a 2-factor experiment (where both factors are "fixed" factors):

$$\widehat{\omega}_{A\times B}^2 = \frac{SS_{A\times B} - (a-1)(b-1)\left(MS_{S(AB)}\right)}{SS_T + MS_{S(AB)}} \tag{7}$$

All the terms in equation (7) are provided as entries in an ANOVA table, so it is easy to compute!

As a special case of an "effect," let us consider a "contrast" (more in Chapter 5). Contrasts are associated with a single degree of freedom (df=1). So, the ω^2 for a 1 df contrast would be:

$$\omega_{contrast}^2 = \frac{SS_{contrast} - (MS_{error})}{SS_T + MS_{error}} \tag{8}$$

Effect Sizes other than ω^2

The ω^2 is a popular effect size index for ANOVA, in part because some research claims it is less biased than alternatives. However, it is important to be aware of alternatives.

Confidence Intervals. First, let's not underestimate the helpfulness of simple solutions. The solution of providing confidence intervals is tried and true. A parameter estimate is surrounded by a range of values which are equally likely, for the given confidence level. While a confidence interval is not an estimate of effect-size per se, its calculation includes the parameter and the standard error. The standard error conveys the precision of the parameter estimate.

Eta-squared, η^2. A large family of effect sizes are like r^2 from regression. In regression, we compute a correlation coefficient, r, or in multiple regression, R, and squaring either is the "coefficient of determination." We speak of r^2 or R^2 as the amount of explained variance in the dependent variable accounted for by the relationship with the independent variable(s):

---

[5] In SAS, adding the option "effectsize" to the model statement produces a number of indices. The ones labeled "semipartial omega-square" are the closest to equation (6).

$$R^2 = \frac{SS_{regression}}{SS_{total}} = 1 - \frac{SS_{error}}{SS_{total}}$$

The coefficient of determination, r^2 or R^2, range from 0 to 1, thus we can speak of the proportion of total variance accounted for (VAF).

In discussions of effect sizes, we speak of eta-squared, $\eta^2 = \frac{SS_{between}}{SS_{total}} = \frac{SS_{effect}}{SS_{total}}$. Eta-squared is very appealing, given its familiarity in the regression context, and it is simpler than ω^2. Compare:

$$\eta^2 = \frac{SS_A}{SS_T} \qquad \omega_A^2 = \frac{SS_A - (a-1)MS_{S(A)}}{SS_T + MS_{S(A)}}$$

Unfortunately η^2 is not a great effect size index (none of them are perfect). Say we have run a 2×2×3 ANOVA, and we wish to isolate an effect size for the A×B interaction. We would calculate η^2 with SS_{AB} in the numerator, and SS_{total} in the denominator. The thing is, all the effects in the model (the main effects of factors A, B, and C, and their 2-way interactions and their 3-way interaction) contribute to SS_{total}. All other things being equal, a 2×2×2×2 design will likely have a greater SS_{total} than a 2×2 design (so the effect would appear proportionally smaller in the larger design). That's goofy. That doesn't quite capture what we're hoping to isolate.

As a solution, a ***"partial" eta-squared*** has been proposed, and it's defined as: $\eta^2_{partial} = \frac{SS_{effect}}{SS_{effect} + SS_{error}}$. In this equation, the effect we're trying to isolate, like the SS_{AB} would serve in the numerator and as part of the denominator, and instead of SS_{total} in the denominator, we'd have $SS_{S(AB)}$ (or SS_{within}), a much smaller term, and one in which the effects of other factors in the model have been partialled out or statistically controlled for.

That seems better. Still, in early research, scholars worried about special designs. If a study was run that involved a 2-factor factorial where factor B was a "blocking" factor (like a covariate, intended to reduce SS_{within}), that might enhance the apparent effect size compared to a 2-factor design in which neither factor was a blocking factor (more on blocking factors in Chapter 7 and covariates in Chapter 9). It has also been shown that η^2 can be sensitive to the number of levels of a factor ("a" and "b") as well as (a×b×n), the overall sample size.[6]

Cohen's d. Other effect size indices are intended to directly capture mean differences, as when comparing 2 means in a t-test or more groups in an ANOVA. The best known is Cohen's "*d*." For 2 independent samples: Cohen's $d = \frac{\bar{x}_1 - \bar{x}_2}{s_{pooled}}$, $s_{pooled} = \sqrt{\frac{(n_1-1)s_1^2 + (n_2-1)s_2^2}{n_1+n_2-2}}$.
Cohen suggested the assessment of the d effect sizes as follows: 0.2 is "small," 0.5 is "medium" and 0.8 is a "large" effect size (Cohen 1992, p.157; 1988, pp.25-26). Thus if the 2 groups' means differ only by a magnitude of 0.2 standard deviations, that's a small effect (whether it's significant or not).

Odds ratios are helpful for categorical data—they're computed as the odds of one outcome occurring compared to another outcome. The "logit" in a logit model is the natural log of odds, so the transformation to this information is straightforward. When outcomes are expressed in terms of probabilities instead, the comparison is referred to as a risk ratio, or an index of ***relative risk***.

Many other effect size indices have been proposed. Great references in these matters include Ellis (2010) and Hedges and Olkin (1985).

---

[6] For more on these issues, see Olejnik and Algina, 2003, pp.435, 441.

Evaluating the Magnitude of Effect Sizes

Recall that we had said that a t-test for a correlation might be significant and if the correlation itself was r = 0.9, we'd probably feel comfortable concluding that the effect is large, the relationship is strong. In another scenario with a slightly smaller correlation coefficient, the t might be significant for r = 0.7, and in real (i.e., noisy) data we'd probably draw the same conclusion. Consider other scenarios with still smaller correlations: around r = 0.3 we might think, ok, that explains about 10% of the variance, and in some contexts that might be quite a lot. A correlation of r = 0.3 is also certainly large enough to cause multicollinearity problems in regression models. But what about a correlation of r = 0.1 or r = 0.01—those seem really small at first blush, but again, in some contexts, perhaps a quality control improvement in plant operations or an efficacy indicator of some new medical treatment, then all of a sudden, r = 0.1 or r = 0.01 might be huge. And therein lies the effect size rub… In test statistics, there is a known distribution, and if we all agree (admittedly by a sociology of science convention) that we'll use a cutoff of 0.05, then we have rules, and playing by those decision rules, everyone would agree: yes, it's significant, or no, it's not. With effect sizes, there are not distributions, so we cannot know if an effect of size 0.3 or 0.7 or whatever is large or small.

Most researchers argue as by legal precedence, citing the one researcher who has offered an opinion of effect sizes: Cohen (1992, p.157; 1988, pp.79-80) has suggested that a correlation, r, around 0.1 is "small," 0.3 is "medium-sized," and 0.5 or larger is "large." His reasoning is sensible—that the small effect should be a bit beyond zero, the medium-sized effect should be a 'just noticeable difference' beyond that, and the large effect is as much larger than the medium-sized effect as the medium was above the small. Furthermore, Cohen does acknowledge that the evaluations of small, medium, and large effect sizes depend on the research contexts and methods.

Nevertheless, his characterization is subjective, and not based on any objective criterion like a statistical distribution. One person's opinion is not good science. What if Einstein had said, "I declare: $E = mc$." Some scared physics doctoral student might want to present a paper at a conference that suggested another model, "Um, Professor Al, cool and all, respect you, dig the hair, but I have data that indicate that the model is more likely: $E = mc^2$." That is, even the opinion of Einstein, a scientist we all hold in highest reverence, would be trumped by data and a better theory (the complication of course being that data and theories are imperfect; it's all relative, which brings us back around to Einstein!). That example might seem silly, but various journals in various fields swing the pendulum back and forth as to whether manuscripts require stated effect sizes, some journals sometimes proposing the elimination of the test statistics. Test statistics (t's, F's, etc.) might be imperfect, but effect sizes offer no simple solution, and can in fact be deeply more problematic.

That's not to say don't use effect sizes. Effect sizes can offer information that is complementary to test-statistics. However, effect sizes raise other issues like interpretation.

Other Uses of Effect Sizes

In this chapter, we've discussed the idea of the possible inclusion in journal articles and research reports of an effect size along with a test statistic to complement the information provided about research findings. Two other arenas in which effect sizes arise are in the computation of power and in meta-analyses.

Power calculations are used to try to estimate requisite sample sizes for studies we're planning to run. Power calculations are simply derived by taking a test statistic and solving for n, the sample size. For example, we've seen the t-test for a one-sample test of means is $t = \frac{(\bar{x}-\mu)}{s_{\bar{x}}}$, which we rearranged to see the role of sample size, $t = \frac{\sqrt{n}(\bar{x}-\mu)}{s_x}$. Power implies that we'd like to find a significant result, so let's say we want our t-test to be at least 2.0 (somewhat arbitrary, but roughly like the 1.96 value of a z-test). Rearranging the equation still further and solving for n, we'd obtain: $n = \frac{t^2 {s_x}^2}{(\bar{x}-\mu)^2} = \frac{4{s_x}^2}{(\bar{x}-\mu)^2}$. That result says that we could obtain a significant t-statistic if we gathered data on a sample of size "n" for the given values of ${s_x}^2$ and $(\bar{x} - \mu)^2$. Power calculations seem like an awesome idea—use an equation and find out how large the sample size needs to be. Yet in practice they are challenging to implement. There are a lot of unknowns in that equation that we must plug in to solve for n. Before collecting data, we must have an estimate of variability (the ${s_x}^2$) and of the effect size itself (the $(\bar{x} - \mu)$). Usually researchers look to past studies for analogous values. We could also do "what if" scenario planning—what if ${s_x}^2$ is smaller or larger, what if $(\bar{x} - \mu)^2$ is smaller or larger, and aim for one of the larger projected n's recommended from among those scenarios.

It is worth noting that in Cohen's book, *Statistical Power Analysis in the Behavioral Sciences* (1988, 1992), he was not considering effect sizes for effect sizes' sake. Rather, he examined various effect sizes (e.g., d, η^2, and several other indices) as a way to enable conducting power estimations. That is, if we knew that $(\bar{x} - \mu)^2$ was "small," we would need a larger n, whereas if we knew the effect $(\bar{x} - \mu)^2$ was "large," we could use a small n. One clear indicator that Cohen was treating effect sizes as integral to power calculations is that, for example, his (1988) Chapter 2 on d describes the index on pp.19-27 and then he follows that presentation with 12 pages of power tables (pp.28-39). He does the same thing in his Chapter 3 on correlation coefficients, following the introduction of the index on pp.75-83 with 12 pages of power tables (pp.84-95). There is a separate chapter for each effect index (e.g., for tests of differences between correlation coefficients, tests of proportions, chi-squares, etc.) and they all follow that same pattern. In the Appendix B to this chapter, we examine another index, f, proposed to handle the ANOVA case, and for it, Cohen's (1988) Chapter 8 begins with the presentation of the index on pp.273-288 followed with 66 pages (!) of power tables (so many pages presumably due to the complexity of multiple degrees of freedom for the tables, pp.289-354). It is not a stretch to conclude that Cohen was interested in effect size indices primarily for the purpose of improving power estimation (per his book's title). The coupling of effect sizes and power co-occur in other references as well, such as Rosenthal and Rosnow (1991, Chapter 19). For more on these issues, see also Kraemer and Blasey (2015), and Murphy, Myors, and Wolach (2008).

A **meta-analysis** is like a literature review on steroids, or a quantitative version of a lit review. Meta-analyses take effect sizes as data points to be synthesized to assess the current standing in a field of certain phenomena. In a meta-analysis, the effect size index is the dependent variable, and the explanatory (or design) factors (or independent variables) would be things we believe explain effects across studies—their commonalities and their differences. For example, an explanatory factor might be whether the sample of study participants was taken in the U.S. or in Brazil, or in person or online, or whether the dependent variables measured in the original studies were actual sales or purchase intentions, or whether recall vs. recognition was used as a measure of memory in an advertising and brand awareness study, etc. Obviously if the intent of the study is to make summary statements, it's important for the contributing elements—the effect size estimates—to be good estimates. For more, see Borenstein et al. (2009), Farley and Lehmann (1986), and Lipsey and Wilson (2000).

SUMMARY

This chapter described effect sizes as information that can complement significance tests. The chapter focused on ω^2 and closed with the mention of a few other indices. Common sense should be used when interpreting any of them.

REFERENCES

1. Borenstein, Michael, Larry V. Hedges, Julian P. T. Higgins, and Hannah R. Rothstein (2009), *Introduction to Meta-Analysis*, Wiley.
2. Chow, Siu L. (1988), "Significance Test or Effect Size?," *Psychological Bulletin*, 103 (1), 105-110.
3. Cohen, Jacob (1992), "A Power Primer," *Psychological Bulletin*, 112 (1), 155-159.
4. Cohen, Jacob (1988), *Statistical Power Analysis for the Behavioral Sciences*, 2nd ed., Routledge.
5. Cohen, Jacob (1969), *Statistical Power Analysis for the Behavioral Sciences*, New York: Academic Press.
6. Cumming, Geoff (2012), *Understanding the New Statistics: Effect Sizes, Confidence Intervals, and Meta-Analysis*, NY: Routledge.
7. Ellis, Paul D. (2010), *The Essential Guide to Effect Sizes: Statistical Power, Meta-Analysis, and the Interpretation of Research Results*, Cambridge University Press.
8. Farley, John and Donald Lehmann (1986), *Meta-Analysis in Marketing: Generalization of Response Models*, Lexington, MA: Lexington Books.
9. Grissom, Robert J. and John J. Kim (2012), *Effect Sizes for Research: Univariate and Multivariate Applications*, 2nd ed., NY: Routledge.
10. Hays, William L. (1994), *Statistics*, 5th ed., Fortworth, TX: Harcourt Brace.
11. Hays, William L. (1963), *Statistics*, (first edition), New York: Holt, Rinehart and Winston.
12. Hedges, Larry V. and Ingram Olkin (1985), *Statistical Methods for Meta-Analysis*, Orlando, FL: Academic Press.
13. Iacobucci, Dawn (2005), "On p-Values," *Journal of Consumer Research*, 32 (1), 6-11.
14. Iacobucci, Dawn, Steve S. Posavac, Frank R. Kardes, Matthew J. Schneider, and Deidre L. Popovich (2015), "The Median Split: Robust, Refined, and Revived," *Journal of Consumer Psychology*, 25 (4), 690-704.
15. Keppel, Geoffrey (1991), *Design and Analysis: A Researcher's Handbook*, 3rd ed., Englewood Cliffs, NJ: Prentice-Hall.
16. Keppel, Geoffrey and Thomas D. Wickens (2004), *Design and Analysis: A Researcher's Handbook*, 4th ed., Upper Saddle River, NJ: Pearson Prentice-Hall.
17. Kraemer, Helena Chmura and Christine M. Blasey (2015), *How Many Subjects? Statistical Power Analysis in Research*, 2nd ed., Sage.
18. Lipsey, Mark W. and David Wilson (2000), *Practical Meta-Analysis*, Sage.
19. Murphy, Kevin R., Brett Myors, and Allen Wolach (2008), *Statistical Power Analysis: A Simple and General Model for Traditional and Modern Hypothesis Tests*, 3rd ed., Routledge.

20. Olejnik, Stephen and James Algina (2003), "Generalized Eta and Omega Squared Statistics: Measures of Effect Size for Some Common Research Designs," *Psychological Methods*, 8 (4), 434-447.
21. O'Grady, Kevin E. (1982), "Measures of Explained Variance: Cautions and Limitations," *Psychological Bulletin*, 92 (3), 766-777.
22. Popper, Karl R. (2002), *Conjectures and Refutations: The Growth of Scientific Knowledge*, New York: Routledge Classics.
23. Rosenthal, Robert and Ralph L. Rosnow (1991), *Essentials of Behavioral Research: Methods and Data Analysis*, 2nd ed., Boston: McGraw-Hill.
24. Rutherford, Andrew (2011), *ANOVA and ANCOVA: A GLM Approach*, 2nd ed., Hoboken, NJ: Wiley.
25. Tversky, Amos and Daniel Kahneman (1983), "Extensional versus Intuitive Reasoning: The Conjunction Fallacy in Probability Judgment," *Psychological Review*, 90 (4), 293-315.

Also related to power and sample size:

1. Holland, Burt S. and Margaret DiPonzio Copenhaver (1988), "Improved Bonferroni-Type Multiple Testing Procedures," *Psychological Bulletin*, 104 (1), 145-149.
2. Keselman, H. J., Paul A. Games, and Joanne C. Rogan (1980), "Type I and Type II Errors in Simultaneous and Two-Stage Multiple Comparison Procedures," *Psychological Bulletin*, 98 (2), 356-358.
3. Levine, Douglas W. and William P. Dunlap (1982), "Power of the *F* Test With Skewed Data: Should One Transform or Not?," *Psychological Bulletin*, 92, 272-280.
4. Maxwell, Scott E. and David A. Cole (1991), "A Comparison of Methods for Increasing Power in Randomized Between-Subjects Designs," *Psychological Bulletin*, 110 (2), 328-337.
5. Ryan, T. A. (1980), "Comment on 'Protecting the Overall Rate of Type I Errors for Pairwise Comparisons With an Omnibus Test Statistic'," *Psychological Bulletin*, 98 (2), 354-355.
6. Wahlsten, Douglas (1991), "Sample Size to Detect a Planned Contrast and a One Degree-of-Freedom Interaction Effect," *Psychological Bulletin*, 110 (3), 587-595.

Appendix A: Omega Squared Definitional and Computational Formulae

We begin with the ***definitional*** formula specified in the chapter in equation (1):

$$\omega_A^2 = \frac{\frac{\sum(\alpha_i)^2}{a}}{\frac{\sum(\alpha_i)^2}{a} + \sigma_\epsilon^2}$$

We know the two expected means squares of the components of the F-test:

$$E(MS_A) = \sigma_\epsilon^2 + \frac{n\sum(\alpha_i)^2}{(a-1)} \qquad \text{(A1)}$$

and

$$E(MS_{S(A)}) = \sigma_\epsilon^2 \tag{A2}$$

We need to isolate the term $\frac{\sum(\alpha_i)^2}{a}$ in (A1) to plug it into the ω_A^2 formula in equation (1). So, in equation (A1), we'll multiply both sides by $\frac{(a-1)}{na}$:

$$\left(\frac{a-1}{na}\right)(MS_A) = \left(\frac{a-1}{na}\right)\sigma_\epsilon^2 + \left(\frac{a-1}{na}\right)\frac{n\sum(\alpha_i)^2}{(a-1)} \tag{A3}$$

Skooch the epsilon term to the left hand side of the equation:

$$\left(\frac{a-1}{na}\right)(MS_A) - \left(\frac{a-1}{na}\right)\sigma_\epsilon^2 = \left(\frac{a-1}{na}\right)\frac{n\sum(\alpha_i)^2}{(a-1)} \tag{A4}$$

Notice in the right hand side of equation (A4), the $(a-1)$'s and n's cancel, so:

$$\left(\frac{a-1}{na}\right)(MS_A) - \left(\frac{a-1}{na}\right)\sigma_\epsilon^2 = \frac{\sum(\alpha_i)^2}{a} \tag{A5}$$

Plug the left hand side of the equation (A5) into the definitional formula (1) (once in the numerator, once in the denominator):

$$\omega_A^2 = \frac{\left(\frac{a-1}{na}\right)(MS_A) - \left(\frac{a-1}{na}\right)\sigma_\epsilon^2}{\left(\frac{a-1}{na}\right)(MS_A) - \left(\frac{a-1}{na}\right)\sigma_\epsilon^2 + \sigma_\epsilon^2} \tag{A6}$$

Next, multiply ω_A^2 by na/na:

$$\omega_A^2 = \frac{(a-1)(MS_A) - (a-1)\sigma_\epsilon^2}{(a-1)(MS_A) - (a-1)\sigma_\epsilon^2 + (na)\sigma_\epsilon^2} \tag{A7}$$

and use $MS_{S/A}$ as an estimate of σ_ϵ^2:

$$\omega_A^2 = \frac{(a-1)(MS_A) - (a-1)MS_{S(A)}}{(a-1)(MS_A) - (a-1)MS_{S(A)} + (na)MS_{S(A)}} \tag{A8}$$

and we know that $MS_A = \frac{SS_A}{(a-1)}$, right? So:

$$\omega_A^2 = \frac{(a-1)\left(\frac{SS_A}{a-1}\right) - (a-1)MS_{S(A)}}{(a-1)\left(\frac{SS_A}{a-1}\right) - (a-1)MS_{S(A)} + (na)MS_{S(A)}} \tag{A9}$$

A couple of (a-1)'s cancel:

$$\omega_A^2 = \frac{SS_A - (a-1)MS_{S(A)}}{SS_A + (na - a + 1)MS_{S(A)}} \tag{A10}$$

Note that $(na - a + 1) = (a(n-1) + 1)$ and replace:

$$\omega_A^2 = \frac{SS_A - (a-1)MS_{S(A)}}{SS_A + (a(n-1) + 1)MS_{S(A)}} \tag{A11}$$

Expand the bottom right:

$$\omega_A^2 = \frac{SS_A - (a-1)MS_{S(A)}}{SS_A + a(n-1)MS_{S(A)} + MS_{S(A)}} \tag{A12}$$

$MS_{S(A)} = \frac{SS_{S(A)}}{a(n-1)}$, so in the denominator, $a(n-1)MS_{S(A)} = SS_{S(A)}$:

$$\omega_A^2 = \frac{SS_A - (a-1)MS_{S(A)}}{SS_A + SS_{S(A)} + MS_{S(A)}} \tag{A13}$$

For a simple one-way ANOVA, $SS_T = SS_A + SS_{S(A)}$,

$$\omega_A^2 = \frac{SS_A - (a-1)MS_{S(A)}}{SS_T + MS_{S(A)}} \tag{A14}$$

Et voilà! This is the ***computational*** formula for omega-squared (for a one-way ANOVA) in equation (4) in the chapter.

Appendix B: Omega Squared Confusion and Magnitude

There is a little bit of confusion about ω^2 in one cite and another. Some mention Cohen's characterizations of effect sizes of "low," "medium," and "large" for ω^2 but Cohen (1969, 1988, 1992) doesn't actually mention ω^2. Curious.

For the ANOVA model, Cohen uses an effect size index called "*f*" (Cohen 1988, pp.275, 285-287; 1992, p.157). In Cohen (1988), this index is defined as $f^2 = \frac{\sigma_m^2}{\sigma^2}$ where σ_m^2 stands for the "variance" of the means between-groups, and σ is the within-group or error variability. When I see "σ^2" or "variance" in the ANOVA context, I think these terms are $MS_{between}$ (or generally MS_{effect}) and MS_{within} (or MS_{error}). There is a little bit of ambiguity as to whether his "variance" means technically "variance" or more loosely, "variability" in particular meaning variability conceptually or at the population level and in computation perhaps meaning sums of squares. Let's see.

1) Perhaps he means "variance" literally, technically. Why?:
   a) He certainly uses the term "variance."
   b) He defines $\sigma_m^2 = \frac{1}{a}(\sum(m_i - m)^2)$, which certainly looks like a typical variance.
   c) If $f^2 = \frac{\sigma_m^2}{\sigma^2}$ were defined as a ratio of MS terms (our unbiased estimates of variances), $f^2 = \frac{MS_{between}}{MS_{within}}$, then $f^2 = F$, which would be consistent with Cohen's $d = \frac{\bar{x}_1 - \bar{x}_2}{s_p}$ which is the same as a 2-sample t-test (on $n_1 + n_2 - 2$ df) and he refers to the *f* index as a generalization of *d* for 2 or more groups (as the F-test is for the t-test). This consistency across his 2 indices is a pretty compelling reason to believe that Cohen meant "variance" literally.
2) On the other hand, by "variance," he might mean more loosely "variability" and more specifically "sums of squares," for two reasons:
   a) He says $\sigma_{total}^2 = \sigma_m^2 + \sigma_{within}^2$ (p.281), an additive relationship which we know holds for SS terms but not MS terms (or it could hold in a Venn diagram sense of "variability" or for population variances).
   b) He relates f^2 to η^2 via the equation $f^2 = \frac{\eta^2}{1-\eta^2}$ (or $\eta^2 = \frac{f^2}{(1+f^2)}$) in which we know η^2 to be defined as $\eta^2 = \frac{SS_{between}}{SS_{total}}$, a ratio of SS terms, not MS terms (he uses his sigma terms, $\eta^2 = \frac{\sigma_m^2}{\sigma_m^2 + \sigma^2} = \frac{\sigma_m^2}{\sigma_{total}^2}$). Using that equation, we can take f^2 and fill in $\frac{SS_{between}}{SS_{total}}$ for the η^2 terms, per: $f^2 = \frac{\eta^2}{1-\eta^2} = \frac{\frac{SS_{between}}{SS_{total}}}{1-\frac{SS_{between}}{SS_{total}}} = \frac{\frac{SS_{between}}{SS_{total}}}{\frac{SS_{total}}{SS_{total}} - \frac{SS_{between}}{SS_{total}}} = \frac{\frac{SS_{between}}{SS_{total}}}{\frac{SS_{total} - SS_{between}}{SS_{total}}} = \frac{\frac{SS_{between}}{SS_{total}}}{\frac{SS_{within}}{SS_{total}}} = \frac{SS_{between}}{SS_{within}}$, just as f^2 was defined ($f^2 = \frac{\sigma_m^2}{\sigma^2}$), i.e., all derived via SS terms.
   c) It would be good if $f^2 = \frac{\sigma_m^2}{\sigma^2}$ were defined as a ratio of SS terms, rather than MS terms (as above in "1c"), because if $f^2 = F$, it would be a little at odds with the intentions of effect size indices. The F-test is the test-statistic for the hypothesis of the effect being tested, and the notion of an effect-size is usually sought to be information that is complementary to the test statistic, not identical to it. So perhaps by the term "variance" he meant "variability" conceptually, such that operationally we'd use SS not MS.
3) On the third hand, his divisor in $\sigma_m^2 = \frac{1}{a}(\sum(m_i - m)^2)$ of "a" rather than "a-1" may offer one clue that he may well be speaking about population values because he's presenting conceptual relationships, whereas I'm thinking about data and so we must divide SS estimates by their df to arrive at unbiased variance estimators.
   a) →So let's proceed using the term "variance" as a population and conceptual idea, i.e., meaning simply "variability," as in: "the total 'variability' may be decomposed into between-group and within-group variability" (see the parenthetical above in "2a").

b) While that issue might be loosely resolved with our taking it to the population or conceptual level, for practical purposes, if we compute η^2 we'll use SS, as most sources indicate, thus, $\eta^2 = \frac{SS_{between}}{SS_{total}}$, and for f^2, we'll use MS per 1c, $f^2 = \frac{MS_{between}}{MS_{within}}$. (Except we won't be using either. We'll use ω^2, calculated as in the chapter or in Appendix A above.)

c) Analogously, Hays (cf., 1963, p.382) speaks of the definitional formula for ω^2 at the population level as well.

Moving on! ...Aside from those ambiguities, the relationship between the index f and η^2 is important because while Cohen does not offer guidelines for η^2 (or ω^2) directly, he offers them for f which can be used, through f^2 and then $\frac{f^2}{(1+f^2)} = \eta^2$ to obtain rules of thumb for η^2. The small, medium, and large effects that are offered in terms of f at 0.1, 0.25, 0.4, respectively (Cohen 1992, p.157; 1988, pp.285-287; 1969, pp.278-280), translate to the case of η^2 being small at 0.0099=0.01, medium is .0588=0.06 and large is 0.1379=0.14. Cohen says that the "medium"-sized value of $f = 0.25$, for example, indicates that 1 standard deviation of the "a" group population means is 0.25 as large as the standard deviation within the groups.

Keppel and Wickens (2004, p.163, footnote 7) provide equations wherein they substitute η^2 with ω^2 and state: $f = \sqrt{\frac{\omega^2}{1-\omega^2}}$ and $\omega^2 = \frac{f^2}{(1+f^2)}$. On the face of it, these substitutions don't seem proper, given that $\eta^2 \neq \omega^2$, not by a long shot. Curiouser. However, in making the substitution, they are perhaps relying less upon the 2 calculated indices, and more upon the conceptual definition common to η^2 and ω^2 as both being interpretable as the proportion of variance or variability in the dependent variable explained by the between-groups effects whose size is being estimated in the index (i.e., variance accounted for, or VAF), again, particularly with an appeal to the population conceptualization. That VAF interpretation is obvious in the equation for $\eta^2 = \frac{SS_{between}}{SS_{total}}$. The interpretation of VAF for ω^2 is not obvious from its computational formula ($\omega^2_{effect} = \frac{SS_{effect} - df_{effect}(MS_{error})}{SS_T + MS_{error}}$) but it is more apparent from its definitional formula,[7] $\omega^2_A = \frac{\frac{\Sigma(\alpha_i)^2}{a}}{\frac{\Sigma(\alpha_i)^2}{a} + \sigma^2_\epsilon}$. Thus, perhaps the substitution of ω^2 in place of η^2 is understandable (acceptable?) from this conceptual point of view, as both indices are characterized as indicators of proportion of explained variance (see also Cumming, 2012; Grissom and Kim, 2012; Rutherford, 2011).[8] If the substitution is acceptable, then so would be the appropriations of the cutoffs for the small, medium, and large effect size thresholds from η^2 to ω^2.

In his truly terrific text (not because of this), Keppel (1991, p.66) used to endorse Cohen's cutoffs for small, medium, and large ω^2 at 0.01, 0.06, and 0.15. However, the more recent edition offers a thoughtful presentation about not using "hard-and-fast rules" and

---

[7] For example, as Hedges and Olkin (1985, p.103) put it, ω^2 is the relative reduction in variance in the dependent variable due to the predictor variable (the effect being represented in ω^2).

[8] And grazie mille to Profs. Don Lehmann of Columbia U, David Brinberg of Virginia Tech, and Jim Jaccard of NYU, for serving as a reality check for me on this.

instead Keppel and Wickens (2004, pp.166-167) offer guidelines. For example, instead of reporting a generic term of "effect size," they encourage precision in specifying which effect size index is being reported, e.g., ω^2 or d or R^2, etc. They encourage presenting basic descriptive statistics (means, standard deviations, etc.) so that readers can see the sizes of group differences for themselves. And they say that obviously the F-tests (and df and p-values) should be reported, because the "effect sizes you report are adjuncts to this information, not substitutes for it" (p.167).

As mentioned in the current chapter, effect size indices like ω^2 have no statistical distributions, so any assessment of small, medium, and large is somewhat arbitrary and subjective. As Keppel and Wickens state (2004, p.167), Cohen's division into small, medium, and large may be helpful "in thinking about how readily an effect stands out from the background [i.e., error] variability"; however, such a classification "says nothing about the practical importance of an effect nor its usefulness to …theory." That's a great plug for common sense. And to be thorough, indeed, going back to the source, Cohen (1969, 1988) is very thoughtful in his presentation and careful to encourage not overgeneralizing the characterizations of "small," "medium," and "large." Cohen (1969, p.278) says "…these qualitative adjectives are relative, and, being general, may not be reasonably descriptive in any specific area." Such careful qualifications are repeated in numerous places throughout his books. Unfortunately, his broader, thoughtful perspective and caveats were left behind even as readers took the cutoffs for small, medium, and large forward.

One last consideration regarding equations for ω^2; we have seen several varieties of equations for ω^2 and still more abound. For example, we had seen in equation (5) the relationship stated between the effect size index and the F-test: $\omega^2 = \frac{F-1}{F-1+\left(\frac{an}{a-1}\right)}$ and that relationship is also given in various alternative forms. It has also been stated as $\omega^2 = \frac{(a-1)(F-1)}{(a-1)(F-1)+an}$ which is also an attractive form, simplifying the calculation of ω^2 from the F-values that the ANOVA table would have produced. It may be translated vis-à-vis equation (5) as follows (here as an example, the "effect" is factor A in a 1-way ANOVA). From our equation (5), $\omega^2 = \frac{F-1}{F-1+\left(\frac{an}{a-1}\right)}$ take the denominator terms without the F and rearrange them: $-1+\left(\frac{an}{a-1}\right) = \left(\frac{an}{a-1}\right) - \left(\frac{a-1}{a-1}\right) = \frac{an-a+1}{a-1} = \frac{a(n-1)+1}{a-1} = \frac{df_{error}+1}{df_{effect}}$. Put that back in the equation to get $\omega^2 = \frac{F-1}{F-1+\left(\frac{an}{a-1}\right)} = \frac{F-1}{F+\frac{df_{error}+1}{df_{effect}}}$. Next multiply the F in the denominator by a term that =1, $\omega^2 = \frac{F-1}{\frac{F(df_{effect})}{df_{effect}}+\frac{df_{error}+1}{df_{effect}}} = \frac{F-1}{\frac{F(df_{effect})+df_{error}+1}{df_{effect}}}$. Put the denominator of the denominator up to the numerator $\omega^2 = \frac{(df_{effect})(F-1)}{F(df_{effect})+df_{error}+1}$. Now for a 1-way ANOVA, the numerator is $(a-1)(F-1)$. The denominator is $F(df_{effect})+df_{error}+1 = F(a-1)+a(n-1)+1 = Fa-F+an-a+1$, which simplifies to $(a-1)(F-1)+an$, so the fuller equation can take the form $\omega^2 = \frac{(a-1)(F-1)}{(a-1)(F-1)+an}$. At least that mystery is resolved.

So what is the bottom line? Certainly we should report the F-test, df, and p-value for any effect being tested. (If we did not report hypothesis test statistics, the journals would be reduced to reports of "so and so says it's a big effect," which doesn't sound like science.) If an effect size is sought to provide complementary information, the ω^2 is considered unbiased relative to other effect size indices. Rather than classifying its size as small, medium, or large, it simply can be compared to reported ω^2 indices in the same data set or in published articles on similar phenomena.

CHAPTER 5

CONTRASTS AND SIMPLE EFFECTS

Questions to guide your learning:
Q_1: What is a contrast?
Q_2: Why or when do we need them?
Q_3: What is a simple effect, a.k.a. how do we do contrasts for interactions?

In this chapter, we'll come to understand contrasts. We need to conduct contrasts if we get a significant result, main effect or interaction, and we want to understand the precise nature of the group mean differences. The only exception (when we don't have to do contrasts) is if there are only a=2 groups, then obviously, when we reject the null $H_0: \mu_1 = \mu_2$ we conclude $H_A: \mu_1 \neq \mu_2$, and all we have to do is look at the means, $\bar{x}_1$ and $\bar{x}_2$ and it will be obvious which mean is larger. However, in a 1-way ANOVA, or for main effects in a two-way or three-way or higher-order ANOVA, if there are more than 2 groups, then simply knowing that we can reject the null $H_0: \mu_1 = \mu_2 = \cdots = \mu_a$ isn't super conclusive. The alternative hypothesis isn't stated very precisely; it's H_A: "H_0 is not true" but we don't know how H_0 is incorrect. Thus, we need to conduct "contrasts" (also called "comparisons") among group means if there are more than 2 groups, which is also to say, if there are more than 1 df. (As a heuristic, think: 1 df means there's only 1 way to interpret how the 2 means differ.)

> Contrasts help tell us why we rejected H_0.

Here's an example from Keppel (1991, pp.112-114). He describes a 1-factor experiment testing study participants' recall of words, a very simple cognitive task. The factor A has a=5 levels:

> "Consider an example of an experiment designed with some explicit comparisons in mind. Suppose that subjects are given a list of 40 common English words to learn. ...The list is presented for six trials, each trial consisting of a study portion, in which the words are presented to the subjects, and a test portion, in which the subjects attempt to recall the words. Thus, each subject sees the list 6 times and is tested 6 times. Subjects are randomly assigned to [1 of 5] different conditions of training [summarized here]."
>
> "For the 1st 2 groups, the words are presented all at once on a piece of paper for 2 minutes; different orderings of the 40 words are used on the 6 trials. [These 2] groups differ with regard to the arrangement of the words on the sheet of paper—for 1 group the words appear in a column and for the other group they are scattered around on the paper. The remaining 3 groups also study the list for a total of 2 minutes, but the words are presented one at a time, at a constant rate [on a laptop or tablet]. For groups 3 and 4, each word is presented once for 3 seconds; the total presentation time for the whole list of words is 120 seconds (3 seconds × 40 words). [These 2] groups differ with regard to the presentation of the words on successive trials. Group 3 receives the same presentation order on all 6 trials, whereas group 4 receives different presentation orders. The final group also receives the [words on the tablet], but at a

faster rate of presentation (1 second per word). However, to equate total study time per word, the words are presented 3 times before the recall test is administered."

Ok, let's paraphrase and summarize:

- We present a list of 40 words to subjects, who study the words and are tested for recall. So, the dependent variable is the number of words recalled. There are 6 trials and 6 test periods (i.e., we present 40 words and test for recall each of 6 times). Let's keep things simple: rather than modeling all 6 trials, let's say that the first 5 trials and tests are just practice, and that all we care about is performance (#words recalled) in the 6th round. (Modeling all 6 trials would require "repeated measures" analyses, something we'll see in Chapter 8.)
- The independent variable or experimentally manipulated factor is how we present the 40 words to subjects. The group definitions follow:
  1) In group 1, all 40 words are presented in a column on a piece of paper. The subjects study the list for 2 minutes (120 seconds).
  2) In group 2, all 40 words are presented on a piece of paper, but not in a column array, rather they are scattered all over the page. The study participants in this condition also try to learn the words in 2 minutes.

  For groups 3, 4, and 5, each word is presented one at a time on a tablet. The entire list of words is still presented for a total of 2 minutes, thus the overall study time is constant across all 5 groups.
  3) In group 3, each word is up on the screen for 3 seconds (3 seconds × 40 words = 120 seconds). The words are presented in the same order over all 6 trials.
  4) Group 4 is like group 3, but the words are presented in different orders over the 6 trials.
  5) In group 5, each word is up on the screen for l second and it comes around 3 times (1 second × 3 times × 40 words = 120 seconds).

So that's the study. We have 1 factor with 5 levels. The overall, omnibus F-test will test the null, H_0: $\mu_1 = \mu_2 = \mu_3 = \mu_4 = \mu_5$. That is, the null states that the presentation style of the words doesn't matter, and that we expect memory recall (the number of words remembered) to be the same in all 5 groups. If we reject the null H_0, what groups might we compare? One set of possible comparisons is charted below. The 5 groups comprise the columns, and 4 contrasts are presented in the rows.

group:	1	2	3	4	5
modality:	paper	paper	tablet	tablet	tablet
order:	varied	varied	same	varied	varied
array:	column	scattered	•	•	•
rate:	•	•	3 sec.	3 sec.	1 sec.

Contrast#	$\bar{x}_1$	$\bar{x}_2$	$\bar{x}_3$	$\bar{x}_4$	$\bar{x}_5$
(1)	1	1	0	-1	-1
(2)	1	-1	0	0	0
(3)	0	0	0	1	-1
(4)	0	0	1	-1	0

Before statistically testing these contrasts, let's look at the logic and the substantive interpretation underlying each of these 4 contrasts and their sets of coefficients:

1. The first row will compare and contrast the medium—paper (l&2) vs. tablet (4&5) (these 4 groups see a varied order of word presentation, so group 3 doesn't enter in—the 0 coefficient says we're ignoring that group for now). The question underlying this contrast is whether the mode of presentation might matter in recall. Perhaps subjects can control their memorization strategy when all words presented at once (on paper; hence, we'd expect groups l&2 to recall more than groups 4&5). Or perhaps presenting all words at once would overwhelm the subjects, thus resulting in better recall in groups 4&5.
2. The 2nd row or contrast indicates that we're comparing only groups 1 and 2, and they vary in whether the words are presented in a column or a scattered array. Perhaps we have a theory about either in terms of how the structures differ in facilitating the efficiency of memory organization.
3. The 3rd contrast compares groups 4 and 5, and they differ in whether a word is presented for l second (three times) vs. 3 seconds (once). Perhaps we'll expect $\bar{x}_4 > \bar{x}_5$ if time exposure per word helps build a memory network; alternatively, perhaps we'd expect $\bar{x}_5 > \bar{x}_4$ if practice and repetition are important.
4. The 4th contrast compares groups 3 and 4, testing the effectiveness on recall of the same vs. different serial orders of words presented on the tablet. We might wonder whether predictability is helpful in memory organization; if so, we'd expect $\bar{x}_3 > \bar{x}_4$.

Note that those 4 contrasts are not the only patterns that might describe how the 5 groups means differ. But the 4 contrasts describe 4 logical or theoretically interesting ways that the groups might differ. If the 5 means differ (we reject the null), but we do not find support for any of the 4 contrasts (by rejecting any of their respective hypotheses), then we have simply not been clever enough to consider how and why the groups differ. That's ok too; we'll test additional contrasts. Exploratory statistics are part of scientific discovery.

Conceptually speaking, those 4 contrasts give a flavor of how we'll try to break down the overall SS_A into pieces that are more specific about the nature of group mean differences. To actually test the contrasts, we'll need more information.

Recall that there is no ambiguity in interpreting a finding when we reject the null for only 2 groups; only 1 df. That clarity is the key to how we proceed. We'll take what is a 5-level 1-way ANOVA and break down the 4 df into a series of tests, each of which has only 1 df, and as a result, there will be no confusion in interpreting the results of each test. Every contrast will have 1 df, so we can think of each comparison as a comparison between 2 means. Sometimes the contrast is defined literally as a comparison of the mean of one group to the mean of another. For example, each of comparisons 2, 3, and 4 involved only 1 pair of means:

comparison #2 $H_0: \mu_1 = \mu_2$ or $H_o: \mu_1 - \mu_2 = 0$ (1)

#3 $H_0: \mu_4 = \mu_5$

#4 $H_0: \mu_3 = \mu_4$

However, even contrast #1 can be seen as a comparison of 2 means, they're just defined in a slightly more complicated way:

$$H_0: (\mu_1 + \mu_2)/2 = (\mu_4 + \mu_5)/2 \text{ or } H_o: \frac{(\mu_1+\mu_2)}{2} - \frac{(\mu_4+\mu_5)}{2} = 0$$

Algebraically, that comparison can also be written:

$$H_0: \; .5\mu_1 + .5\mu_2 + (0 \times \mu_3) - .5\mu_4 - .5\mu_5 = 0 \qquad (2)$$

/ \ / \

means μ_i's contrast coefficients

A contrast is defined to look at the population means, the μ_i's, in equation (2), by using "contrast coefficients" that we'll denote c_i, one c for each group. We'll call each contrast a ψ:

$$\text{Contrast}_1 = \psi_1 = .5\mu_1 + .5\mu_2 + 0\mu_3 - .5\mu_4 - .5\mu_5 \qquad (3)$$

$$= \text{a linear combination of group means, } \psi = \textstyle\sum_{i=1}^{a} c_i \mu_i \qquad (4)$$

Naturally, our estimate of ψ will be: $\hat{\psi} = \sum_{i=1}^{a} c_i \bar{x}_i$ (5)

Constraints on Contrast Coefficients

In defining contrasts, the only rule we must follow is that the c_i contrast coefficients must sum to 0: $\sum_{i=1}^{a} c_i = 0$ (6)
(but not all c_i's can be 0). So in ψ_1 in equation (3), $\Sigma c_i = .5 + .5 + 0 + (-.5) + (-.5) = 0$, so we're ok! But for example, the following is not a contrast: $\psi = \frac{1}{2}\mu_1 + \frac{1}{3}\mu_2 - 1\mu_3$ because $\sum c_i = \frac{1}{2} + \frac{1}{3} - 1 \neq 0$.

If we are examining contrasts in a 2-way (or higher) effect, that constraint, $\Sigma c_i = 0$ becomes a little more restrictive. The sum of contrast coefficients must be 0 "in all directions." In a 1-way design, that's easy: $\sum_{i=1}^{a} c_i = 0$. And obviously when conducting contrasts on the main effects in 2-way or higher-way designs, the situation is similar:

for a contrast on factor A: $\sum_{i=1}^{a} c_i = 0$

$$c_1\bar{x}_{A1} + c_2\bar{x}_{A2} + \cdots + c_a\bar{x}_{Aa} \qquad (7)$$

for a contrast on factor B: $\sum_{i=1}^{b} c_i = 0$

$$c_1\bar{x}_{B1} + c_2\bar{x}_{B2} + \cdots + c_b\bar{x}_{Bb} \qquad (8)$$

But for a contrast in an interaction, here's how the notion of "summing over all subscripts" plays out:

$$\sum_{i=1}^{a} c_{ij} = \sum_{j=1}^{b} c_{ij} = 0 \qquad (9)$$

Take a table of means for a 2×3 design:

	b_1	b_2	b_3
a_1	μ_{11}	μ_{12}	μ_{13}
a_2	μ_{21}	μ_{22}	μ_{23}

There are 6 means, so there will be 6 contrast coefficients, and they have to sum to 0 for each row and for each column. So we define

$$\psi = c_{11}\mu_{11} + c_{12}\mu_{12} + c_{13}\mu_{13} + c_{21}\mu_{21} + c_{22}\mu_{22} + c_{23}\mu_{23} \qquad (10)$$

$$\hat{\psi} = \sum_i \sum_j c_{ij} \bar{x}_{ij}$$

For example, say we are considering the following coefficients:

	b_1	b_2	b_3
a_1	.5	0	-.5
a_2	-.5	0	.5

↑ the table entries are the c_{ij}'s

$\hat{\psi} = .5\bar{x}_{11} + 0\bar{x}_{12} - .5\bar{x}_{13} - .5\bar{x}_{21} + 0\bar{x}_{22} + .5\bar{x}_{23}$. This $\hat{\psi}$ is an appropriately defined contrast because $\sum_i c_{ij} = \sum_j c_{ij} = 0$ (sums over i and sums over j).

So this is a contrast:

½	-¼	-¼
-½	¼	¼

but this is not:

½	-¼	-¼
½	-¼	-¼

and this is not:

1	-1	0
-1	0	1

The constraint that the coefficients sum to zero is also important in statistical computing packages. In SAS, we enter "proc glm;" and we specify the factors with "class a b;" and the model with "model x = a b a*b / ss3;" (more details are provided in Chapter 11). We obtain means "means a b a*b;" and we run contrasts with the code: "contrast 'factor A contrast 1' A 1 -1 0;" for example. In the contrast statement, the characters between the single quotes serve as a short title, then the factor is specified, then the contrast coefficients are listed. If the contrast coefficients do not sum to 0, the computer will generate an error message, saying the contrast is "not estimable."

How Many Contrasts Can We Run, and What is "Orthogonal"?

So that's how we define a contrast. In a moment, we'll get to the SS, F-tests, etc. The logic in contrasts is to take some df>2 and break them down, using them to make finer inquiries into how group means might differ. A natural question is just how many of these contrasts can we run to follow-up an inquiry into the situation of having rejected a H_0? The answer is: well, any number really. However, only "df" of them can be "mutually orthogonal." Huh?

The definition and test is that 2 contrasts are ***orthogonal*** if the sum of their cross products is zero. That is, for 2 contrasts c_i and c_i', we'll test:

$$\sum_{i=1}^{a} c_i \times c_i' = 0. \qquad (11)$$

Here is an example where 2 contrasts (the rows) defined to compare a=3 groups are orthogonal:

group mean:	a=1	a=2	a=3	
contrast #1	1	-½	-½	
contrast #2	0	1	-1	
their product:	0	-½	½	sum=0, so ψ_2 is orthogonal to ψ_1.

In this example, the contrasts are not orthogonal:

group mean:	a=1	a=2	a=3	
contrast#1	1	0	-1	
contrast#2	0	1	-1	
their product:	0	0	1	sum=1 (i.e., $\Sigma \neq 0$), so ψ_2 is not orthogonal to ψ_1.

With a set of more than 2 contrasts, k contrasts are ***mutually orthogonal*** if all pairs of the k contrasts are orthogonal. Thus for factor A, there can be at most (a-1) mutually orthogonal contrasts; i.e., (a-1) pieces of non-redundant, independent information, and for factor B, there could be at most (b-1) mutually orthogonal ψ's.

Let's apply these criteria to the memory experiment. We had 1-factor with 5-levels and 4 contrasts. Each contrast and its coefficients form the rows of this table:

group:	1	2	3	4	5
ψ_1	1	1	0	-1	-1
ψ_2	1	-1	0	0	0
ψ_3	0	0	0	1	-1
ψ_4	0	0	1	-1	0

To test the mutual orthogonality of the 4 ψ's, we test each pair (the symbol for orthogonal is $\perp$):

$\psi_1 \perp \psi_2$? (1)(1) + (1)(-1) + (0)(0) + (-1)(0) + (-1)(0) = 0; yes
$\psi_1 \perp \psi_3$? (1)(0) + (1)(0) + (0)(0) + (-1)(1) + (-1)(-1) = 0; yes
$\psi_1 \perp \psi_4$? (1)(0) + (1)(0) + (0)(1) + (-1)(-1) + (-1)(0) = 1; no
$\psi_2 \perp \psi_3$? (1)(0) + (-1)(0) + (0)(0) + (0)(1) + (0)(-1) = 0; yep
$\psi_2 \perp \psi_4$? (1)(0) + (-1)(0) + (0)(1) + (0)(-1) + (0)(0) = 0; yep
$\psi_3 \perp \psi_4$? (0)(0) + (0)(0) + (0)(1) + (1)(-1) + (-1)(0) = -1; nope

In this example, not all pairs are orthogonal, so the set of 4 ψ's are not mutually orthogonal. Does that matter? Not necessarily. The theoretical questions raised in a study are more important than their statistical properties. These 4 contrasts were the ones that interested us, so these are the contrasts we should run. Still, let's see what we're giving up if a set of contrasts are not orthogonal.

Look at ψ_1 and ψ_2 (or ψ_1 and ψ_3). This is one way to find orthogonal sets—when 2 groups are lumped together in 1 contrast, create a contrast between them for another, orthogonal contrast.

The (Un)Importance of Mutually Orthogonal Contrasts

If we can find (a-1) orthogonal contrasts, that's got the nice quality that when we add the SS for each ψ, that sum will equal SS_A. That property fits nicely with how we think about SS, and breaking down the df; we'd be apportioning some SS to each conceptual question in each contrast. If the contrasts are orthogonal, we can imagine a Venn diagram of SS_{total} that is nice and clean.

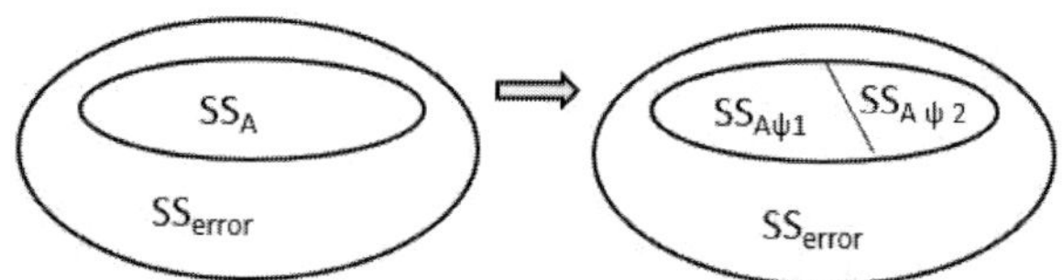

When a set of ψ's are not mutually orthogonal, this nice summing property doesn't hold. In the 5-group example, $SS_{\psi 1} + SS_{\psi 2} + SS_{\psi 3} + SS_{\psi 4} \neq SS_A$. If contrasts are not orthogonal, the sum might not meet SS_A (below, left), there might be some overlap or redundancy in the variance explained between the 2 contrasts (below, middle), and the sum might exceed SS_A (below, right).

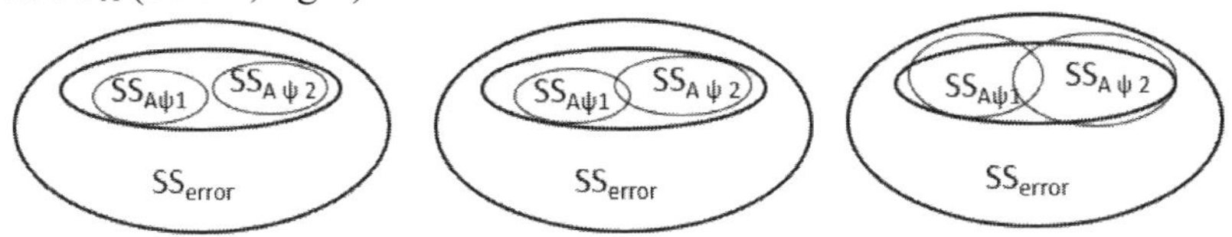

When $\sum SS_\psi = SS_A$, it's easy to interpret the contrast ψ's—we're partitioning SS_A into (a-1) independent components of variability. However, if $\sum SS_\psi \neq SS_A$, we still interpret the contrasts, but the notion that they don't sum is a little harder to interpret. Still, as in the previous example, these ψ's might still be the ones we're most interested in for theoretical reasons. We'd prefer them to a set of contrasts that are orthogonal but not as theoretically meaningful.

Contrast F-tests

At this point, let's say we've selected some theoretically interesting contrasts, and we'd like to test each. The null hypothesis is H_0: $\psi = 0$. We compute SS:

$$SS_{Acomp} = \frac{(\hat{\psi})^2}{\sum(c_i)^2/n} = \frac{n(\hat{\psi})^2}{\sum(c_i)^2} = \frac{n\sum(c_i\bar{x}_i)^2}{\sum(c_i)^2} \qquad (12)$$

where "comp" stands for "comparison" (or we could say $SS_{A\psi}$).

Each contrast is defined to have only 1 df, so $MS_{Acomp} = SS_{Acomp}/1 = SS_{Acomp}$ and we compare that MS to the same error term as we had compared the overall effect, in this case, the error term for the main effect for factor A.

For factor A, we had used the F-test: $F = \frac{MS_A}{MS_{S(A)}} = \frac{SS_A/df_A}{MS_{S(A)}}$ (13)

So for a contrast involving factor A, to test ψ_A, we use the F-test:

$$F = \frac{MS_{Acomp}}{MS_{S(A)}} = \frac{SS_{Acomp}/df_{Acomp}}{MS_{S(A)}} = \frac{SS_{Acomp}/1}{MS_{S(A)}} \qquad (14)$$

and we compare our observed F-value to the critical value for F with 1 and a(n-l) df.

We can include the contrasts in the ANOVA table:

Source	df	SS	MS	F-test
A	a-1	SS_A	MS_A	$MS_A/MS_{S(A)}$
Acomp1	*1*	SS_{Acomp1}	MS_{Acomp1}	$MS_{Acomp1}/MS_{S(A)}$
Acomp2	*1*	SS_{Acomp2}	MS_{Acomp2}	$MS_{Acomp2}/MS_{S(A)}$
Acomp3	*1*	SS_{Acomp3}	MS_{Acomp3}	$MS_{Acomp3}/MS_{S(A)}$
Acomp4	*1*	SS_{Acomp4}	MS_{Acomp4}	$MS_{Acomp4}/MS_{S(A)}$
Error	a(n-1)	$SS_{S(A)}$	$MS_{S(A)}$	
Total	an-1	SS_{total}		

Presenting the contrasts in an ANOVA table might make sense, especially if the contrasts are selected to be mutually orthogonal. If they are not, we might just present them in text, because the ANOVA table format just encourages readers to add up the SS in his/her head (and they won't add up in the non-orthogonal case).

A Priori vs. Post Hoc Contrasts/Comparisons

There are different kinds of contrasts or comparisons. One way to distinguish among contrasts is to recognize them as "*a priori*" ("planned comparisons") vs. "*post hoc*" comparisons. *A priori* or planned contrasts are specified at the design stage of the experiment, long before any data are collected, explicitly or implicitly (e.g., if we weren't interested in contrast #4 in the memory experiment, we probably wouldn't have included group 3 in design because that's the only contrast in which that group appears).

Alternatively, "*post hoc* comparisons" are contrasts that we test once we've seen the data because something just looks interesting. There is nothing wrong with exploring one's data—that is a legitimate part of the discovery of science. However, when one examines the data, such as the 5 means in the memory study, in a sense doing so, just eye-balling the data, implicitly "tests" (though not statistically officially) a number of comparisons. For example, in the memory experiment, say the group means had been:

$$\bar{x}_1 = 25.0,\ \bar{x}_2 = 26.1,\ \bar{x}_3 = 24.9,\ \bar{x}_4 = 26.3,\ \bar{x}_5 = 34.9$$

It looks like, for whatever reason, group 5 recalled the most words. We hadn't planned a contrasts with coefficients: .25, .25, .25, .25, -1, but seeing those means, that's the impression we get, and we'd want to test to see if that subjective impression is supported by an objective statistical test. So we'll run that contrast.

The biggest difference between *a priori* and *post hoc* comparisons is that the latter makes scientists a little nervous that perhaps the statistical test will be significant because it's somehow capitalizing on chance in the data. It's like "data fishing," if we fish (do some analysis) and the fish (F-test) isn't big enough, we'd throw it back and cast again. The typical treatment is to use a more conservative Type I error rate for *post hoc* tests (e.g., take α from .05 to .01) to help counter any effect of capitalizing on chance.

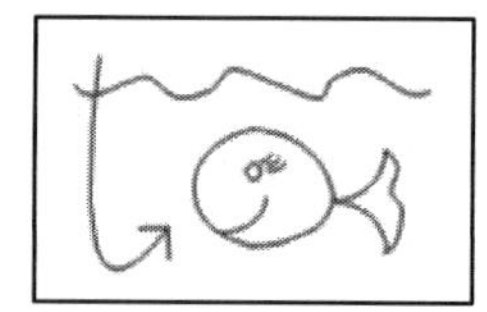

Another difference is that running a *post hoc* contrast, to compare 2 or more treatment means, is conditional on having a significant overall F-test. That is, *if* the omnibus F-test (e.g., for the main effect of factor A) is significant, *then* we can conduct *post hoc* contrasts on that factor.

There is a perspective (but mixed opinions) that we can test a planned comparison whether the omnibus F-test is significant or not. This pursuit seems a little inconsistent. The overall, omnibus F-test for the main effect of factor A tests the null hypothesis H_0: $\mu_1 = \mu_2 = \cdots = \mu_a$ vs. the alternative H_A: at least one μ is different, which seems to imply that at least one contrast, some combination of the means, will be significant. So if we don't reject H_0, it would seem that there would be no contrast that would be significant, so why would we go ahead and test the planned comparisons? It's possible in some situations that a planned comparison test might sometimes be more powerful than the overall F-test (i.e., sometimes the contrast might be significant when the overall F-test was not). For example, we assume the within-cell σ^2's are homogeneous (that's what allows us to pool in $MS_{S(A)}$), yet perhaps the σ^2's for the 2 groups involved in the contrast are less than those in the other groups, then we might compute a test using only the 2 relevant groups' data (with a modified $MS_{S(A)}$ term). Or, we might choose some α like .05 for the planned comparison (because it's planned, there is no possibility for capitalizing on chance), and choose a smaller α for the overall F-test (e.g., .01) because we know we're likely to also be conducting additional *post hoc* contrasts. Both of these situations seem a tad squirrely to me, or at least a bit of a stretch, so I'm suggesting it would be better practice to simply let the overall F-test (e.g., for factor A) indicate whether we should investigate any contrasts (*a priori* or *post hoc*). If the main effect for A (or B) is significant (or the interaction), get in there with contrasts and find out what's going on. If an effect is not significant, logic would suggest that we can mine our data for all it is worth and we'd likely come up with nothing.

There is another suggestion, also debated, that we don't need to adjust α (e.g., from .05 to .01) for the number of contrasts tested for planned comparisons, but that α does need to be adjusted for *post hoc* comparisons. That logic is somewhat sensible if the point of the adjustment is to make more conservative tests that are conducted after the data are viewed. Yet most research reports and articles don't list excessive numbers of tested contrasts. Furthermore, one might imagine jotting down an *a priori* list of many, many contrasts, and just because we have them on paper prior to collecting data doesn't mean we're not being opportunistic. So use common sense; if there won't be too many tests, *a priori* or *post hoc*, perhaps α doesn't need to be adjusted, but if there will be many: a) reduce α, and b) be more thoughtful when planning future studies!

Family-wise and Per-comparison Error Rates

If we conduct "k" number of contrasts, how should we adjust our α? The Type I error rate for a single test is α:

α = p(make type I error). (15)

The likelihood of not making a type I error would be:

1- α = p(not make type I error).

The likelihood of not making a single type I error across all "k" of our tests would be

$(1-\alpha)^k$ = p(NOT make a type I error in ANY of the "k" comparisons).

Accordingly, the probability that we make at least 1 type I error across the "k" contrasts is:

=1-p(not doing so)

$= 1-(1-\alpha)^k$.

For example, say we use a standard $\alpha = 0.05$. Note that as "k" goes up, the likelihood that we make at least 1 Type I error goes up quickly, well past .05:

k	$1-(1-\alpha)^k$	
3	$1-(1-.05)^3$	=.14
5	$1-.95^5$	=.23
10	$1-.95^{10}$	=.40 (40% chance of make ≥1 Type I error)

This relationship is what connects the so-called family-wise error rate to the error rate of each contrast:

Error rates: $\alpha_{FW} = 1-(1-\alpha_{PC_i})^k$ (16)

/ \

FW = familywise PC = per comparison

To adjust the Type I error rate, we are usually more sophisticated than merely dropping α from .05 to .01. It is traditional to distinguish between the "family-wise" error rate and the error rate "per comparison." Here, "family-wise" means across all the tests we conduct for one effect, e.g., the main effect for A, it's not a correction across the entire study, all effects including A, B, A×B, etc. (The main effect for B gets its own family-wise rate, etc.)

The correction we apply is to simply divide our α by the number of contrasts, "k," based on the Bonferonni inequality:

$\alpha_{FW} \leq \sum_{i=1}^{k} \alpha_{PC_i}$ (17)

We'll choose some α_{FW} (often $\alpha = 0.05$) and spread it (usually evenly) over the "k" contrasts:

$$\alpha_{PC_i} = {}^{\alpha_{FW}}/_{k} \qquad (18)$$

So for example, if we want α_{FW} = .05 and we have k = 5 contrasts:

α_{PC_i} = .05/5 = .01.

This reduction in α requires that our observed F-test surpasses a larger critical value, as protection for a possibly inflated Type I error rate. Doing so necessarily leads to a more conservative, less powerful, F-test. This figure depicts the rejection region in the F-distribution; the critical F-value for $\alpha = 0.01$ will simply be larger than the F-value for $\alpha = 0.05$.

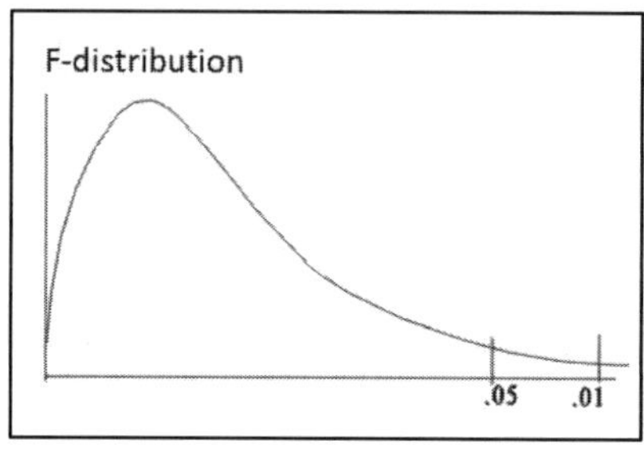

Analysis of "Trends" (Just a Special Kind of Contrasts)

Occasionally, researchers read about or wish to conduct analysis of trends. In the ANOVA context, a trend is just a special kind of contrast, a comparison that is made among group means where the groups are defined along a quantitative factor.

A "qualitative" factor is one in which the groups are defined by some nominal or categorical distinction. For example, factor A might be "teaching method," and its a=3 levels might be operationalized as: a=1 lessons on YouTube, a=2 self-guided, a=3 lectures. For such factors, the labels "a=1" through "a=3" are arbitrary; we could instead call the YouTube group #2 or #3. Wc could plot means using any order (and any spacing); on the left, the effect looks linear, on the right, non-monotonic. Yet these trends are not real because the ordering of the factor levels is arbitrary.

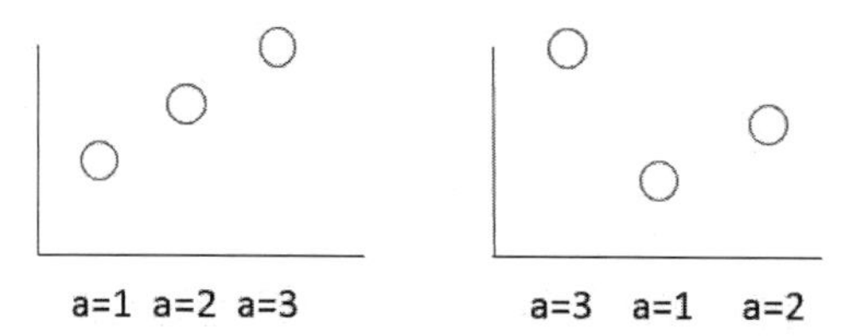

For a "quantitative" factor, the groups are defined by an underlying numeric ordering. For example, factor A might be "time allotted" to complete some task, and its levels might be operationalized as: a=1 60 seconds, a=2 90 seconds, and a=3 120 seconds. For such a factor, it is less arbitrary to order the groups from 1 to 3 (or 3 to 1, but not 2, 1, 3). If factor A is ordinal, then we would plot the groups in order:

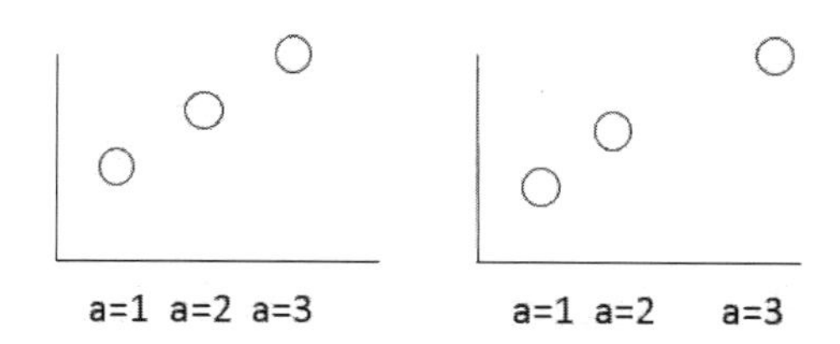

If factor A is "quantitative," say interval or ratio (e.g., with equally spaced intervals like 30-60-90), we could connect the dots and look for trends:

linear: quadratic:

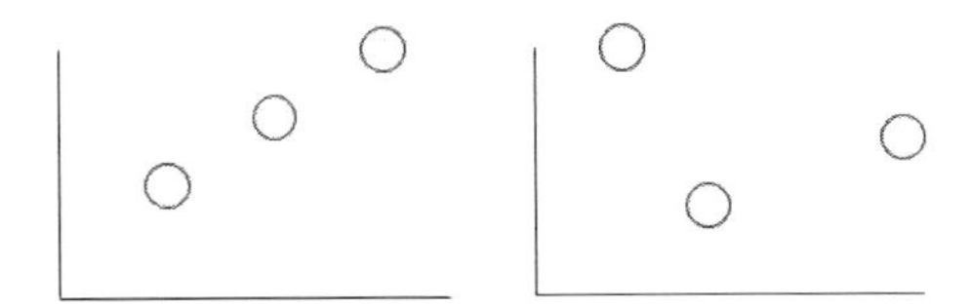

Specifically, for factor A with "a" levels, we can estimate a polynomial of (a-1) power. For example, if a=2, we can study a linear trend: $Y = b_0 + b_1 X$.

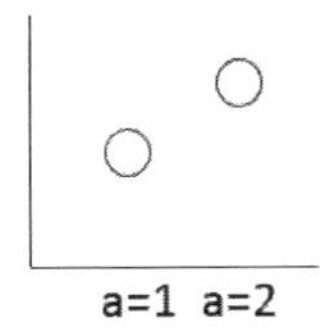

If a=3, we can study a linear and a quadratic trend: $Y = b_0 + b_1 X + b_2 X^2$.

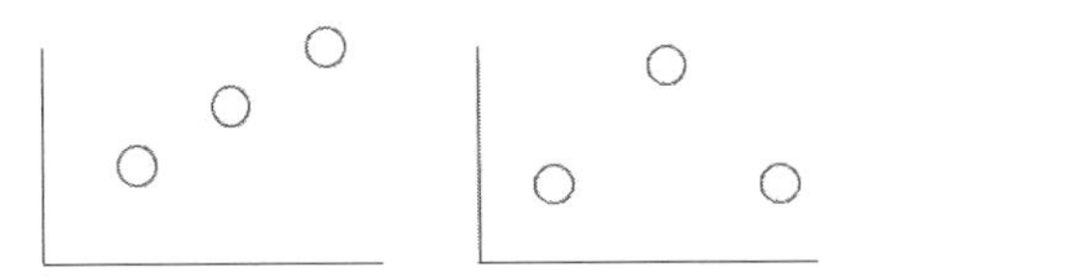

If, say, a=5, we can study a linear, quadratic, cubic, and quartic trend. The names describe the number of directions the curves takes, e.g., quartic:

We have to select trend coefficients carefully, otherwise, we proceed as we do for any other contrast. We'll define:

$$H_0\colon \psi_{trend} = 0 \qquad \hat{\psi} = \sum_{i=1}^{a} c_i \bar{x}_i \tag{19}$$

For example, for a=3, to test a linear trend and a quadratic trend, we'd use contrast coefficients:

		$\bar{x}_1$	$\bar{x}_2$	$\bar{x}_3$
linear trend	c_i's:	-1	0	1
quad		-1	2	-1

For a=5, the coefficients to create (a-1) = 4 orthogonal contrasts would be:

		$\bar{x}_1$	$\bar{x}_2$	$\bar{x}_3$	$\bar{x}_4$	$\bar{x}_5$
linear trend	c_i's:	-2	-1	0	1	2
quad		2	-1	-2	-1	2
cubic		-1	2	0	-2	1
quartic		1	-4	6	-4	1

We compute the SS as for other contrasts:

$$SS_{Atrend} = \frac{n(\hat{\psi}_{trend})^2}{\Sigma(c_i)^2} \tag{20}$$

In fitting a linear component and a quadratic component and any other, we are estimating the form of our data curve, but do not extrapolate beyond the curve (as if we had knowledge of what is going on in the data for points less than a=1 or greater than a=a).

The main point is, there's no mystery in trends; they're just another kind of contrast. We can test trends up to the (a-1) degree for a factor with "a" levels.

More Types of Comparisons

There are numerous types of contrasts. As a benchmark, recall that for a planned contrast, we calculated: $SS_{Acomp} = \frac{n(\hat{\psi})^2}{\Sigma(c_i)^2}$ and $F = \frac{SS_{Acomp}}{MS_{S(A)}}$ on 1, a(n-1) df.

A ***Dunnett*** test is applicable if we have "a-1" experimental groups and 1 "control" group and all we want to do is compare each treatment group to the standard base. We obtain the means, and we'd conclude that an experimental group mean is statistically significantly different from the control group mean if the difference between these two means exceeds this value (the bar represents means, the "d" is for the difference between two means, and the subscript of "D" is for Dunnett): $\bar{d}_D = q_D\sqrt{\frac{2 \times MS_{S(A)}}{n}}$. There are tables of critical values, q_D's (Dunnett 1964) but we'll obtain the tests easily from statistical computing packages like SAS.

A ***Tukey*** test is constructed to test all possible pairwise comparisons between the study's group means (Tukey 1949). This test is also referred to as Tukey's "honestly significant difference" (HSD) test (there must be a story to that name). When comparing each pair of means, we conclude that 2 group means are statistically significantly different if their mean difference exceeds: $\bar{d}_T = q_T\sqrt{\frac{MS_{S(A)}}{n}}$. Once again, tables of critical values, q_T's, exist, but we'll rely on SAS to do the work for us.

A ***Scheffé*** test allows us to test comparisons of any sort, not just pairs of means, but say, an average of two groups contrasted against a third, say. It is extremely flexible because it allows us to test a contrast of any kind, and it does so by being extremely conservative (Scheffé 1953). To use this approach, we'd calculate an F-test for a contrast, as usual, but rather than comparing to the standard critical F-value, call it F_{crit} (on 1 and a(n-1) df), we'd compare our observed F-test to a critical F-value defined as (a-1)F_{crit}. That is, for significance, our computed F-value must exceed: F_S = (a-1) F_{crit} (now on (a-1) and a(n-1) df). To see what a huge difference that can make, imagine a study with a=5 levels and n=9 subjects per cell. The regular critical F-value would be $F_{4,40,.05} = 2.61$, and the Scheffé adjustment takes it to F_S = (5-1)*2.61 = 10.44. This F-test is conservative (and fairly robust), but the most appropriate test in many circumstances, and many prefer it ("conservative" tests are macho in statistics).

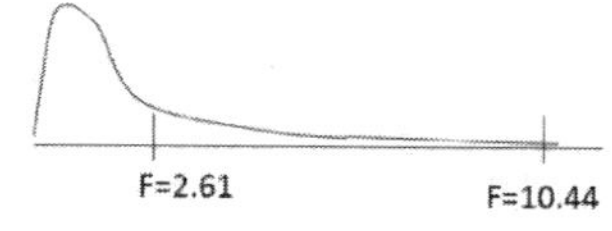

In comparing these 3 types of contrasts, Dunnett is the most sensitive (its critical difference value will be the smallest), because the test is designed only for one type of comparison—the scenario of comparing each of several experimental treatments to a control group. Tukey's test allows for more comparisons, e.g., all experimental treatments could also be compared to each other. The Scheffé test is the least sensitive (or the most conservative), meaning our F-value will have to be very large, because it's a test that can be used for all kinds of comparisons. Each of these tests may be obtained from packages like SAS (in SAS's proc glm, the means statement would end with the option "/ Dunnett Scheffe Tukey;").

IN HIGHER-ORDER FACTORIALS

Thus far, we have considered contrasts for a 1-way ANOVA, and the contrasts for main effects in a 2-way or higher-order ANOVA are run similarly. To run follow-up tests on a significant main effect for factor A, we'd define the contrast, make sure the coefficients sum to zero, compute the SS and MS and F-test, where the error term in the F-test is the same error term that was used to test the significant of that overall main effect:

$$\hat{\psi}_A = \sum c_i \bar{x}_i \qquad \left(\sum_{i=1}^{a} c_i = 0\right) \tag{21}$$

$$SS_{Acomp} = \frac{bn(\psi_A)^2}{\sum(c_i)^2}$$

$$MS_{Acomp} = \frac{SS_{Acomp}}{df_{Acomp}} = \frac{SS_{Acomp}}{1}$$

$$F_{Acomp} = \frac{MS_{Acomp}}{MS_{S(AB)}}$$

Similarly, for factor B: $\hat{\psi}_B = \sum c_j \bar{x}_j \qquad \left(\sum_{j=1}^{b} c_j = 0\right)$ (22)

$$SS_{Bcomp} = \frac{an(\psi_B)^2}{\sum(c_j)^2}$$

$$MS_{Bcomp} = \frac{SS_{Bcomp}}{df_{Bcomp}} = \frac{SS_{Bcomp}}{1}$$

$$F_{Bcomp} = \frac{MS_{Bcomp}}{MS_{S(AB)}}$$

Here too, $MS_{S(AB)}$ is the same error term that was used for the overall test of the main effect for B, in $F_B = \frac{MS_B}{MS_{S(AB)}}$.

"Simple Effects" are Contrasts for Interactions

What's new in a 2-way or higher-order factorial is how to conduct follow-up tests when the effect that is significant is an interaction term. Contrasts or follow-up tests on interactions in ANOVA are called "simple effects."

To understand simple effects, consider the following example. The table contains cell means from a 2×2 factorial. The means are then plotted, where one plot has factor A as the horizontal axis and factor B forms the profiles, and the other plot has the roles reversed. To facilitate the translation back and forth across the table and two forms of the plots, the cell means have been labeled.

	a=1	a=2
b=1	X = 4	Y = 6
b=2	Q = 3	W = 2

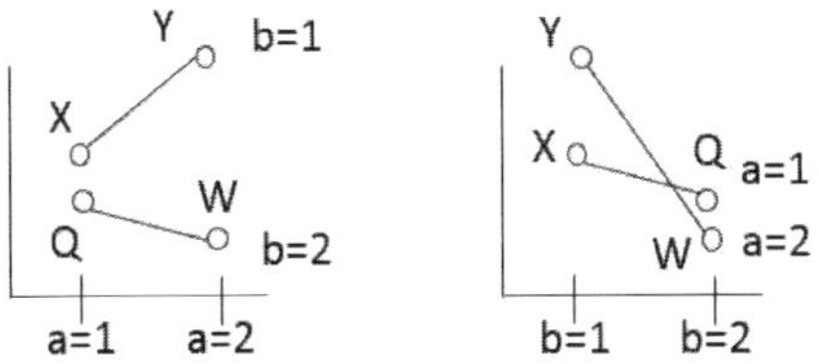

If we wish to test whether point X is different from point Y, notice it involves whether there is an effect of factor A but only in the b=1 conditions. This is called the simple effect of A at level 1 of B:

	a=1	a=2
b=1	4	6

If we wish to compare means Q and W, this is the simple effect of A at level 2 of B:

	a=1	a=2
b=2	3	2

Alternatively, we can study the simple effects of B at different levels of A, first comparing means X and Q, the simple effect of B at level 1 of A:

	a=1
b=1	4
b=2	3

Next, comparing means Y and W is the simple effect of B at level 2 of A:

	a=2
b=1	6
b=2	2

How to Test Simple Effects

To test the simple effects, we calculate:

$$SS_{A\ at\ b_j} = \frac{1}{n}\sum_i\left(\sum_{k=1}^{n} x_{ijk}\right)^2 - \frac{1}{an}\left(\sum_i\sum_k x_{ijk}\right)^2$$

Note the first piece sums over k so it's like the sums for each AB cell, and the second piece sums the data over i and k so it's the sums for each level j of B. That SS has $df_{A\ at\ b_j} = (a-1)$, so the F-test is computed as: $F = \frac{MS_{A\ at\ b_j}}{MS_{S(AB)}}$.

In our example, the cell means are:

	a=1	a=2
b=1	4	6
b=2	3	2

And say n=5, then the cell sums would be $\sum_{k=1}^{n} x_{ijk}$, that is, a sum for each (i,j) cell where we've summed over k (the n subjects within each i,j cell), so $20 = \sum_{k=1}^{n} x_{11k}$, $30 = \sum_{k=1}^{n} x_{21k}$, etc.:

	a=1	a=2
b=1	20	30
b=2	15	10

Then we'll use the SS equation above, but note the tiny notation differences. That equation had been written for the general case of some level of factor B "j" and here, we're specifically looking at j=1 and j=2. Thus, instead of x_{ijk} there are x_{i1k} and x_{i2k} terms in this first pair of SS (and x_{1jk} and x_{2jk} terms in the next pair of SS for specific levels of factor A). For the simple effects of factor A at each level of factor B:

$$SS_{A\ at\ b_1} = \frac{\sum_i(\sum_{k=1}^{n} x_{i1k})^2}{n} - \frac{(\sum_i \sum_k x_{i1k})^2}{an} = \frac{20^2+30^2}{5} - \frac{(20+30)^2}{2\times5} = 260 - 250 = 10.0$$

$$SS_{A\ at\ b_2} = \frac{\sum_i(\sum_{k=1}^{n} x_{i2k})^2}{n} - \frac{(\sum_i \sum_k x_{i2k})^2}{an} = \frac{15^2+10^2}{5} - \frac{(15+10)^2}{2\times5} = 65 - 62.5 = 2.5.$$

Similarly, for the simple effects of factor B at each level of factor A, we compute:

$$SS_{B\ at\ a_i} = \frac{1}{n}\sum_j \left(\sum_{k=1}^{n} x_{ijk}\right)^2 - \frac{1}{bn}\left(\sum_j \sum_k x_{ijk}\right)^2$$

Specifically, we'll calculate:

$$SS_{B\ at\ a_1} = \frac{\sum_j(\sum_{k=1}^{n} x_{1jk})^2}{n} - \frac{(\sum_j \sum_k x_{1jk})^2}{bn} = \frac{20^2+15^2}{5} - \frac{(20+15)^2}{2\times5} = 125 - 122.5 = 2.5$$

$$SS_{B\ at\ a_2} = \frac{\sum_j(\sum_{k=1}^{n} x_{2jk})^2}{n} - \frac{(\sum_j \sum_k x_{2jk})^2}{bn} = \frac{30^2+10^2}{5} - \frac{(30+10)^2}{2\times5} = 200 - 160 = 40.$$

df for Simple Effects

Each simple effect of factor A at each level of b, $SS_{A\ at\ b_j}$ has (a-1) df. Thus, for both simple effects of factor A at b=1 and b=2, together use 2 df in total. Analogously, together, the 2 simple effects of factor B at a=1 and a=2, $SS_{B\ at\ a_i}$, each has (b-1) df, again, for a total of 2 df.

How can this be? These tests are follow-ups to an interaction effect which has (a-1)(b-1) df, specifically 1 df in the 2×2 factorial example. Unlike contrasts in a 1-way ANOVA or contrasts for the main effects in a 2-way or higher-order ANOVA, where the SS_A are broken into pieces, when we study simple effects, we're not just breaking down df (or SS) of the interaction term alone.

In the example, looking at both simple effects of factor A for each level of factor B, at b=1 and b=2 sums up to the SS for the interaction but also in that aggregate is the SS for factor A: $\sum_j SS_{A\ at\ b_j} = SS_{AB} + SS_A$. Similarly, $\sum_i SS_{B\ at\ a_i} = SS_{AB} + SS_B$.

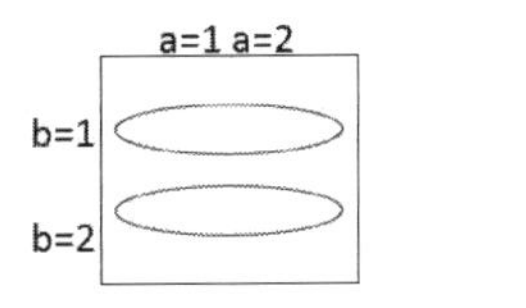

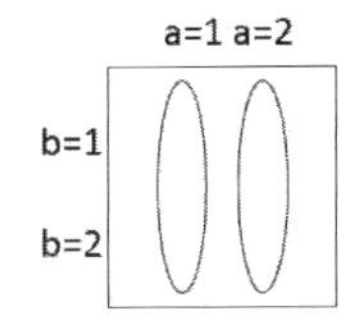

That's just stating the relationships, but here's how to understand why those equations hold. Imagine we had a data pattern like this parallel plot. Even if the interaction term was not significant ($SS_{AB} = 0$), and it doesn't look like it is (the lines are parallel), the simple effects could be significant if SS_A is significant (e.g., the SS for factor A might well be significant at b=1 and at b=2). That's what the equation means when it says that the 2 $SS_{A\ at\ b_j}$ add up to, or reflect, both SS_{AB} (which we expected) and also SS_A (which perhaps we hadn't).

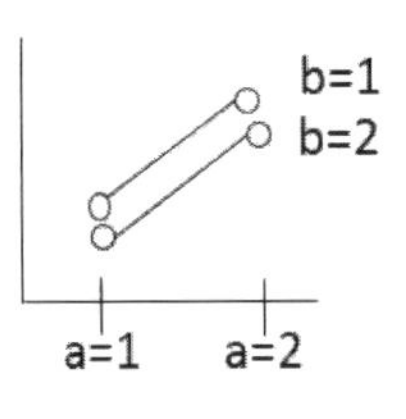

Accordingly, it's not a good idea to test for simple effects unless the interaction is significant. For interactions, we may well have expectations and explanations for why an interaction should be significant *a priori*, but we are not supposed to test the simple effects unless the interaction has been demonstrated empirically (*post hoc*) to be significant.

Computations on Single df Comparisons

Occasionally, if a factor involved in an interaction term has more than 2 levels, we'd have more than 1 df for the interaction term, so conducting the simple effect contrasts becomes slightly more complicated. Here is an example of a 3×2 factorial. We'll look at the simple effect of a contrast within a factor (say A) at each level of the other factor (B).

$$\hat{\psi}_{A\ at\ b_j} = \sum c_i \bar{x}_{ij} \qquad SS_{Acomp\ at\ b_j} = \frac{n\left(\psi_{A\ at\ b_j}\right)^2}{\Sigma(c_i)^2}$$

The contrast coefficients might be something like these c_i's:

	a=1	a=2	a=3
b=1	0	1	-1
b=2	0	0	0

Say we have cell means:

	a=1	a=2	a=3
b=1	5	4	6
b=2	8	3	2

Then the simple effect contrasts within factor A at level b=1 would be:

$$\hat{\psi}_{Acomp\ at\ b_1} = (0)(5) + (1)(4) + (-1)(6) = 0 + 4 - 6 = -2$$

$$SS_{Acomp\ at\ b_1} = \frac{5(-2)^2}{(0)^2+(1)^2+(-1)^2} = \frac{5\times 4}{0+1+1} = \frac{20}{2} = 10$$

$$F_{Acomp\ at\ b_1} = \frac{SS_{Acomp\ at\ b_1}}{MS_{S(AB)}}.$$

Alternatively, on the flip side, we might examine a contrast effect of factor B for each level of factor A (in the example, with only 2 levels of factor B, these tests would be the same as the simple effects already discussed):

$$\hat{\psi}_{B\ at\ a_i} = \sum c_j \bar{x}_{ij} \qquad SS_{Bcomp\ at\ a_i} = \frac{n\left(\psi_{B\ at\ a_i}\right)^2}{\Sigma\left(c_j\right)^2}.$$

Say:

	a=1	a=2
b=1	-1	0
b=2	1	0

Or

	a=1	a=2
b=1	0	1
b=2	0	-1

(These coefficients don't have to sum in all directions because a simple effect is not actually examining all cells, only those in 1 level of A, in this case.)

EXTEND TO THREE-WAY ANOVA

In this final section, let's extend the logic and math to a 3-factor ANOVA. We have factors: A, B, and C. We know how to test contrasts on the main effects (A, B, C) and the 2-way interactions (A×B, A×C, B×C). The new term is the 3-way interaction, and the question is how to test for simple effects when the A×B×C term is significant.

There are several patterns that might give rise to a significant 3-way interaction. Let's say it appears that maybe the nature of the A×B interaction seems to differ depending on the level of factor C. (This investigation would be a little like running a 2-factor ANOVA and looking at A×B for the c = 1 data and another A×B for the c = 2 data, but not quite because we need to use the proper error term, which is $MS_{S(ABC)}$ not $MS_{S(AB)}$.

First, $df_{AB\ at\ c_k} = (a-1)(b-1)$.

Next, $SS_{AB\ at\ c_k} = \frac{\Sigma\left(\Sigma_{\ell=1}^{n} x_{ijk\ell}\right)^2}{n} - \frac{\Sigma\left(\Sigma_j \Sigma_\ell x_{ijk\ell}\right)^2}{bn} - \frac{\Sigma\left(\Sigma_i \Sigma_\ell x_{ijk\ell}\right)^2}{an} + \frac{\left(\Sigma(\Sigma_\ell x_{ijk\ell})\right)^2}{abn}$.

In that SS equation, the first term sums over ℓ, so it's the squared sums in each cell for ABC (call it $\frac{\Sigma\left(ABC_{ijk}\right)^2}{n}$). The second term sums over j and ℓ, so those are squared sums for AC (say $\frac{\Sigma(AC_{ik})^2}{bn}$). The third term is for BC (say $\frac{\Sigma\left(BC_{jk}\right)^2}{an}$), and the final term takes the ABC sums and then squares the whole (call the term $\frac{\left(ABC_{ijk}\right)^2}{abn}$).

The sum of these simple effects equal the SS for the 3-way interaction term that we're trying to understand, as well as the SS_{AB} term (remember: the key is always to look at the subscripts, and here we are examining the simple effects of "AB at c_k" hence it's the A×B sums of squares that are added into the 3-way interaction SS): $\sum_{k=1}^{C} SS_{A\times B\ at\ c_k} = SS_{ABC} + SS_{AB}$.

An Example

This is getting a little hairy, so an example might help. The following data come from a 3×2×2 factorial, that is, a=3, b=2, c=2, with cell size n=4 (these data are modified from a subset of data in Keppel and Wickens, p.484).

	a1		a2		a3	
	b1	b2	b1	b2	b1	b2
c1	7	2	10	4	13	9
	4	4	7	6	10	8
	5	3	6	3	13	9
	6	3	8	5	8	10
	7	7	6	5	12	4
c2	5	6	5	8	13	6
	5	5	5	7	11	4
	6	5	6	6	12	5

The ANOVA table indicates that the main effects for factors A and B are significant, as are the A×B, A×C, and A×B×C interactions:

Source	df	SS	MS	F	p-value	big?
A	2	151.625	75.812	47.88	<.0001	←*
B	1	65.333	65.333	41.26	<.0001	←*
C	1	0.083	0.083	0.05	0.8198	
A×B	2	31.542	15.771	9.96	0.0004	←*
A×C	2	19.542	9.771	6.17	0.0050	←*
B×C	1	0.750	0.750	0.47	0.4957	
A×B×C	2	51.125	25.562	16.14	<.0001	←*
error	36	57.000	1.583			
total	47	377.000				*p<.05

Per a standard analysis, if these were our data, let's look at the means and contrasts for all the significant effects. Beginning with the main effects, for factor A, the means are:

$$\text{A:}\ \bar{x}_{a1} = 5.000, \qquad \bar{x}_{a2} = 6.062, \qquad \bar{x}_{a3} = 9.187$$

It looks like $\bar{x}_{a3} >>> \bar{x}_{a1}$ and $\bar{x}_{a2}$, so let's define an appropriate contrast:

$$H_0\colon \psi = \frac{\mu_1 + \mu_2}{2} - \mu_3 = 0$$

Test it:

$$\hat{\psi} = \frac{1}{2}(5.000) + \frac{1}{2}(6.062) - (9.187) = -3.656$$

$$SS_{Acomp1} = \frac{nbc(\psi)^2}{\Sigma(c_i)^2} = \frac{4 \times 2 \times 2(-3.656)^2}{\left(\frac{1}{2}\right)^2 + \left(\frac{1}{2}\right)^2 + (-1)^2} = \frac{213.861}{1.5} = 142.57$$

$$F_{Acomp1} = \frac{SS_{Acomp1}/1}{MS_{S(ABC)}} = \frac{142.57}{1.583} = 90.07$$

Our F-value is big, so we reject H_0 and conclude that indeed, the 3rd mean is significantly larger than the 1st and 2nd means.

Continuing with factor A, now we know that $\bar{x}_{a3} > \frac{\bar{x}_{a1}+\bar{x}_{a2}}{2}$ and the next step is to determine whether $\bar{x}_{a2} > \bar{x}_{a1}$; the 2nd mean is larger than the first, but is it significantly larger? We test the contrast hypothesis:

$$H_0\colon \psi = \mu_1 - \mu_2 = 0$$

By computing:

$$\hat{\psi} = 1(5.000) - 1(6.062) + 0(9.187) = -1.062$$

$$SS_{Acomp2} = \frac{4 \times 2 \times 2(-1.062)^2}{(1)^2 + (-1)^2 + (0)^2} = \frac{18.046}{2} = 9.023$$

$$F_{Acomp2} = \frac{SS_{Acomp2}/1}{MS_{S(ABC)}} = \frac{9.023}{1.583} = 5.70$$

This F-value is also significant, so we reject this H_0 as well. Thus, we know that all 3 means on factor A are significantly different.

Next, the main effect for factor B was significant, so we would proceed to run contrasts on it as we just did for factor A. However, in this case we don't need to because there are only 2 groups, b=2, or there is only 1 df, so we're done. We see the means are: $\bar{x}_{b1} = 7.917$, $\bar{x}_{b2} = 5.583$, so we can conclude that the b=1 group mean is significantly larger than the b=2 group mean (well, strictly speaking, we conclude that $\bar{x}_{b1} \neq \bar{x}_{b2}$, but duh, all we have to do is look at the means and if they're different, we can see that 7.92>5.58).

Thus, the following step is unnecessary, but just to verify that in a 1 df effect (in this case, the main effect for B), the 1 df contrast would yield the same information, let's test the null, H_0: $\psi = \mu_1 - \mu_2 = 0$.

$$\hat{\psi} = 1(7.917) - 1(5.583) = 2.334$$

$$SS_{Bcomp1} = \frac{nac(\psi)^2}{\Sigma(c_j)^2} = \frac{4 \times 3 \times 2(2.334)^2}{(1)^2 + (-1)^2} = 65.37$$

$$F_{Bcomp1} = \frac{SS_{Bcomp1}/1}{MS_{S(ABC)}} = \frac{65.37}{1.583} = 41.30$$

The values of SS = 65.37 and F = 41.30 are equal to those in the ANOVA table, within rounding error.

The next significant effect is the A×B interaction. The means are in this table (and we should plot them):

	b=1	b=2
a=1	5.625	4.375
a=2	6.625	5.500
a=3	11.500	6.875

Given those means and knowledge that n=4 and c=2, the cell sums are these:

	b=1	b=2
a=1	45	35
a=2	53	44
a=3	92	55

To examine the simple effects of factor B at each level of factor A:

$$SS_{B\ at\ a1} = \left[\frac{45^2 + 35^2}{2 \times 4}\right] - \left[\frac{(45 + 35)^2}{2 \times 2 \times 4}\right] = 6.25$$

And we divide by the MS_{error}=1.583, to see if the simple effect is significant or not; doing so, we obtain F=3.948. Next:

$$SS_{B\ at\ a2} = \left[\frac{53^2 + 44^2}{8}\right] - \left[\frac{(53 + 44)^2}{16}\right] = 5.0625, \ \ F = 3.198$$

$$SS_{B\ at\ a3} = \left[\frac{92^2 + 55^2}{8}\right] - \left[\frac{(92 + 55)^2}{16}\right] = 85.5625, \ \ F = 54.051$$

Each of those 3 simple effects has (b-1) = 1 df, and 36 error df.

Note that the sum of the 3 simple effects indeed add up to the interaction SS along with the SS for B:

$$\sum_{i=1}^{a} SS_{B\ at\ ai} = SS_B + SS_{AB} = 6.25 + 5.06 + 85.56 = 96.87 = 65.33 + 31.54$$

We would test the simple effects on the A×C similarly. Finally, we'll test the simple effects on the 3-way A×B×C interaction. We might be curious as to whether the nature of the A×B relationship is different depending on levels of C (e.g., for c=1 maybe A×B is an interaction, whereas for c=2 maybe A×B is an interaction of a different nature or the interaction there is not significant), or, given that we just looked at the effect of B at each level of a_i, let's look at the B×C interaction at each level of a_i.

The 3-way cell means are:

	b1		b2	
	c1	c2	c1	c2
a1	5.50	5.75	3.00	5.75
a2	7.75	5.50	4.50	6.50
a3	11.00	12.00	9.00	4.75

So the cell sums are (given that n=4):

	b1		b2	
	c1	c2	c1	c2
a1	22	23	12	23
a2	31	22	18	26
a3	44	48	36	19

Those particular simple effects SS, each on (b-1)(c-1) = 1 df are:

$$SS_{B\times C\ at\ a1} = \frac{\Sigma\left(ABC_{1jk}\right)^2}{n} - \frac{\Sigma\left(AB_{1j}\right)^2}{cn} - \frac{\Sigma(AC_{1k})^2}{bn} + \frac{\left(ABC_{1jk}\right)^2}{bcn}$$

$$= \left[\frac{(22^2+23^2+12^2+23^2)}{4}\right] - \left[\frac{(22+23)^2+(12+23)^2}{2\times 4}\right]$$

$$- \left[\frac{(22+12)^2+(23+23)^2}{2\times 4}\right] + \left[\frac{(22+23+12+23)^2}{2\times 2\times 4}\right]$$

$$= 421.5 - 406.25 - 409.0 + 400.0 = 6.25$$

And $SS_{B\times C\ at\ a2} = 18.06$, and $SS_{B\times C\ at\ a3} = 27.56$.

That concludes the analysis of the 3-way factorial. It's a little hairy but the logic is straightforward and we can thank our lucky stars for computers to do the work for us.

Signs and Values of the Contrast Coefficients

Before leaving this chapter, let's close out one final issue. Earlier, I had promised that the scale of the c's didn't matter, right? So check this out.

From the previous example, the means on factor A were:

$$A\!: \ \bar{x}_{a1} = 5.000, \qquad \bar{x}_{a2} = 6.062, \qquad \bar{x}_{a3} = 9.187$$

And we want to compare groups 1 and 2 vs. group 3. In the example, we had selected contrast coefficients of $.5, .5, -1$:

$$\hat{\psi} = \frac{1}{2}(5.000) + \frac{1}{2}(6.062) - (9.187) = -3.656$$

And we got an SS of:

$$SS_{Acomp1} = \frac{nbc(\psi)^2}{\sum(c_i)^2} = \frac{4 \times 2 \times 2(-3.656)^2}{\left(\frac{1}{2}\right)^2 + \left(\frac{1}{2}\right)^2 + (-1)^2} = \frac{213.861}{1.5} = 142.57.$$

What happens if we change the c's? It don't make no nevermind. Say we are still comparing groups 1 and 2 vs. group 3, but instead of using contrast coefficients of $.5, .5, -1$, we used 1, 1, -2. We'd get:

$$\hat{\psi} = 1(5.000) + 1(6.062) - 2(9.187) = -7.312$$

That's different, but in the SS, the coefficients are corrected out in the denominator:

$$SS_{Acomp1} = \frac{nbc(\psi)^2}{\sum(c_i)^2} = \frac{4 \times 2 \times 2(-7.312)^2}{(1)^2 + (1)^2 + (-2)^2} = \frac{855.446}{6} = 142.57.$$

That's the same SS!

And say we went bananas and used coefficients: -50, -50, 100, we'd get:

$$\hat{\psi} = -50(5.000) - 50(6.062) + 100(9.187) = -250 - 303.1 + 918.7 = 365.6$$

That's wildly different, but again, the SS equation compensates:

$$SS_{Acomp1} = \frac{nbc(\psi)^2}{\sum(c_i)^2} = \frac{4 \times 2 \times 2(365.6)^2}{(-50)^2 + (-50)^2 + (100)^2} = \frac{2{,}138{,}613.76}{15{,}000} = 142.57.$$

Those are the same SS. Way cool, eh?

SUMMARY

This chapter presented the logic and math of contrasts for main effects and "simple effects" (contrasts) for interactions. The appendix shows how to run simple effects in SAS.

REFERENCES

- Dunnett, Charles W. (1964), "New Tables for Multiple Comparisons with a Control," *Biometrics*, 20 (3), 482-491.
- Scheffé, H. (1953), "A Method for Judging All Contrasts in the Analysis of Variance," *Biometrika*, 40 (1-2), 87-110.
- Tukey, John W. (1949), "Comparing Individual Means in the Analysis of Variance," *Biometrics*, 5 (2), 99-114.

APPENDIX:
RUNNING SIMPLE EFFECTS FOR 2×2'S AND 2×2×2'S IN SAS

When running simple effects (a.k.a. interaction contrasts) in SAS, we have to modify how we specify the contrast statements. Why? Contrast coefficients don't sum to 0 in all directions across a 2-way table. For example, in this 2-way table, we see the coefficients to specify a simple effect of B with a=1. The column sums are fine, but the row sums do not equal 0.

	a=1	a=2	
b=1	-1	0	$\Sigma = -1$
b=2	1	0	$\Sigma = 1$
	$\Sigma = 0$	$\Sigma = 0$	

When the coefficients don't sum properly, SAS will give an error message saying "non-est" (the contrast specified is "non-estimable"; i.e., SAS is confused), and some programs just won't print the contrasts, with no error message or warning.

Or, say the data look like this plot, where 3 of the means look the same, and 1 looks different, it's tempting to think the contrast coefficients would be .33, .33, .33, -1, but they won't work because they don't sum to 0 in all directions.

	a=1	a=2
b=1	.33	.33
b=2	.33	-1

Instead, in such a scenario, we'd test these simple effects: 1) B@a1 (expecting it to be n.s.) and 2) B@a2 (expecting it to be significant). Or, depending on which pair of simple effects is more interesting, or easier to understand, theoretically, we'd test these simple effects: 1) A@b1 (expecting it to be significant), and 2) A@b2 (expect it to be n.s.).

2×2 Simple Effects in SAS

This figure depicts 4 means, as would be obtained from a 2×2 ANOVA. Across the 4 means, there are several simple effects that we might wish to test. We might wish to compare means 1 and 3, and then 2 and 4. Alternatively, we might wish to contrast means 1 and 2, and then 3 and 4.

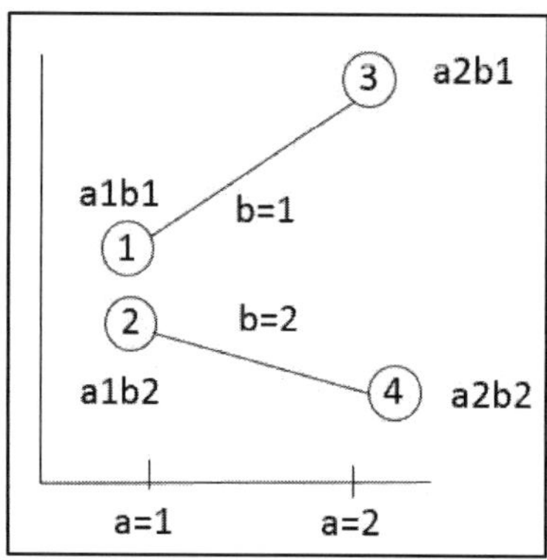

When comparing means 1 and 3, that's the simple effect of the factor A at level b1. To do so in SAS, we'd follow the means or model statement with:

contrast 'a @ b1' a 1 -1 a*b 1 0 -1 0;

To test the simple effect of A at b2 (means 2 vs. 4), we'd say:

contrast 'a @ b2' a 1 -1 a*b 0 1 0 -1;

We could verify: $SS_{A@b1} + SS_{A@b2} = SS_{AB} + SS_A$.

Alternatively, if we sought to contrast means 1 and 2, we'd be testing the simple effect of B at a1: contrast 'b @ a1' b 1 -1 a*b 1 -1 0 0;

To test the simple effect of B at a2 (means 3 vs. 4): contrast 'b @ a2' b 1 -1 a*b 0 0 1 -1;

And again, we can check: $SS_{B@a1} + SS_{B@a2} = SS_{AB} + SS_B$.

Numerical Example

So let's see how to specify contrast coefficients in SAS for simple effects. Imagine a 2×2 factorial design, like the data below (with n=6 subjects per cell):

```
data my2x2; input
subjid  a  b     x;  datalines;
     1  1  1     4
     2  1  1     5
     3  1  1     6
     4  1  1     4
     5  1  1     5
     6  1  1     6
     7  1  2     6
     8  1  2     7
     9  1  2     8
    10  1  2     3
    11  1  2     4
    12  1  2     5
    13  2  1     6
    14  2  1     7
    15  2  1     8
    16  2  1     5
    17  2  1     6
    18  2  1     7
    19  2  2     2
    20  2  2     3
    21  2  2     4
    22  2  2     2
    23  2  2     3
    24  2  2     4
proc glm; class a b; model x = a b a*b /ss3;
means a b a*b; *the contrast statements for testing the simple effects follow;
contrast 'a at b1' a 1 -1 a*b 1 0 -1 0; contrast 'a at b2' a 1 -1 a*b 0 1 0 -1;
contrast 'b at a1' b 1 -1 a*b 1 -1 0 0; contrast 'b at a2' b 1 -1 a*b 0 0 1 -1;
```

For these sample data, the SAS output is as follows.

General Linear Models Procedure

Class Level Information

Class	Levels	Values
A	2	1 2
B	2	1 2

Number of observations in data set = 24

Source	DF	Sum of Squares	Mean Square	F Value	Pr > F
Model	3	39.000	13.000	8.39	0.0008
Error	20	31.000	1.550		
Corrected Total	23	70.000			

R-Square	C.V.	Root MSE	x Mean
0.557	24.8998	1.24499	5.000

Source	DF	Type III SS	Mean Square	F Value	Pr > F
A	1	1.500	1.500	0.97	0.3370
B	1	13.500	13.500	8.71	0.0079
A*B	1	24.000	24.000	15.48	0.0008

Level of A	N	---------------X-------------- Mean	SD
1	12	5.2500	1.4222
2	12	4.7500	2.0505

Level of B	N	---------------X-------------- Mean	SD
1	12	5.7500	1.2154
2	12	4.2500	1.9129

Level of A	Level of B	N	------------X----------- Mean	SD
1	1	6	5.000	0.894
1	2	6	5.500	1.871
2	1	6	6.500	1.049
2	2	6	3.000	0.894

Dependent Variable: X

Contrast	DF	Contrast SS	Mean Square	F Value	Pr > F
a at b1	1	6.750	6.750	4.35	0.0499
a at b2	1	18.750	18.750	12.10	0.0024
b at a1	1	0.750	0.750	0.48	0.4947
b at a2	1	36.750	36.750	23.71	<.0001

2×2×2 Simple Effects in SAS

Next, we'll consider a 2×2×2 factorial design, like the data that follow, with n=3 observations per cell.

```
data my2x2x2; input
subjid  a  b   c   x; datalines;
    1   1  1   1   4
    2   1  1   1   5
    3   1  1   1   6
    4   1  1   2   4
    5   1  1   2   5
```

```
 6  1  1  2  6
 7  1  2  1  6
 8  1  2  1  7
 9  1  2  1  8
10  1  2  2  3
11  1  2  2  4
12  1  2  2  5
13  2  1  1  6
14  2  1  1  7
15  2  1  1  8
16  2  1  2  5
17  2  1  2  6
18  2  1  2  7
19  2  2  1  2
20  2  2  1  3
21  2  2  1  4
22  2  2  2  2
23  2  2  2  3
24  2  2  2  4
proc glm; class a b c; model x = a b c a*b a*c b*c a*b*c /ss3; means a b c a*b a*c b*c a*b*c;
*the contrast statements for testing the simple effects follow;
```

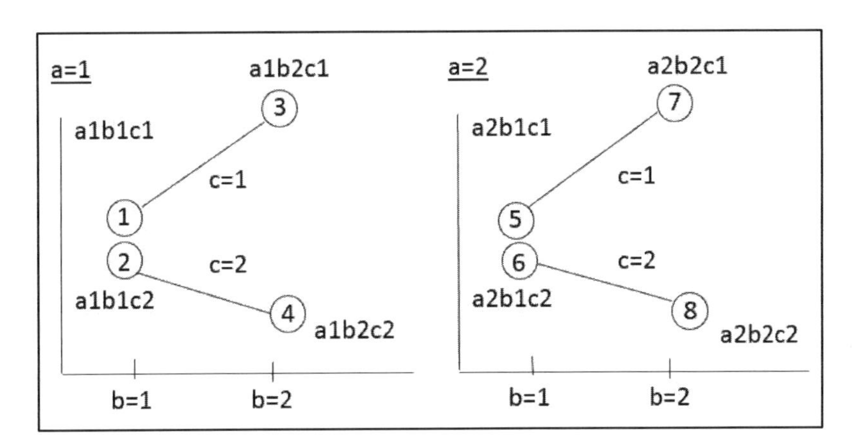

As this figure reflects, things are more complicated. We have 8 means in a 3-way interaction for a 2×2×2 factorial. In it, we can test 2 different kinds of simple effects.

First, we can test whether the effect of A×B varies at different levels of Ck. To test the simple effect of A×B at C1 (means 1, 3, 5, 7):
contrast 'axb @c1' a*b 1 -1 -1 1 a*b*c 1 0 -1 0 -1 0 1 0;
To test the simple effect of A×B at C2 (means 2, 4, 6, 8):
contrast 'axb @c2' a*b 1 -1 -1 1 a*b*c 0 1 0 -1 0 -1 0 1;
And we can verify that $SS_{AB@C1} + SS_{AB@c2} = SS_{AB} + SS_{ABC}$.

To test the simple effect of A×C at B1 (means 1, 2, 5, 6):
contrast 'axc @b1' a*c 1 -1 -1 1 a*b*c 1 -1 0 0 -1 1 0 0;
To test the simple effect of A×C at B2 (means 3, 4, 7, 8):
contrast 'axc @b2' a*c 1 -1 -1 1 a*b*c 0 0 1 -1 0 0 -1 1;

And we'd see that $SS_{AC@b1} + SS_{AC@b2} = SS_{AC} + SS_{ABC}$.

As the last option, to test the simple effect of B×C at A1 (means 1, 2, 3, 4):
contrast 'bxc @a1' b*c 1 -1 -1 1 a*b*c 1 -1 -1 1 0 0 0 0;
To test the simple effect of B×C at A2 (means 5, 6, 7, 8):
contrast 'bxc @a2' b*c 1 -1 -1 1 a*b*c 0 0 0 0 1 -1 -1 1;
And we'd see that $SS_{BC@a1} + SS_{BC@a2} = SS_{BC} + SS_{ABC}$.

Alternatively, we can test a different kind of simple effect, namely, we can see whether a main effect (for A, B, or C) varies depending on the values of the other two factors. To test the simple effect of A at b1c1 (means 1 vs. 5):
contrast 'a@ b1c1' a 1 -1 a*b 1 0 -1 0 a*c 1 0 -1 0 a*b*c 1 0 0 0 -1 0 0 0;
To test the simple effect of A at b1c2 (means 2 vs. 6):
contrast 'a@ b1c2' a 1 -1 a*b 1 0 -1 0 a*c 0 1 0 –1 a*b*c 0 1 0 0 0 -1 0 0;
To test the simple effect of A at b2c1 (means 3 vs. 7):
contrast 'a@ b2c1' a 1 -1 a*b 0 1 0 -1 a*c 1 0 -1 0 a*b*c 0 0 1 0 0 0 -1 0;
To test the simple effect of A at b2c2 (means 4 vs. 8):
contrast 'a@ b2c2' a 1 -1 a*b 0 1 0 -1 a*c 0 1 0 -1 a*b*c 0 0 0 1 0 0 0 -1;
And they would sum properly:
$SS_{A@b1c1} + SS_{A@b1c2} + SS_{A@b2c1} + SS_{A@b2c2} = SS_A + SS_{AB} + SS_{AC} + SS_{ABC}$.

Analogously, to test the simple effect of B at a1c1 (means 1 vs. 3):
contrast 'b @ a1c1' b 1 -1 a*b 1 -1 0 0 b*c 1 0 -1 0 a*b*c 1 0 -1 0 0 0 0 0;
To test the simple effect of B at a1c2 (means 2 vs. 4):
contrast 'b @ a1c2' b 1 -1 a*b 1 -1 0 0 b*c 0 1 0 -1 a*b*c 0 1 0 -1 0 0 0 0;
To test the simple effect of B at a2c1 (means 5 vs. 7):
contrast 'b @ a2c1' b 1 -1 a*b 0 0 1 -1 b*c 1 0 -1 0 a*b*c 0 0 0 0 1 0 -1 0;
To test the simple effect of B at a2c2 (means 6 vs. 8):
contrast 'b @ a2c2' b 1 -1 a*b 0 0 1 -1 b*c 0 1 0 -1 a*b*c 0 0 0 0 0 1 0 -1;
They would sum:
$SS_{B@a1c1} + SS_{B@a1c2} + SS_{B@a2c1} + SS_{B@a2c2} = SS_B + SS_{AB} + SS_{BC} + SS_{ABC}$.

As the last permutation, to test the simple effect of C at a1b1 (means 1 vs. 2):
contrast 'c @ a1b1' c 1 -1 a*c 1 -1 0 0 b*c 1 -1 0 0 a*b*c 1 -1 0 0 0 0 0 0;
To test the simple effect of C at a1b2 (means 3 vs. 4):
contrast 'c @ a1b2' c 1 -1 a*c 1 -1 0 0 b*c 0 0 1 -1 a*b*c 0 0 1 -1 0 0 0 0;
To test the simple effect of C at a2b1 (means 5 vs. 6):
contrast 'c @ a2b1' c 1 -1 a*c 0 0 1 -1 b*c 1 -1 0 0 a*b*c 0 0 0 0 1 -1 0 0;
To test the simple effect of C at a2b2 (means 7 vs. 8):
contrast 'c @ a2b2' c 1 -1 a*c 0 0 1 -1 b*c 0 0 1 -1 a*b*c 0 0 0 0 0 0 1 -1;
And: $SS_{C@a1b1} + SS_{C@a1b2} + SS_{C@a2b1} + SS_{C@a2b2} = SS_C + SS_{AC} + SS_{BC} + SS_{ABC}$.

Source	DF	Type III SS	Mean Square	F Value	Pr > F
A	1	1.500	1.500	1.50	0.2384
B	1	13.500	13.500	13.50	0.0021
C	1	6.000	6.000	6.00	0.0262
A*B	1	24.000	24.000	24.00	0.0002
A*C	1	1.500	1.500	1.50	0.2384
B*C	1	1.500	1.500	1.50	0.2384
A*B*C	1	6.000	6.000	6.00	0.0262
Error	16	16.000	1.000		

Contrast	DF	Contrast SS	Mean Square	F Value	Pr > F
A at b1	1	6.750	6.750	6.75	0.0194
A at b2	1	18.750	18.750	18.75	0.0005
B at a1	1	0.750	0.750	0.75	0.3993
B at a2	1	36.750	36.750	36.75	<.0001
AxB @c1	1	27.000	27.000	27.00	<.0001
AxB @c2	1	3.000	3.000	3.00	0.1025
B @ a1c1	1	6.000	6.000	6.00	0.0262
B @ a1c2	1	1.500	1.500	1.50	0.2384
B @ a2c1	1	24.000	24.000	24.00	0.0002
B @ a2c2	1	13.500	13.500	13.50	0.0021

Simple Effects for More Complicated Designs

While the previous SAS commands are the official way of testing for simple effects in 2×2 and 2×2×2 factorials, when all else fails, there is a mechanical, if less elegant way, to test contrasts throughout a design. In a pinch, we can simply create a new variable with the requisite number of levels, and compare means across those levels, as if the design had been a big 1-factor design. So for a 2×2, we'll create a new variable with 4 levels. For example, here is a 2×2.

```
data my2x2; input subjid a b x;
if a=1 and b=1 then newvar=1;
if a=1 and b=2 then newvar=2;
if a=2 and b=1 then newvar=3;
if a=2 and b=2 then newvar=4;
datalines;
        1  1  1    4
        2  1  1    5
        3  1  1    6
        4  1  1    4
        etc., the rest of the data go here…
```

Then continuing in SAS: proc glm; class newvar; model x = newvar;
contrast 'a at b1' 1 0 -1 0; *this example tests the simple effect of a at b1;
That contrast command will calculate the SS for the numerator of the contrast F-value. Double check that the error MS for the "newvar" analysis is equal to the $MS_{S(AB)}$ from the ANOVA in which factors A and B were properly treated as a 2×2 (if not, divide the newvar SS by the $MS_{S(AB)}$). For this "new variable" example, SAS generates the following.

Source	DF	Type III SS	Mean Square	F Value	Pr > F
newvar	3	39.000	13.000	8.39	0.0008
Error	20	31.000	1.550		

Level of newvar		N (X)	Mean (X)	SD (X)
1	1	6	5.000	0.894
2	2	6	5.500	1.871
3	1	6	6.500	1.049
4	2	6	3.000	0.894

Contrast	DF	Contrast SS	Mean Square	F Value	Pr > F
a at b1	1	6.750	6.750	4.35	0.0499
a at b2	1	18.750	18.750	12.10	0.0024
b at a1	1	0.750	0.750	0.48	0.4947
b at a2	1	36.750	36.750	23.71	<.0001

Note that these results are identical to those for the 2×2 analysis above. It works!

Analogously for bigger designs, string out a 2×2×2 as a vector:

```
data strung; input x a b c x;
if a=1 and b=1 and c=1 then new=1;
if a=1 and b=1 and c=2 then new=2;
if a=1 and b=2 and c=1 then new=3;
if a=1 and b=2 and c=2 then new=4;
if a=2 and b=1 and c=1 then new=5;
if a=2 and b=1 and c=2 then new=6;
if a=2 and b=2 and c=1 then new=7;
if a=2 and b=2 and c=2 then new=8;
datalines;
1 1 1 1 4
2 1 1 1 5
3 1 1 1 6
4 1 1 2 4 etc…
proc glm; class new; model x = new; means new;
contrast 'b at a1c1' new 1 0 -1 0  0 0 0 0; contrast 'b at a1c2' new 0 1 0 -1  0 0 0 0;
contrast 'b at a2c1' new 0 0 0 0  1 0 -1 0; contrast 'b at a2c2' new 0 0 0 0  0 1 0 -1;
contrast 'axb at c1' new 1 0 -1 0 -1 0 1 0; contrast 'axb at c2' new 0 1 0 -1 0 -1 0 1; run;
```

Contrast	DF	Contrast SS	Mean Square	F Value	Pr > F
B @ a1c1	1	6.000	6.000	6.00	0.0262
B @ a1c2	1	1.500	1.500	1.50	0.2384
B @ a2c1	1	24.000	24.000	24.00	0.0002
B @ a2c2	1	13.500	13.500	13.50	0.0021
AxB @c1	1	27.000	27.000	27.00	<.0001
AxB @c2	1	3.000	3.000	3.00	0.1025

These contrasts also replicate the results above, on the 2×2×2 factorial.

HOMEWORK

This HW covers contrasts.

Problem 1. In a lot of medical experiments, there will be experimental groups comprised of new drugs purporting to be a superior treatment of some condition, and the study also includes a "control group" not in which nothing happens, rather in which the current status quo best currently optimal treatment serves as a benchmark. In business, this situation can also be found, for example, a brand manager for some company can think of "the currently running ad campaign" as a control group, and then investigate whether newly proposed ad campaigns might be more persuasive (i.e., more effective in enhancing brand attitudes or even increasing sales). Say the situation is that in addition to a "control group" (business as usual), the company's ad agency is proposing to use a celebrity endorser (which the brand manager is worried about because celebrities can be both expensive and crazy), and whether the ad appeal is cognitive (our brand is better than the competition for these reasons) or emotional (our brand makes you feel awesome and will improve your life, making you richer and better looking).

This design is referred to as a "2×2 +1" factorial (literally read as "2 by 2 plus one") because it is a 2×2 structure with the addition of another cell. Logically the design is best drawn out as follows (this layout is important for analyses too, given that the "control" group doesn't fit into the factorial and will confuse the computer):

control	No celeb Cognitive	No celeb Emotional	Celebrity Cognitive	Celebrity Emotional
a1	a2	a3	a4	a5

Construct a set of coefficients that will provide the following comparisons:

i) Control vs. the combined new ad options (the experimental groups).
ii) Celebrity endorser vs. no celebrity.
iii) Cognitive vs. emotional for no celebrity.
iv) Cognitive vs. emotional with a celebrity.

Show that the 4 comparisons are mutually orthogonal.

Problem 2. Verify that the 4 trend contrasts for a=5 are mutually orthogonal (we saw those coefficients in the contrasts chapter but they are repeated here for your convenience):

a=5 trend coefficients:

a=5 ci's	$\bar{y}_1$	$\bar{y}_2$	$\bar{y}_3$	$\bar{y}_4$	$\bar{y}_5$
linear	-2	-1	0	1	2
quad	2	-1	-2	-1	2
cubic	-1	2	0	-2	1
quartic	1	-4	6	-4	1

Problem 3. Find a contrast ψ_5 that would be orthogonal to ψ_1, ψ_2, ψ_3, and ψ_4 in the "40 words memory experiment," and describe what the contrast would mean substantively.

modality:	paper	paper	tablet	tablet	tablet
order:	varied	varied	same	varied	varied
array:	column	scattered	-	-	-
rate:			3 sec.	3 sec.	1 sec.
ci's	$\bar{y}_1$	$\bar{y}_2$	$\bar{y}_3$	$\bar{y}_4$	$\bar{y}_5$
ψ_1	1	1	0	-1	-1
ψ_2	1	-1	0	0	0
ψ_3	0	0	0	1	-1
ψ_4	0	0	1	-1	0

CHAPTER 6

FIXED VS. RANDOM FACTORS, EXPECTED MEAN SQUARES

Questions to guide your learning:
Q_1: What are fixed and random factors?
Q_2: What is an Expected Mean Square (EMS)?
Q_3: How are EMS's important in determining F-tests?

In planning a study, it's usually clear what the focal factors and dependent measure(s) will be—they're the motivation for running the study. We're conducting the study because of our interest in how several variables combine to produce an effect—so we know the factors we want to include in the design. The only possible difficulty is deciding when to stop, e.g., we might know that factors A, B, and C are crucial factors, but including D would be interesting yet that would start making the study large and unmanageable, requiring a larger sample, etc.

So let's say we have selected our factors. In the stage of planning the experimental design, we must choose levels for each factor—the values for a=1, a=2, …, a=a. This task is somewhat more difficult—we must operationalize each factor in terms of the number of levels and the definition of each level. For example, if factor A is "the complexity of an ad message," how many messages are necessary? Two, as in "easy" vs. "hard"? Three, as in "easy," "moderately difficult," and "difficult"? To be fair, let's acknowledge that most (99%) of studies published in psych, marketing, and management journals are 2×2's, so we will probably select two factors A and B, and then their levels of "low" and "high" must be defined more precisely so we can create the experimental stimuli materials.

"Fixed" vs. "random" tells us about EMS. EMS tells us how to create the F-test.

Depending on how the levels of a factor are conceptualized and chosen, the factor may be classified as "fixed" or "random." Here are the definitions of those terms:

- A ***fixed factor*** is created when the levels of the factor are systematically chosen and assumed to represent the entire population of treatment conditions of interest. We're not interested in, nor will we be able to, generalize whatever results we find beyond the particular treatments we test and observe. For many studies, this is perfect—we only care about "current med" vs. "our newly proposed med" or "easy" vs. "complicated" ad message (and we're not too particular about how easy and complicated are defined, other than that they are relatively easier or more complicated than each other).
- A ***random factor*** is created when the levels of the independent variable are selected randomly (or unsystematically). Given their random selection, they can be assumed to represent a random sample of the larger population of treatment conditions, and at the end of the study, we are in a logical and statistical position to make claims about that larger population based on our results.

In most applications (and in most published articles), all of the factors will be considered fixed, except subjects because those study participants were indeed randomly selected. Why? The notion of random factors sounds desirable because of the implication that we can generalize results further, yet for theoretical reasons, it is usually more important to select levels of a factor in a way that assures coverage of an appropriate range. Thus, in controlling the levels, we are systematically selecting them, creating a fixed factor.

For example, say we want to vary the length of time a subject has to study and memorize a list of words. The factor is "length of time." What values should this variable take on its various levels (a=1, a=2, ...)? If we select the levels randomly, we might end up selecting something like: 30 seconds, 34 seconds, and 154 seconds. That's just weird but it's a random sample, as likely as any other. For theoretical reasons, we'd be better off selecting "sensibly" (per the literature), such as 30, 60, and 90 seconds, thereby creating a fixed factor.

Why does all this matter? The distinction between random and fixed factors affects the F-tests we use to conduct our hypothesis testing. Fixed vs. random factors affect the expected mean squares (EMS) which determine the appropriate error term for each F-test. Up to this point, we've seen only fixed factors A, B, C..., and for any H_0, we built an $F=MS_{effect}/MS_{S(AB)}$ (e.g., for a 2-factor factorial). In this chapter, we're looking at the rules to generate EMS's, so we know how to select the appropriate error term for the denominator of the F-test of any H_0.

If all factors are fixed (which is what we've seen so far, and which covers most of the literature), the S(AB) term (or S(ABC) in a 3-factor study) will serve as the error term for all F-tests in the ANOVA. The variations discussed in this chapter show that when 1 or more of the factors A, B, C, etc. (any factor in the design other than subjects) are random, then their EMS's will change the nature of the F-tests that must be calculated.

The Rules for Deriving EMSs

Let's introduce the rules with an example.[9] Say we have 2 fixed factors (our typical case). Factors A and B take on "a" and "b" levels, and there are n subjects in each cell. Subjects are always taken to be random.

Step 0. Write the model:

i) list the factors: A, B, S(AB)

ii) form the allowable interactions: A×B, ~~A×S(AB)~~, ~~B×S(AB)~~, ~~A×B×S(AB)~~ (remember, we cannot have a factor that is both nested and crossed, hence the strike-throughs)

iii) write the model statement: $x_{ijk} = \mu + \alpha_i + \beta_j + (\alpha\beta)_{ij} + \epsilon_{k(ij)}$

Step 1. (except for μ) write the terms in the model as rows of a table:

α_i

β_j

$(\alpha\beta)_{ij}$

$\epsilon_{k(ij)}$

[9] Different authors use different rules to derive EMS's, cf., Hasse diagrams. The rules followed in this chapter owe to Hicks (1982). For another approach, see the appendix.

Step 2. Write every subscript in the model as columns of the table. Indicate whether the subscript is associated with a factor that is random (R) or fixed (F), and indicate the number of levels the factor takes:

	a	b	n	← #levels
	F	F	R	Fixed/Random
	i	j	k	subscripts
α_i				
β_j				
$(\alpha\beta)_{ij}$				
$\epsilon_{k(ij)}$				

Step 3. Proceeding row by row: If the column subscript is not one of the row subscripts, write down the number of levels in that column space. For example, for row α_i the 1st column is for 'i' –this is the subscript in the row term, so skip the column. The 2nd column is for 'j,' which is not a subscript for the row term, so write down 'b' in the 2nd space. The 3rd column is for 'k,' which is also not a subscript for the row term, so write down 'n.' So far, we would have:

	a	b	n
	F	F	R
	i	j	k
α_i	blank	b	n
β_j			
$(\alpha\beta)_{ij}$			
$\epsilon_{k(ij)}$			

Similarly, for row $(\alpha\beta)_{ij}$ we would skip both the 1st and 2nd columns, since each of them represent a subscript that is included in the row term. In the last column, we would write an 'n' since the subscript 'k' is not a subscript in the term $(\alpha\beta)_{ij}$.

	a	b	n
	F	F	R
	i	j	k
α_i		b	n
β_j	a		n
$(\alpha\beta)_{ij}$			n
$\epsilon_{k(ij)}$			

Step 4. For any term (any row) that has parentheses in the subscripts (i.e., there is nesting in the term, as in "subjects are nested within AB" in this design), write a "1" under the columns for the subscripts that are inside the parentheses:

	a	b	n
	F	F	R
	i	j	k
α_i		b	n
β_j	a		n
$(\alpha\beta)_{ij}$			n
$\epsilon_{k(ij)}$	1	1	

Step 5. Fill in the remaining blanks as follows. For a column headed by a fixed factor, enter 0's. For a column headed by a random factor, enter 1's:

	a	b	n	
	F	F	R	
	i	j	k	EMS
α_i	0	b	n	$\varepsilon(MS_A)$ =?
β_j	a	0	n	$\varepsilon(MS_B)$ =?
$(\alpha\beta)_{ij}$	0	0	n	$\varepsilon(MS_{AB})$ =?
$\epsilon_{k(ij)}$	1	1	1	$\varepsilon(MS_{error})$ =?

Step 6. Proceeding row by row, compute the EMS for that term in the model:

i) cover the column(s) (like with a pencil) that are associated with subscripts that are included in the model term that are not in parentheses. In this example, when we're working on α_i then cover column i, when we're working on β_j cover column j; for $(\alpha\beta)_{ij}$ cover columns i and j; for $\epsilon_{k(ij)}$ cover column k.

ii) multiply the model term by the values remaining visible in the table. Include the term in the EMS formula if all of the subscripts in the term are included (perhaps with other subscripts). For example, to compute $\varepsilon(MS_A)$, cover column i (note the strike-throughs) and multiply the remaining:

	a	b	n	
	F	F	R	
	i	j	k	
α_i	~~0~~	b	n	←include
β_j	~~a~~	0	n	←do not include (i is not a subscript)
$(\alpha\beta)_{ij}$	~~0~~	0	n	←include (i is one of the subscripts)
$\epsilon_{k(ij)}$	~~1~~	1	1	←include (i is one of the subscripts)

What we have so far: $(\alpha_i \times b \times n) + ((\alpha\beta)_{ij} \times 0 \times n) + \left(\epsilon_{k(ij)} \times 1 \times 1\right) = \alpha_i \text{bn} + \epsilon_{k(ij)}$.

This stuff is tricky, so here's another example. To compute $\varepsilon(MS_{AB})$, cover columns i and j (note the strike-throughs):

	a	b	n	
	F	F	R	
	i	j	k	
α_i	~~0~~	~~b~~	n	←do not include (need both i & j)
β_j	~~a~~	~~0~~	n	←do not include (need both i & j)
$(\alpha\beta)_{ij}$	~~0~~	~~0~~	n	←include
$\epsilon_{k(ij)}$	~~1~~	~~1~~	1	←include (contains i & j)

What we have now: $(\alpha\beta)_{ij}\text{n} + \epsilon_{k(ij)}$.

As a final touch, when doing the multiplication, instead of saying: $\alpha_i \times b \times n$, we use the notation $\phi_A$bn for a fixed factor A (think ϕ, phi, for fixed). For a random factor, we'd use the notation σ_A^2bn (this looks like a variance because it's a random factor and so the effect would have a distribution).

	a	b	n	
	F	F	R	
	i	j	k	EMS
α_i	0	b	n	$bn\phi_A + \sigma_\epsilon^2$
β_j	a	0	n	$an\phi_B + \sigma_\epsilon^2$
$(\alpha\beta)_{ij}$	0	0	n	$n\phi_{AB} + \sigma_\epsilon^2$
$\epsilon_{k(ij)}$	1	1	1	σ_ϵ^2

These EMS's should look familiar. Each EMS contains the "effect" denoted in the row that we're interested in testing, plus σ_ϵ^2. So the F-test for each H_0: is $MS_{effect}/MS_{S(AB)}$, as we have been using for our fixed-effects designs (i.e., all experimental factors are fixed, except subjects are always random). Having confirmed the form of the F-tests in that design, let's consider alternative designs. What follows are the derivations of EMSs for three more examples: a mixed-design (one factor is random and one is fixed), a random-effects design (i.e., both factors are random), and a three-factor mixed design (with 1 factor fixed, 2 factors random).

EMS for 1 Fixed Factor and 1 Random Factor

In this next example, we'll keep the design simple to only 2 factors, but we'll add the complication that while factor A is a fixed factor, factor B is a random factor. The number of levels for the factors are still "a" and "b" and there are n subjects per cell, and they're always random. Just proceeding through the same steps laid out previously:

Step 0. Model: $X_{ijk} = \mu + \alpha_i + \beta_j + (\alpha\beta)_{ij} + \epsilon_{k(ij)}$
So far, nothing looks different.

Steps 1, 2, 3.

	a	b	n	
	F	R	R	Step 2
Step 1	i	j	k	
α_i		b	n	
β_j	a		n	Step 3
$(\alpha\beta)_{ij}$			n	
$\epsilon_{k(ij)}$				

Steps 4, 5:

	a	b	n
	F	R	R
	i	j	k
α_i	0	b	n
β_j	a	1	n
$(\alpha\beta)_{ij}$	0	1	n
$\epsilon_{k(ij)}$	1	1	1

Note that in this table, the column for factor B, the random factor, has some "1"s in it that used to be "0"s and that will end up making the difference.

Step 6:

	a	b	n	
	F	R	R	
	i	j	k	EMS
α_i	0	b	n	$bn\phi_A + n\sigma_{AB}^2 + \sigma_\epsilon^2$
β_j	a	1	n	$an\sigma_B^2 + \sigma_\epsilon^2$
$(\alpha\beta)_{ij}$	0	1	n	$n\sigma_{AB}^2 + \sigma_\epsilon^2$
$\epsilon_{k(ij)}$	1	1	1	σ_ϵ^2

The EMS for the main effect of factor A is now more complicated than when both factors A and B were fixed factors (in the previous example). It is interesting that while it is factor B that is the fixed factor, the EMS it affects is that for factor A. Anyway, now the new F-tests follow:

To test H_0: no main effect for A, we would compare MS_A to MS_{AB}:

$F = MS_A/MS_{AB}$ on (a-1), (a-1)(b-1) df, this is different!

To test H_0: no main effect for B:

$F=MS_B/MS_{S(AB)}$ on (b-1), ab(n-1) df, as we've seen before

To test H_0: no interaction:

$F=MS_{AB}/MS_{S(AB)}$ on (a-1)(b-1), ab(n-1) df, as we've seen

All 3 F-tests have the same numerators as before, naturally, we are trying to test the same 3 effects. What is different is that we now have 2 different terms that serve as error terms. For the tests of the main effect of B and the interaction, we use the error term we've always used, $MS_{S(AB)}$. However, for the test of the main effect for factor A, the denominator in the F-test is actually the interaction MS! So, when testing for the presence of an interaction, MS_{AB} serves as the numerator effect, and when testing for the presence of the main effect of factor A, it serves as the denominator error term.

Also, a technical point of clarification: when testing whether there is a fixed effect, the null is as we've seen, H_0: $\alpha_1 = \alpha_2 = \cdots \alpha_a = 0$. Strictly speaking, when testing whether there is a random effect, the null is: H_0: $\sigma_b^2 = 0$. Remember, a fixed effect is a factor where we care about the particular points, μ_1, μ_2 ... or their differences from μ, hence the α_i's, whereas a random factor is one where we are drawing from a population that has a number of different effects, some σ^2 variance of effects.

EMS for 2 Random Factors

We've seen a 2-factor factorial with both factors being fixed, and we've seen what happens when one of those factors is random. In this example, let's see what happens when both factors A and B are random.

As an (admittedly odd) example (Hicks, 1982, p.225), a company might be reviewing the efficacy of its plant operations, with a special focus on, say, its Zorro machines. Factor A is comprised of a random selection of some number ("a") of Zorro machine operators (so however the Zorro machines work, we'll be able to generalize to all operators, not just those selected systematically, in a "fixed" factor sense, such as the best operators, or something). Factor B is a random selection of some number ("b") of the Zorro machines themselves (again, so we can generalize the results to the population of all Zorro machines, regardless of what plant they're in, or which operator is running it, etc.). In Hicks's example, the replications per cell (n) were multiple tests on the same machine by the same operator.

Step 0. Model: $X_{ijk} = \mu + \alpha_i + \beta_j + (\alpha\beta)_{ij} + \epsilon_{k(ij)}$
The model statement does not look different.

Steps 1, 2, 3, 4:

Step 1	a R i	b R j	n R k	Step 2
α_i		b	n	
β_j	a		n	Step 3
$(\alpha\beta)_{ij}$			n	
$\epsilon_{k(ij)}$	1	1		
	Step 4			

Step 5:

	a R i	b R j	n R k
α_i	1	b	n
β_j	a	1	n
$(\alpha\beta)_{ij}$	1	1	n
$\epsilon_{k(ij)}$	1	1	1

Step 6:

	a R i	b R j	n R k	EMS
α_i	1	b	n	$bn\sigma_A^2 + n\sigma_{AB}^2 + \sigma_\epsilon^2$
β_j	a	1	n	$an\sigma_B^2 + n\sigma_{AB}^2 + \sigma_\epsilon^2$
$(\alpha\beta)_{ij}$	1	1	n	$n\sigma_{AB}^2 + \sigma_\epsilon^2$
$\epsilon_{k(ij)}$	1	1	1	σ_ϵ^2

With both factors A and B being random, the F-tests of both main effects are affected (note the EMS for the interaction has remained the same throughout these examples; this is true of the highest-order interaction term in a model). The new F-tests follow:

To test H_0: no main effect for factor A, we would compare MS_A not to $MS_{S(AB)}$, but to MS_{AB}:

$F = MS_A, / MS_{AB}$, on (a-1), (a-1)(b-1) df

Analogously, to test H_0: no main effect for B:

$F = MS_B, / MS_{AB}$, on (b-1), (a-1)(b-1) df

To test H_0: no interaction,

$F = MS_{AB}, / MS_{S(AB)}$, on (a-1)(b-1), ab(n-1) df

It should be coming clear that any time even 1 factor in a model is "random" (other than subjects, which is always a random factor), then the error terms for the F-test will be different from those in the fully fixed-effects model. Furthermore, the error terms for the F-tests will be different from H_0 to H_0 within the same study (e.g., we've used both MS_{AB} and $MS_{S(AB)}$ as error terms in different F-tests).

To sum up so far: in examining the EMS and F-tests for all permutations of 2-factor factorials (both factors are fixed, one is fixed and one is random, or both factors are random), we've covered the majority of scenarios given the prevalence of 2-factor designs over designs with more factors. This table shows how these 3 scenarios compare:

	A and B are fixed		A fixed, B random		A and B are random	
effect	EMS	F-tests	EMS	F-tests	EMS	F-tests
A	$bn\phi_A + \sigma_\epsilon^2$	$\frac{MS_A}{MS_{S(AB)}}$	$bn\phi_A + n\sigma_{AB}^2 + \sigma_\epsilon^2$	$\frac{MS_A}{MS_{AB}}$	$bn\sigma_A^2 + n\sigma_{AB}^2 + \sigma_\epsilon^2$	$\frac{MS_A}{MS_{AB}}$
B	$an\phi_B + \sigma_\epsilon^2$	$\frac{MS_B}{MS_{S(AB)}}$	$an\sigma_B^2 + \sigma_\epsilon^2$	$\frac{MS_B}{MS_{S(AB)}}$	$an\sigma_B^2 + n\sigma_{AB}^2 + \sigma_\epsilon^2$	$\frac{MS_B}{MS_{AB}}$
AB	$n\phi_{AB} + \sigma_\epsilon^2$	$\frac{MS_{AB}}{MS_{S(AB)}}$	$n\sigma_{AB}^2 + \sigma_\epsilon^2$	$\frac{MS_{AB}}{MS_{S(AB)}}$	$n\sigma_{AB}^2 + \sigma_\epsilon^2$	$\frac{MS_{AB}}{MS_{S(AB)}}$
S(AB)	σ_ϵ^2		σ_ϵ^2		σ_ϵ^2	

EMS for A Bigger Mixed Model: 2 Random Factors, 1 Fixed Factor

The logic and rules for deriving EMS's are easily extended, so let's look at one more example. In the following study, factors A and B are random, and factor C is fixed.

Step 0. $X_{ijk} = \mu + \alpha_i + \beta_j + \gamma_k + (\alpha\beta)_{ij} + (\alpha\gamma)_{ik} + (\beta\gamma)_{jk} + (\alpha\beta\gamma)_{ijk} + \epsilon_{l(ijk)}$

Steps 1, 2, 3, 4:

	a	b	c	n	
	R	R	F	R	
Step 1	i	j	k	ℓ	Step 2
α_i		b	c	n	
β_j	a		c	n	Step 3
γ_k	a	b		n	
$(\alpha\beta)_{ij}$			c	n	
$(\alpha\gamma)_{ik}$		b		n	
$(\beta\gamma)_{jk}$	a			n	
$(\alpha\beta\gamma)_{ijk}$				n	
$\epsilon_{l(ijk)}$	1	1	1		←Step 4

Steps 5, 6:

	a	b	c	n	
	R	R	F	R	
	i	j	k	ℓ	EMS
α_i	1	b	c	n	$bcn\sigma_A^2 + cn\sigma_{AB}^2 + \sigma_\epsilon^2$
β_j	a	1	c	n	$acn\sigma_B^2 + cn\sigma_{AB}^2 + \sigma_\epsilon^2$
γ_k	a	b	0	n	$abn\phi_C + bn\sigma_{AC}^2 + an\sigma_{BC}^2 + n\sigma_{ABC}^2 + \sigma_\epsilon^2$
$(\alpha\beta)_{ij}$	1	1	c	n	$cn\sigma_{AB}^2 + \sigma_\epsilon^2$
$(\alpha\gamma)_{ik}$	1	b	0	n	$bn\sigma_{AC}^2 + n\sigma_{ABC}^2 + \sigma_\epsilon^2$
$(\beta\gamma)_{jk}$	a	1	0	n	$an\sigma_{BC}^2 + n\sigma_{ABC}^2 + \sigma_\epsilon^2$
$(\alpha\beta\gamma)_{ijk}$	1	1	0	n	$n\sigma_{ABC}^2 + \sigma_\epsilon^2$
$\epsilon_{l(ijk)}$	1	1	1	1	σ_ϵ^2

Well, those EMSs have gone bananas. To formulate F-tests, we put the MS for the effect we're testing in the numerator. That's the easy part. The denominator requires an error term that logically cancels out all of the terms of the numerator's EMS except the effect we're trying to isolate. So, for the first EMS, for the main effect of factor A, the EMS is $bcn\sigma_A^2 + cn\sigma_{AB}^2 + \sigma_\epsilon^2$. We know the main effect is captured by the first term $bcn\sigma_A^2$, so we need to find another EMS that has the other terms $cn\sigma_{AB}^2 + \sigma_\epsilon^2$ to serve as the comparative error term. Looking down the table a little further, we can see that that's the EMS for the AB interaction. We just use this relative comparison logic for each F-test:

to test for:	F
main effect A	MS_A/MS_{AB}
main effect B	MS_B/MS_{AB}
main effect C	$MS_C/$?
interaction AB	$MS_{AB}/MS_{S(ABC)}$
interaction AC	MS_{AC}/MS_{ABC}
interaction BC	MS_{BC}/MS_{ABC}
interaction ABC	$MS_{ABC}/MS_{S(ABC)}$

First note that we used 3 different error terms: $MS_{S(ABC)}$ of course, but also MS_{AB} and MS_{ABC}. Next, note that we don't have a great comparison for testing the main effect for factor C. Yet presumably we wish to test it… If we could assume $\sigma_{AC}^2 = 0$, then we could test MS_C vs. MS_{BC}. Or, if we could assume $\sigma_{BC}^2 = 0$, then we could test MS_c vs MS_{AC}. We might assume $\sigma_{AC}^2 = 0$ (or $\sigma_{BC}^2 = 0$), if we tested the null hypothesis that there was no significant interaction between factors A and C (or B and C) and were unable to reject H_0. However, even if the A×C (or B×C) interaction wasn't significant, its effect might still be sufficiently nonzero (these are variances, after all), as to make a rather insensitive test of C.

So a better alternative is to create a "Pseudo F-test," denoted F' by finding some comparison for MS_C based on a linear combination of other MS terms. For example consider:

$MS_{AC} + MS_{BC} - MS_{ABC}$

The EMS for this combination =

$$(bn\sigma_{AC}^2 + n\sigma_{ABC}^2 + \sigma_\epsilon^2) + (an\sigma_{BC}^2 + n\sigma_{ABC}^2 + \sigma_\epsilon^2) - (n\sigma_{ABC}^2 + \sigma_\epsilon^2)$$
$$= bn\sigma_{AC}^2 + an\sigma_{BC}^2 + n\sigma_{ABC}^2 + \sigma_\epsilon^2$$

That term is just what we need. It differs from the $E(MS_C)$ by only the term we'd like to isolate and test: $abn\phi_C$.

So we construct $F' = \frac{MS_C}{MS_{AC}+MS_{BC}-MS_{ABC}}$.

Now the only remaining trick is what will the df be? The numerator df are still (c-1). The denominator df are as follows. The MS combination itself appears in the numerator, and the pieces and their respective df appear in the denominator. The "$(1)^2$" and "$(-1)^2$" terms in the deonominator come from the values in the linear combination (add MS_{AC}, add MS_{BC}, and subtract MS_{ABC}):

$$\frac{(MS_{AC} + MS_{BC} - MS_{ABC})^2}{(1)^2 \frac{(MS_{AC})^2}{(a-1)(c-1)} + (1)^2 \frac{(MS_{BC})^2}{(b-1)(c-1)} + (-1)^2 \frac{(MS_{ABC})^2}{(a-1)(b-1)(c-1)}}$$

This df might not be an integer, but the tables of critical values in the back of stats books (and Excel's functions for that matter) require integers, so round down to the next integer to be conservative. For example, for an F' on 4 and 12.7 df, we'd use 4 and 12 df.

As a final note, SAS, SPSS, and other statistical computing packages don't handle random factors well. By default, they will calculate all the F-tests based on the same error term in the denominator as if the factors were all fixed ($MS_{S(ABC)}$, for example).[10]

So! The bottom line: if you have at least 1 random factor (not including subjects), derive the EMS's by these rules, get the MS from SAS, and then compute the F-tests by hand.

SUMMARY

This chapter explained the difference between fixed and random factors. The concept of expected mean squares (EMS) was presented, and it was shown how they determine the forms of F-tests. Rules were presented and several examples illustrated how to derive the EMSs and therefore the proper forms of F-tests.

---

[10] In SAS, there is an optional statement "random;" but do not use it; it doesn't correct the F's. In proc glm, after the model statement, use "test" statements, e.g., test h=a e=a*b;.

REFERENCES

Hicks, C. R. (1982), *Fundamental Concepts in the Design of Experiments*, 3rd ed., NY: Holt, Rinehart, and Winston, pp.210-226, 406-407.

APPENDIX

I recently came across another approach to deriving EMS's that seems so nice and simple. I haven't investigated the extent to which this approach generalizes to many experimental designs, but thus far it works on the simple designs on which I've 'tested' it. I don't know the ultimate source, so I will attribute it to online materials currently posted at: www.ndsu.edu/ndsu/horsley/Expected_Mean_Squares_(HZAU).pdf. In case the web address changes, note that this resource was written by Professor Richard Horsley, of the Plant Sciences department at North Dakota State University. That should be enough identifying information to search and find the resource if the website changes.

In the first step, we build a table with the "sources" from the ANOVA model as the rows and their variance components serving as the columns. All sources/rows will have the error variance as a part of its formula so we enter it as well. Note the order in the columns proceed first with the basic error term, then the most complicated higher-order term, then the main effects. For a 2-factor factorial, that table follows:

Source	σ^2	$n\sigma^2_{AB}$	$nb\sigma^2_A$	$na\sigma^2_B$
A	σ^2			
B	σ^2			
A×B	σ^2			
error	σ^2			

Next, going row by row, we add the term that serves as the column label if the column label has the source letter(s) perhaps with other letters in its subscript. (Said the other way, if (all) the row source letter(s) is(are) in the subscript at the top of the column, perhaps with other letters, then we include the column term.) The filled in table follows:

	EMS when factors A and B are random			
Source	σ^2	$n\sigma^2_{AB}$	$nb\sigma^2_A$	$na\sigma^2_B$
A	σ^2	$+n\sigma^2_{AB}$	$+nb\sigma^2_A$	
B	σ^2	$+n\sigma^2_{AB}$		$+na\sigma^2_B$
A×B	σ^2	$+n\sigma^2_{AB}$		
error	σ^2			

Those are the EMS for the 2-factor factorial when both factors A and B are random. If factors A and B are fixed, then keep the first term (σ^2) and the last term from the row in the table above. And to denote that they are fixed, replace the σ^2 with ϕ. Thus, when both factors A and B are fixed, the EMS table looks like this:

Source	Before	→	EMS when A and B are fixed After
A	$\sigma^2 + n\sigma_{AB}^2 + nb\sigma_A^2$	→	$\sigma^2 + nb\phi_A$
B	$\sigma^2 + n\sigma_{AB}^2 + na\sigma_B^2$	→	$\sigma^2 + na\phi_B$
A×B	$\sigma^2 + n\sigma_{AB}^2$	→	$\sigma^2 + n\phi_{AB}$
error	σ^2	→	σ^2

For the intermediate case, when one factor, say A is fixed, and the other factor B is random, begin with the fully random case on the left, and on the right, keep the first and last terms. The new twists: keep the term if any letter(s) in the variance component is used in naming the source but if any remaining letter(s) correspond to a fixed effect, drop that term:

Source	Before	→	EMS for A fixed, B random After
A	$\sigma^2 + n\sigma_{AB}^2 + nb\sigma_A^2$	→	$\sigma^2 + n\sigma_{AB}^2 + nb\phi_A$
B	$\sigma^2 + n\sigma_{AB}^2 + na\sigma_B^2$	→	$\sigma^2 + na\sigma_B^2$
A×B	$\sigma^2 + n\sigma_{AB}^2$	→	$\sigma^2 + n\sigma_{AB}^2$
error	σ^2	→	σ^2

HOMEWORK

This HW covers EMS (expected mean squares).

Problem 1. This ANOVA table provides the results for a 3×5 factorial experiment testing the effects of 3 randomly selected test markets (cities) and 5 ads (a fixed factor) on the sales of a new product. Two measures of monthly sales are made for each city-ad combination (i.e., n=2). The following ANOVA table is compiled.

source	df	SS	MS
Test markets (cities)	2	100	50
Ads	4	120	30
TM×Ad interaction	8	160	20
within cell error	15	225	15
total	29	605	

Given that (at least) one of the experimental factors is a random factor, determine the EMS, and indicate the appropriate significance tests.

Problem 2. Time to get complicated. Imagine a 4-factor experiment, an A×B×C×D factorial, with n observations in each cell. Take the first two factors (A and B) as fixed and the other two factors (C and D) as random (i.e., the levels of the factors C and D were selected randomly to enable conclusions generalizing beyond the actual levels chosen). Derive the EMS and construct the proper forms for the F-tests.

Problem 3. Write out the SAS proc glm statements necessary to run problems 1 and 2.

Problem 4 extra credit. Derive the EMS for a 3 factor factorial in which the levels of all three factors, A, B, and C, are selected randomly.

CHAPTER 7

EXPERIMENTAL DESIGNS

Questions to guide your learning:
Q_1: What kinds of experimental designs exist besides factorials?
Q_2: What are completely randomized designs, randomized block designs, Latin Squares, nested factors, split-plot designs, and fractional factorials?

Thus far, we have focused on ANOVA models for factorial designs, wherein we manipulate 1 or more (though usually 2) factors, each with 2 or more levels (though usually 2) to determine their effects on a dependent variable of interest. The 2×2 factorial accounts for easily 99% of published research designs. In this chapter, we'll open horizons and see more options, coming to understand randomized block designs, Latin Squares, nested designs, split-plot designs, and fractional factorials.

Types of Experimental Designs

In the jargon of experiments, the design we've relied upon is called a "completely randomized design" (CRD) because we randomly assign each subject to 1 and only 1 treatment condition.[11] (And we had drawn a random sample of subjects, to generalize to the population.) Due to the random assignment, "good" subjects (e.g., smart or friendly) are just as likely to be assigned to any given condition as "bad" subjects (e.g., dumb or unfriendly). Thus, our assumption and hope is that various individual differences average out across the experimental groups. We might indeed end up with some confounding (e.g., all "good" subjects in one condition) due to the nature of randomness, but we don't expect systematic bias. Given that the groups were approximately the same at the start of the experiment, any group differences at the end of the experiment may be attributed to our manipulations.

Randomized Block Designs (RBD)

In a "randomized block design" (RBD), we form blocks of subjects, matching them on some relevant characteristic, then we randomly assign subjects within the blocks to the treatment conditions. We form blocks to account for a variable that we suspect might matter but one which we don't care about, a so-called "nuisance" factor.

For example, say we draw a random sample of 60 subjects, and we have a = 4 levels of factor A in our experiment. Say the experiment involves some cognitive task, puzzle-solving or something. We would probably expect a subject's "intelligence" to influence their problem-solving ability or related behaviors, but we're not interested in that effect—that effect seems obvious. Our theoretical interest is in how our 4 levels of factor A (whatever it is) might

---

[11] In studies of human behavior, units in experiments were traditionally called "subjects." That term has fallen out of favor in recent years, perhaps because it connotes royalty and status differences. Contemporary terminology includes "study participants" or "respondents" or "observations," etc. However, there are several phrases in experimentation that use the term "subjects" in a technical manner, so in this chapter, I'm going to forgo PC worries and use the term "subjects."

differentially affect the problem-solving ability regardless of the subjects' innate intellectual levels. If that interested us, we could run a study in 1 of 2 ways:

- In a CRD, we would simply randomly assign 15 subjects to each level of A.
- In a RBD, we would block the subjects by some kind of intelligence measure (an IQ type of pre-test, students' G.P.A., etc.). We might form 3 groups: 20 low, 20 medium, and 20 high "intellectual ability" subjects. Then we would randomly assign 5 of the 20 subjects with low ability to each of the 4 levels of A, and same for the subjects of medium and high ability.

The procedures for these 2 designs illustrate the extra step (of forming blocks) in the randomized block design (RBD):

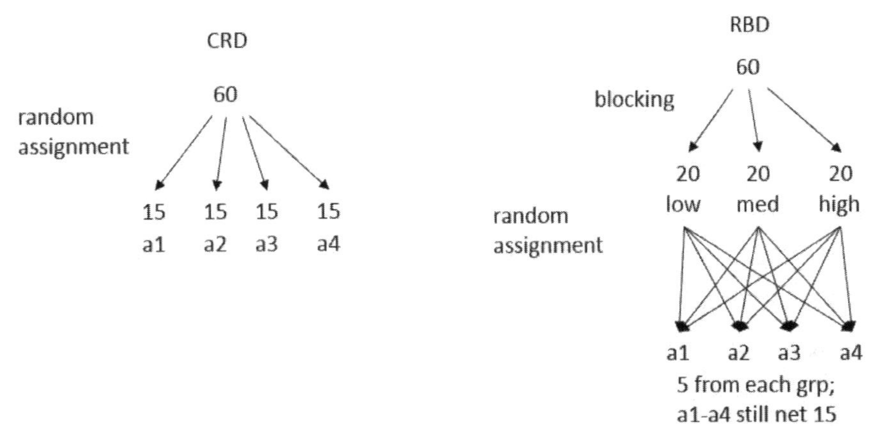

After blocking, the proportion of subjects (Ss) in each condition of the design would be:

	Treatment factor A, a = 1, 2, 3, 4			
	a=1	a=2	a=3	a=4
blocking factor (ability):	5 Ss low 5 medium 5 high	5 Ss low 5 medium 5 high	5 Ss low 5 medium 5 high	5 Ss low 5 medium 5 high

If the blocking has been done well, we're guaranteed that cognitive ability won't be confounded with factor A because there will be subjects of all 3 types (low, medium, and high ability) in every treatment condition a = 1, 2, 3, 4. (A confound would be something like many "low" ability subjects being in a=1 and a=2 and many "high" ability subjects being in a=4. Then, if there are group differences, we might assume they're due to the 4 operationalizations of Factor A, but they might instead be due to the pre-existing group differences on ability, and we wouldn't even know it.)

That procedure was how to run RBDs. The next question is how to analyze data from RBDs. We'll continue calling factor A our independent variable, and we'll call factor B the blocking variable (like cognitive ability). The ANOVA table would look like this:

source	df	SS, MS as usual…
A	a-1	
B	b-1	#1
A×B	(a-1)(b-1)	#2
S(AB)	ab(n-1)	
total	abn-1	

Two notes in that ANOVA table: #1) we hope the blocking effect is significant, even if it is not particularly interesting theoretically. If it's not significant, we either didn't do a very good job at blocking, or the population we sampled from was already homogeneous and we didn't need to block.

Note #2) We hope and expect the interaction term should not be significant. We don't want blocks to interact with factor levels—in a sense, they shouldn't. If there are 5 "low," "medium" and "high" subjects in a=1 and the same in a=2, etc., it would seem that they couldn't interact. If the interaction is significant, we probably hadn't done a good job at blocking. It's probably worth testing for the interaction effect; we'd want to verify that it's not significant (or if our total sample size is large, with such power that the interaction is significant, then we hope that the effect is small). If the interaction is significant, we should worry, though there isn't any advice as to what to do then), other than rerun the study and do a better job at blocking (but at least the blocking main effect and interaction have been partialled out of the main effect for A, the central focus of the study.

Some stats types argue that we can assume the interaction effect to be negligible (insignificant) and therefore we'd aggregate it into the error term. Thus, while the full ANOVA model is: $X_{ijk} = \mu + \alpha_i + \beta_j + (\alpha\beta)_{ij} + \epsilon_{k(ij)}$, if we were to assume that factor A (α_i) and the blocking factor B (β_j) wouldn't interact, we'd posit a reduced model with just the main effects: $X_{ijk} = \mu + \alpha_i + \beta_j + \epsilon_{k(ij)}$. The hope is that this RBD model is more powerful because we've added df to the error term (but we also better hope that those additional df aren't offset by having added SS_{AB}, even if small, to SS_{error}).

Comparing the models in the table below, we see the df_{error} for the RBD reduced model are indeed > df_{error} for the full RBD model. But the df_{error} for the CRD are even more. Does that mean the CRD is "more powerful"? Not necessarily. We expect the RBD to more than compensate for the loss of a few df by the effect of blocking—we've reduced a source of error. We thought cognitive ability might impact the dependent variable. If we hadn't blocked on some intelligence measure, then that variability would be in the SS_{error}. Having blocked, the cognitive ability variance is taken out of SS_{error} and is in SS_B. The smaller our SS_{error}, the larger our F-test (for A) will be. Thus, if we blocked well, the RBD should be more sensitive.

CRD (a=4)		RBD full (4×3)		RBD reduced (4×3)	
source	df	source	df	source	df
A	(a-1)=3	A	(a-1)=3	A	(a-1)=3
		B	(b-1)=2	B	(b-1)=2
		A×B	(a-1)(b-1)=6		
S(A)	a(n-1)=**56**	S(AB)	ab(n-1)=**48**	$SS_{error} = SS_{AB}+SS_{S(AB)}$	(abn-a-b+1)=**54**
total	an-1=59	total	abn-1=59	total	abn-1=59

Using the notion of expected mean squares (EMS) from Chapter 6, we can construct the appropriate F-tests. Factor A could be fixed or random (e.g., say it's fixed). Factor B is usually considered to be random because the blocks represent groups of subjects (who are random creatures and are certainly considered random in the S(AB) term).

	a	b	n	
	F	R	R	
	i	j	k	EMS
α_i	0	b	n	$bn\theta_A^2 + n\sigma_{AB}^2 + \sigma_\epsilon^2$
β_j	a	1	n	$an\sigma_B^2 + \sigma_\epsilon^2$
$(\alpha\beta)_{ij}$	0	1	n	$n\sigma_{AB}^2 + \sigma_\epsilon^2$
$\epsilon_{k(ij)}$	1	1	1	σ_ϵ^2

Here we see a little more on the notion of ignoring an A×B interaction and lumping it into error. The central effect we care about in this study is the main effect for A. Normally we would say that the F-test would be: $F=MS_A/MS_{AB}$ (and $F=MS_B/MS_{S(AB)}$). However in this RBD world, in constructing the F-test for the main effect we're interested in, factor A, we could test H_0: $(\alpha\beta)_{ij} = 0$ (via the usual $F = MS_{AB}/MS_{S(AB)}$) and hope it is not significant, or we could just assume $\sigma_{AB}^2 = 0$ (given that conceptually it is essentially 0 because people in the blocks are spread out evenly across levels of factor A). (I'll always test; I want to know what's going on in my data.) If the F-test for A×B is not significant (or we assume it away as n.s.), we would combine SS_{AB} and $SS_{S(AB)}$ to get SS_{error}. Then in effect, we'd have:

source	EMS
α_i	$bn\theta_A^2 + \sigma_\epsilon^2$
β_j	$an\sigma_B^2 + \sigma_\epsilon^2$
error	σ_ϵ^2

From this structure, we would test $F=MS_A/MS_{error}$ and $F=MS_B/MS_{error}$.

If the F-test for the A×B interaction effect is significant, then we probably shouldn't assume $\sigma_{AB}^2 = 0$, and aggregating the SS_{AB} into $SS_{S(AB)}$ would be incorrect and counter-productive to us given that it would make the SS_{error} larger, working against the likelihood of finding a significant F-test.

That's it on RBDs—we've seen how they're done and how they're analyzed. They're not difficult to execute or analyze. Sometimes the biggest challenge is simply determining which characteristic is relevant and should be used as a blocking factor.

As a last few words about RBD's, consider that again, the notion of creating blocks is to create homogeneous groups (pulling the heterogeneity out of the error SS). Increasing the number of blocks allows the blocks to be more homogeneous within the block but having "too many" blocks can get a little cumbersome.

A small block, of group 2, would be like a matched t-test (e.g., studies on twins). If factor A has a=3 levels, then we need to have blocks of size 3 (or multiples of 3) matched closely on the blocking factor so that the subjects within the blocks are homogeneous. Then for each of the b blocks, we'd assign 1 of the 3 subjects to conditions a_1, a_2, and a_3. Or if we have 2 experimental factors, A and C, and we seek to add B as a blocking factor, and say a = 4 and c = 3, then each of the b blocks must contain a×c = 4×3 = 12 subjects who are very similar on the blocking factor (the block must be of size 12 or multiples of 12). As the number of factors and the number of levels per factor increase, we need bigger blocks, but as block size increases, it becomes more difficult to keep each block homogeneous.

One form of an RBD can be thought of is when each subject is a block (a very homogeneous block) who is exposed to all treatment conditions. This is a "within-subjects design" or a "repeated measures design." In a CRD, differences between 2 groups, say a_2 and

a_3, reflect both treatment differences (i.e., factor A) and group differences due to subjects' individual difference variables. When each subject is his or her own block, then those individual differences are "controlled for." We'll see more in Chapter 8.

When researchers debate about sampling vis-à-vis internal and external validity, some might argue that using students as subjects is like choosing subjects from an already fairly homogeneous "block" (at least for such variables as intelligence, SES, etc.). No additional blocking might be necessary. Questions remain about generalizing results to the larger, more heterogeneous population. In the journals for most of the social sciences, a focus on internal validity seems to dominate. In applied research, external validity may loom larger.

There are variants on blocking designs. Imagine the following example, run on retail stores as "subjects" (not humans). We might wish to study the effect on sales in test markets of 2 factors: promotional display and price. Factor A would be display 1 or display 2, and factor B would be price levels b_1, b_2, and b_3. A blocking factor C might be the size (e.g., sales volume) of the store, which we might use to classify each store as "big" or "small." The size of a store (the blocking factor) would surely be expected to influence the dependent variable of sales, but it is not a very interesting effect. It's a "nuisance" variable, so we'll control for it, but it is not of much theoretical interest. To run such a study, we would: 1) randomly select a sample of stores, 2) classify each store to the "big" or "small" block based on the store's past sales data, 3) for all big" stores, randomly assign each to 1 of the A×B combinations (and the same is done for all "small" stores). Since there are $a \times b = 2 \times 3 = 6$ combinations, each block would need to be of size 6, so we could sample 6 big and 6 small stores, or 12 + 12 or 18 + 18 and so on. Numbers like those would be easily achieved if the experimental units were people, but to get 12 stores or 24 etc. might be rather demanding. If all we could get was a sample of say 15 big stores (and 15 small stores), we would form 2 full blocks (6 + 6) and have 3 stores left over to fill an "incomplete block."[12]

Finally, a related topic is the analysis of covariance. In that model, we attempt to statistically control for some nuisance variable rather than mechanically blocking on it. We'll see the analysis of covariance in Chapter 9.

Latin Squares

A Latin Square (LS) design is an extension of RBD, in that 2 blocking factors are used in a very particular way. Imagine we have 3 factors: factor A is the treatment factor, the focus of the study, and now we have 2 blocking factors, factors B and C. To compare the designs CRD, RBD, and LS, imagine drawing a random sample of 40 subjects, and assigning them to 1 of a = 2 levels of a treatment factor.

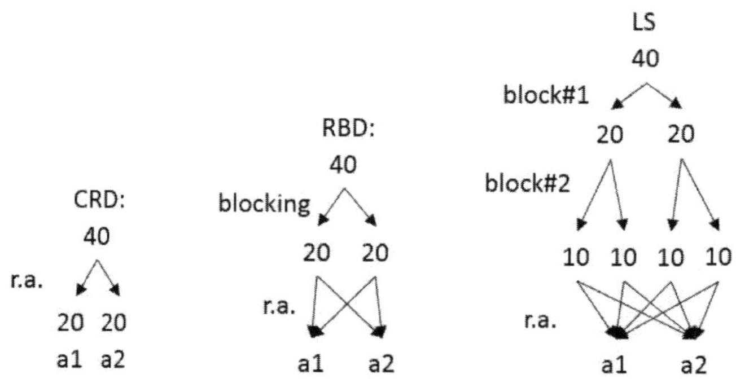

[12] See John (1980) and Dey (2010), Lawson (2010), Moen, Nolan, and Provost (2012).

The goal in RBDs was to control for a nuisance variable, and hope that the SS_{block} was sufficiently large to compensate for having lost some df for the blocking. If we use 2 blocking factors in a LS in an attempt to further reduce error variability by controlling 2 nuisance variables, we'd lose more df. Still, the loss of some df may well be compensated by the further reduction in SS_{error}. The df across the 3 designs depicted in the diagram would compare as follows:

CRD
$df_{total} = df_A + df_{S(A)}$ an – 1 = (a-1) + a(n-1) (2×20)-1 = (2-1) + 2(20-1) 39 = 1 + 38
RBD
$df_{total} = df_A + df_B + df_{error}$ abn-1 = (a-1) + (b-1) + (abn-a-b+1) (2×2×10)-1 = (2-1) + (2-1) + (2×2×10-2-2+1) 39 = 1 + 1 + 37
LS
$df_{total} = df_A + df_B + df_C + df_{error}$ abcn-1 = (a-1) + (b-1) + (c-1) + (abcn-a-b-c+2) = (2-1) + (2-1) + (2-1) + (2×2×2×5-2-2-2+2) 39 = 1 + 1 + 1 + 36

The thing that is a little peculiar about LS's is that all factors must have the same number of levels (which is a rather restrictive requirement). For example, let's say a=4, then it must also be the case that b=c=4. If we were to run a full factorial, it would require $4^3 = 64$ cells, so we would need many subjects. In a Latin Square, we would run a subset of cells, so fewer subjects would be needed. For example, we would run these 16 cells:

	b1	b2	b3	b4
c1	a1	a2	a3	a4
c2	a2	a4	a1	a3
c3	a3	a1	a4	a2
c4	a4	a3	a2	a1

Note that the 2 blocking factors, B and C are completely crossed, like a traditional 2-factor factorial. What is different is that only 1 level of factor A is run for each B×C cell. Also note that each of the 4 levels of A exist only once in each row and in each column. When the combinations are strung out in vector form, as when we enter data, we'd see the 16 conditions clearly (note there are not 64 lines for all possible combinations):

a	b	c
1	1	1
1	2	3
1	3	2
1	4	4
2	1	2
2	2	1
2	3	4
2	4	3

3 1 3
3 2 4
3 3 1
3 4 2
4 1 4
4 2 2
4 3 3
4 4 1

As an example, one nuisance factor might be the time of day an experiment is run. If we are measuring some cognitive performance as the dependent variable, we might expect scores to diminish later in the day (perhaps not, but the point is, we could control for it).

A second nuisance factor might be the experimenter who oversaw the data collection in that time slot; after all, if we have 4 research assistants helping to run the study, we wouldn't want them to influence the outcomes in anyway (e.g., if they gave hints or odd body language or made too much noise). We could control for them by spreading them over treatment conditions.

The experimental treatment factor that interests us might be something like the number of seconds allotted to study a list of words for subjects to memorize (so the dependent variable might be the number of words recalled). Taking those 2 blocking factors over the 16 cells, we'd have the following mapping:

		Experimenter (C)			
		Matthew	Mark	Luke	John
Time of Day (B)	10am	60 sec	90	120	150
	12	90	150	60	120
	2pm	120	60	150	90
	4pm	150	120	90	60

Again, notice that each treatment level of factor A occurs once in each level of B (row) and once in each level of C (column).

So Latin Square designs are a little peculiar in form but they allow us to study 2 blocking factors and do so efficiently. Naturally that efficiency induces a cost: factor A is not completely crossed with B or C, so we cannot study higher-order interactions. As we have seen in RBDs, users of LS designs usually assume the blocking factors don't interact with each other (and they cannot interact with the treatment factor). As a result, the model is fairly simple: $X_{ijk\ell} = \mu + \alpha_i + \beta_j + \gamma_k + \epsilon_{pooled}$.

In the model, α_i represents the treatment factor A, β_j reflects blocking factor 1, and γ_k reflects the blocking factor 2. The ANOVA table would have these sources:

source	df
A (treatment)	(a-1)
B (block 1)	(b-1)
C (block 2)	(c-1)
error (pooled)	(abcn-a-b-c+2)
total	abcn - 1

Just as Latin Squares are an extension of RBD's, there are extensions of LS designs. If we blocked on 3 factors, these designs are called Graeco-Latin Squares, and if we were to run an incomplete LS, it would be called a Youden Squares design.

Nested Factors

It is very frequently the case that 2 factors are crossed, as in factorial designs. However it might instead be the case that one factor is nested in another. Thus far, the only kind of nesting we've seen is subjects being nested in 1 level of factor A, S(A), or in 1 cell in a 2-way design, S(AB). Yet, an entire experimental factor factor (B) might be nested in another (A). If B were nested within A, we'd denote it as B(A).

For example, we might study the effect of schools and their teachers and the variety of philosophies and training they represent on their students' scores on the dependent variable of standardized achievement tests. The teachers only work at one school, so teachers 1 and 2 are nested in school 1, teachers 3 and 4 are nested in school 2, etc.

Factor A (school)	1	2	3
Factor B (teacher)	1,2	3,4	5,6

Drawn another way, to emphasize that the factors A and B are not crossed, the "X" marks the spot of where we have data, again, teachers 1 and 2 at school 1, etc.

Teacher \ School:	1	2	3
1	X		
2	X		
3		X	
4		X	
5			X
6			X

To model such data, we normally think of crossing factors A and B. However, in this case, B is already nested in A, that is, B(A), and it makes no sense to have a factor be crossed with and nested in the same other factor. The model term would look like: $(\alpha\beta)_{ij(i)}$, it can't happen. The model would be:

$$x_{ijk} = \mu + \alpha_i + \beta_{j(i)} + \epsilon_{k(ij)}$$

With the ANOVA table:

Source	df	F
A	(a-1)	$MS_A/MS_{S(AB)}$
B(A)	a(b-1)	$MS_{B(A)}/MS_{S(AB)}$
S(AB)	ab(n-1)	
total	abn-1	

Those F-tests are conditioned on both factors being fixed. But we might wish to draw random samples of schools and of teachers within their schools, making factors A and B random.

Thus, as a brief side bar, let's consider all permutations of 2 factors, A and B(A), being fixed or random to see the proper F-tests.

I) A is fixed, and B(A) is fixed:

	F	F	R		
	i	j	k	EMS	F-tests
α_i	0	b	n	$bn\theta_A^2 + \sigma_\epsilon^2$	$MS_A/MS_{S(AB)}$
$\beta_{j(i)}$	1	0	n	$n\theta_{B(A)}^2 + \sigma_\epsilon^2$	$MS_{B(A)}/MS_{S(AB)}$
$\epsilon_{k(ij)}$	1	1	1	σ_ϵ^2	

II) A is fixed, and B(A) is random:

	F	R	R		
	i	j	k	EMS	F-tests
α_i	0	b	n	$bn\theta_A^2 + n\sigma_{B(A)}^2 + \sigma_\epsilon^2$	$MS_A/MS_{B(A)}$
$\beta_j(i)$	1	1	n	$n\theta_{B(A)}^2 + \sigma_\epsilon^2$	$MS_{B(A)}/MS_{S(AB)}$
$\epsilon_{k(ij)}$	1	1	1	σ_ϵ^2	

III) A is random, and B(A) is fixed:

	R	F	R		
	i	j	k	EMS	F-tests
α_i	1	b	n	$bn\sigma_A^2 + \sigma_\epsilon^2$	$MS_A/MS_{S(AB)}$
$\beta_{j(i)}$	1	0	n	$n\sigma_{B(A)}^2 + \sigma_\epsilon^2$	$MS_{B(A)}/MS_{S(AB)}$
$\epsilon_{k(ij)}$	1	1	1	σ_ϵ^2	

IV) A and B(A) both random:

	R	R	R		
	i	j	k	EMS	F-tests
α_i	1	b	n	$bn\sigma_A^2 + n\sigma_{B(A)}^2 + \sigma_\epsilon^2$	$MS_A/MS_{B(A)}$
$\beta_{j(i)}$	1	1	n	$n\sigma_{B(A)}^2 + \sigma_\epsilon^2$	$MS_{B(A)}/MS_{S(AB)}$
$\epsilon_{k(ij)}$	1	1	1	σ_ϵ^2	

So, in sum (for two factor designs), B(A) is always tested by $MS_{B(A)}/MS_{S(AB)}$. If B is random, then A is tested by: $MS_A/MS_{B(A)}$ but if B is fixed, then A is tested by: $MS_A/MS_{S(AB)}$.

Another Nesting Example: C Nested in B

Here is a second example with nesting. The factors are: A, B, C(B), S(ABC), or in terms of their subscripts: A_i, B_j, $C_{k(j)}$. The model statement is:

$$X_{ijk\ell} = \mu + \alpha_i + \beta_j + \gamma_{k(j)} + (\alpha\beta)_{ij} + (\alpha\gamma)_{ik(j)} + \epsilon_{\ell(ijk)}$$

Notice that if C is nested in B, there cannot be any interaction with BC, so there will be no B×C and no A×B×C interaction term. There can still be an A×B interaction and even an A×C interaction (specifically it would be A×C(B)). Diagrammatically, this design (with a=2, b=3, c=6) would look like this:

	a_1	a_2
b_1	c_1 c_2	c_1 c_2
b_2	c_3 c_4	c_3 c_4
b_3	c_5 c_6	c_5 c_6

To run just one set of expected mean squares, say factors A and B are fixed, and C is random:

	F	F	R	R		
	i	j	k	ℓ	EMS	F-tests
α_i	0	b	c	n	$bcn\theta_A^2 + n\sigma_{AC(B)}^2 + \sigma_\epsilon^2$	$MS_A/MS_{AC(B)}$
β_j	a	0	c	n	$acn\theta_B^2 + an\sigma_{C(B)}^2 + \sigma_\epsilon^2$	$MS_B/MS_{C(B)}$
$\gamma_{k(j)}$	a	1	1	n	$an\theta_{C(B)}^2 + \sigma_\epsilon^2$	$MS_{C(B)}/MS_{S(ABC)}$
$(\alpha\beta)_{ij}$	0	0	c	n	$cn\theta_{AB}^2 + n\sigma_{AC(B)}^2 + \sigma_\epsilon^2$	$MS_{AB}/MS_{AC(B)}$
$(\alpha\gamma)_{ik(j)}$	0	1	1	n	$n\sigma_{AC(B)}^2 + \sigma_\epsilon^2$	$MS_{AC(B)}/MS_{S(ABC)}$
$\epsilon_{k(ij)}$	1	1	1	1	σ_ϵ^2	

One Final Nesting Example: B Nested within A, C Nested within B ("hierarchical")

Finally, imagine one more nesting example, a hierarchical structure where C is nested in B, and B is in A. The model would be: $x_{ijk\ell} = \mu + \alpha_i + \beta_{j(i)} + \gamma_{k(j(i))} + \epsilon_{\ell(ijk)}$. (I like the notation of $\gamma_{k(j(i))}$ because it is quite explicit. But sometimes it will be written $\gamma_{k(ji))}$ like in SAS where the terms would be A, B(A), C(BA).)

Say all the factors are random, we'd have the F-tests following their nesting:

	R	R	R	R		
	i	j	k	ℓ	EMS	F-tests
α_i	1	b	c	n	$bcn\sigma_A^2 + cn\sigma_{B(A)}^2 + n\sigma_{C(B(A))}^2 + \sigma_\epsilon^2$	$MS_A/MS_{B(A)}$
$\beta_{j(i)}$	1	1	c	n	$cn\sigma_{B(A)}^2 + n\sigma_{C(B(A))}^2 + \sigma_\epsilon^2$	$MS_{B(A)}/MS_{C(B(A))}$
$\gamma_{k(j(i))}$	1	1	1	n	$n\sigma_{C(B(A))}^2 + \sigma_\epsilon^2$	$MS_{C(B(A))}/MS_{S(ABC)}$
$\epsilon_{\ell(ijk)}$	1	1	1	1	σ_ϵ^2	

Nesting in SAS

In SAS, in proc glm, the nested factors are denoted as follows. In the first example, B was nested in A, B(A), and the SAS statements would be: proc glm; class a b; model X = A B(A); *also, if B is random, we'd need to include the following: test h=a e=b(a);

In the second example, with A and B and C(B), we'd say: proc glm; class a b c; model y = A B C(B) A*B A*C(B); *if A and B are fixed and C is random add the test statements; test h=a e=a*c(b); test h=b e=c(b); test h=a*b e=a*c(b); *the default error term is S(ABC), so the other F-tests will be fine;

In the third example, where C is nested in B which is nested within A, and say all 3 factors random, we'd say: proc glm; class a b c; model y = A B(A) C(B A); test h=a e=b(a); test h=b(a) e=c(b a); The default error term and F-test for C is okay.

Nesting and HLMs

Nested designs are similar to "hierarchical linear models" (HLMs). In HLMs, there are (at least) two models. Recall the first nesting example earlier, of teachers being nested in their schools. The "micro" HLM model would reflect the teachers, and the "macro" model would reflect the schools. In the micro, "level 1" model, we'd look at, say teachers' self-reported job satisfactions (x) as a function of the number of students assigned to them (v for volume). The model would be a regression of the form: $x_{ij} = b_{0j} + b_{1j}(v_{ij} - \bar{v}_j) + e_{ij}$. We'd collect all the intercepts (b_{0j}) and slopes (b_{1j}) across all the schools (j) and model those parameter estimates in the macro, "level 2" models, one for the intercepts ($b_{0j} = \gamma_{00} +$

$\gamma_{01}w_j + u_{0j}$), and one for the slopes ($b_{1j} = \gamma_{10} + \gamma_{11}w_j + u_{1j}$) as a function of some variable measured at the school level, such as a school's budget, w_j. Obviously the HLM approach is a bit complicated (for more, see Heck and Thomas, 2000 and Raudenbush and Bryk, 2002).

Split-Plot Designs

A design we don't see much but it's just great—great heritage, great stories—is the so-called "split-plot" design. Split-plot got its name from farming experiments. Long before contemporary genetic farming, farmers wanted to know which seeds were best, which fertilizers were best, which insecticides were best. Agriculture was the priority business, and many of the old-time classic statisticians (e.g., Sir Ronald Fisher's examples abound, and universities like U of Illinois, U of Iowa, and Iowa State have strong statistics traditions) were set on solving problems of how to produce the most, best yields. So, the term "plot" was literally a plot of land, and in an experiment, we'd "split" the plot into sections where we'd try insecticide #1 over here, and insecticide #2 over there, etc.

So here is a bird's eye view of a farm. There are orchards in the NW, cows grazing in the NE, people in the SE. We want the land to be relatively homogeneous before treating it with anything (so we can attribute any yield differences to the various insecticides). Yet the farmland that borders the orchards might do better simply because it will be enriched by the plants' nutrients, and the farmland that borders the cows might also be enriched by, well the cows' nutrients. The farmland that is near the people might be do worse, being a little scruffed up with pollution or lack of full sun exposure, etc.

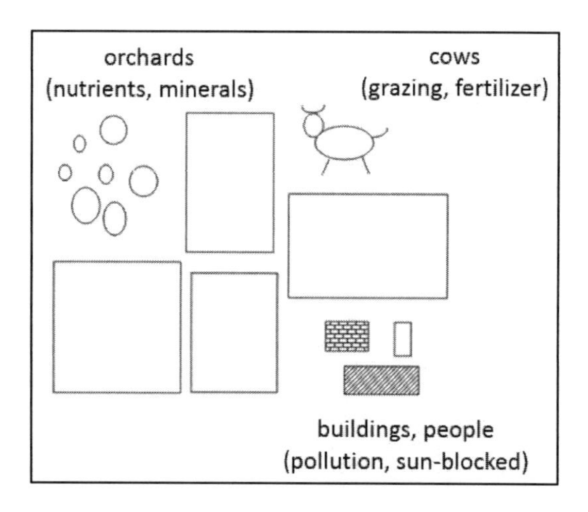

So, we won't apply our 3 different insecticides like this (doing so would create possible confounds).

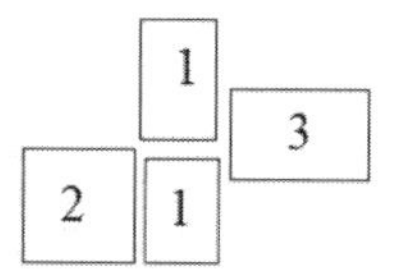

Instead, we'll consider each of those 4 rectangles of land 4 "whole plots," and we'll apply the 3 different experimental treatments (the insecticides) to "split-plots," like this:

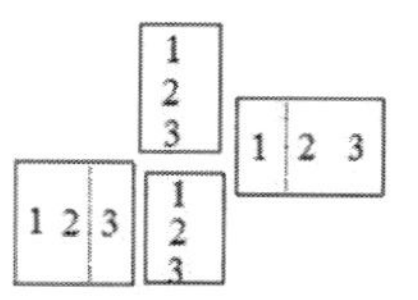

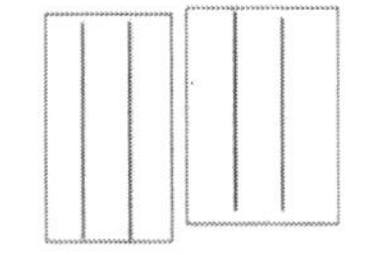

Or consider another agricultural example. For some treatments, we might need a large area for application (e.g., different crop-dusting herbicides), so that would be the "whole plot" factor and another factor (such as different fertilizers) may be applied to the smaller areas, so that would be the "split-plot factor."

As an attempt to offer a non-farming example, let me try one more. Say we are testing the effectiveness of anti-aging skin creams. If we put cream 1 on some people, and cream 2 on other people, we'd have a standard 1-factor, 2-level design. Of course we assume the people in the 2 groups are roughly the same (if they were randomly assigned to 1 cream or the other), but we need reasonably large sample sizes to feel confident that we've controlled for all the skin types. A more sensitive experiment could be run in which we consider one person's "face" as the "whole plot," so there is homogeneity (we truly are treating the same skin), and then the application of the anti-aging cream to the left or right eye is considered the "split-plot" factor, so we could put cream 1 on the left and cream 2 on the right.

If factor A is the "whole plot" factor, and B is the "split-plot" factor, the 2-factor model would be:

$$X_{ijk} = \mu + \alpha_i + \beta_j + (\alpha\beta)_{ij} + \epsilon_{k(ij)}$$

/ \ \

whole plot split-plot (anything involving B)

Split-plot designs have similarities and differences when compared to some other experimental designs. For example, a plot might look a bit like a block. In the "crop dusting" example, each whole plot is a crop treated with a different herbicide. A "whole plot" here (the large crop area) might look like a "block" and in either design, it would be taken to be homogeneous (which it is likely to be). However, recall from the blocking example, while we would assign members within a block to different treatment conditions, we expected that blocks and treatments should not interact. With split-plot designs, we take the whole plot and divvy it up and treat the split-plots with some treatment, and here, we will test for whether the whole plot factor might interact with the split-plot factor. So in a RBD, there is a blocking factor and a treatment factor, whereas in split-plot designs, there is a large treatment factor and a small treatment factor.

Whole plots and split-plots might also look a bit like nesting factors, in that a split-plot is indeed part of the larger whole plot. However, in nesting designs, we'd have some element (like a teacher) nested in a larger factor (the school), and the teacher would be nested in 1 and only 1 school. With whole plots and split-plots, the experimental treatment applied to the split-plots (i.e., b = 1, 2, …, b) would be applied across the split-plots in every whole plot. (If the split-plots were nested in the whole plots, then fertilizers 1, 2, 3 would be applied within whole plot #1, and fertilizers 4, 5, 6 applied to whole plot #2, etc.)

In practice, probably the most likely split-plot design we'll see is to treat a person as a whole plot, and have the person engage in multiple levels of the experimental factor. These designs are called "repeated measures" or "within subjects" designs, and they're important enough and complicated enough that they are discussed in Chapter 8 and again in Chapter 16. Such designs are also sometimes referred to as mixed designs (though "mixed" can also refer to a design with 1 or more fixed factors along with 1 or more random factor). Here is a comparison between a CRD (each subject is assigned to only 1 condition), a fully within-subjects design (each subject is exposed to all experimental conditions), and a mixed design with a between-subjects factor (a whole plot) and a within-subjects factor (the split-plots). Note that in the CRD, the sample size would need to be multiples of 4, vs. the efficiency and sensitivity (issues discussed more in Chapter 8) of the fully within subjects design and the split-plot version of a within-subjects design:

CRD 2 factors A×B

	b_1	b_2
a_1	S_1	S_2
a_2	S_3	S_4

n=4

fully w/in subject (A×B×S)

	b_1	b_2
a_1	S_1	S_1
a_2	S_1	S_1

n=1

Split-plot B×(A×S) (A w/in, B between)

	b_1	b_2
a_1	S_1	S_2
a_2	S_1	S_2

n=2

As a final word on split-plot designs…what if there is a whole plot factor A, and a split-plot factor B, and we wish to cross B with another experimental factor across the split-plots? We divide up the split-plot again and apply the new factor: it's a "split-split-plot" design.

Fractional Factorials

A fractional factorial design is just what it sounds like, a "fraction" of the familiar factorial. Why would we not run the whole thing? In factorial designs, wherein each level of a factor is crossed with each level of all the other factors, as the number of factors increases, the number of treatment combinations increases even more quickly, so we would need many subjects, even if we aimed for only a modest cell sample size n.

For example, for a factorial experiment with "f" factors, each taking on only 2 levels, that would mean 2×2×2... or 2^f cells. In the table below, if f is only 2, then we have the familiar 2×2, and we can handle that, no problemo. Now let's think bigger. When we get caught up in studying some phenomenon, we can typically imagine a good number of factors that might impact our dependent variable(s). Let's say we think up 5 factors. In the table below, we see that f=5 results in 5 main effects, of course, ten 2-way interactions, um that's a lot of plotting, ten 3-ways, good grief, five 4-way interactions and one 5-way. The design necessitates 32 cells, so we'd have to have sample sizes with multiples of 32. Or, in the last line of the table, in an experiment with 8 factors (i.e., with 2^8 cells), even without replications (i.e., even if n=1), we'd need 256 subjects.

f= #factors	2^f= #cells	#main effects	#interactions: 2-way	3-way	4	5	6	7	8
2	4	2	1	-	-	-	-	-	-
3	8	3	3	1	-	-	-	-	-
4	16	4	6	4	1	-	-	-	-
5	32	5	10	10	5	1	-	-	-
6	64	6	15	20	15	6	1	-	-
7	128	7	21	35	35	21	7	1	-
8	256	8	28	56	70	56	28	8	1

Instead of running a full factorial, what about running just a part of it, a fraction of it. If we had only 3 factors, we would no doubt run the full factorial, but we'll use a small example to show the issue that arises in fractional factorials. So, instead of running a full 2×2×2, say we were to run a fraction of the entire factorial experiment. Specifically, we'll run half of the 2^3 (sometimes denoted 2^{3-1}, $2^2 = 4$ cells, or 2^{-1} is half so that's half of a 2^3). In this example, we'll run the $a_1b_1c_1$, $a_1b_2c_1$, $a_2b_1c_2$, and $a_2b_2c_2$ cells (those with *s):

		c1	c2
a1	b1	*	
	b2	*	
a2	b1		*
	b2		*

In that design, it's easy to see that the means we'd use to test $c_1$vs. c_2 are the same means we'd use to test a_1 vs. a_2. The main effects for factor A and C are confounded (yikes!). If there is a significant effect, we wouldn't know whether the variability is attributable to factor A or factor C.

In this second example, we'll run a different "half" of the 2×2×2, a different subset of 4 cells from the full 8, i.e., the cells: $a_1b_1c_2$, $a_1b_2c_1$, $a_2b_1c_1$, and $a_2b_2c_2$:

		c1	c2
a1	b1		*
	b2	*	
a2	b1	*	
	b2		*

In this design, the a_1 vs. a_2 comparison draws on the same cell means that a B×C interaction would, so we'd say A and B×C were "confounded" or "aliased."

To understand the pattern of "confounding" or "aliases" that would result from a proposed design (to see if it's tolerable), we start by confounding the grand mean, GM (usually of no interest) with the highest order interaction (unfortunately too often thought of as too complicated to be of interest). We write all the effects as rows, in this case for a 3-factor factorial. We symbolically "multiply" the confound of GM (in this case, ABC) with each row effect. We then delete any terms that are squared, and the last column is the result, the confounding or aliasing pattern.

effect in full 2^3 GM	GM = ABC	confounds
A	A(ABC) = A^2BC = BC	A and BC are aliases
B	B(ABC) = AB^2C = AC	B and AC are aliases
C	C(ABC) = ABC^2 = AB	C and AB are aliases
AB	AB(ABC) = A^2B^2C = C	
AC	AC(ABC) = A^2BC^2 = B	
BC	BC(ABC) = AB^2C^2 = A	
ABC	ABC(ABC) = $A^2B^2C^2$ =GM	

In the ANOVA table for that design, given the confounding, we would have only the following sources of variation:

source	df
A (or BC)	1
B (or AC)	1
C (or AB)	1

The confounding, and that truncated ANOVA table, should give us pause before using a fractional factorial design. If we've designed an experiment with 3 factors, we usually want more precise knowledge as to what effects have an impact on the dependent variable. Using these designs, we cannot say for certain. Usually a main effect is the larger "effect size" (it almost has to be, given that an interaction is conditional upon 2 factors), so some experimentalists would say that the F-tests are more likely to reflect the main effects than the interactions. Perhaps that is true, that it's more likely, but we wouldn't know. (In addition, in creating any F-tests, we'll need an error term.)

As another way to translate the design into the aliasing patterns, let's look at the "effects codes" for the design depicting all effects in a 3-way factorial. Design matrices may be read somewhat like contrast coefficients. The grand mean (GM) is defined by averaging over all cells (all values in the GM column are "1"). The main effect for factor A contrasts the 1st four rows (where a=1) with the last 4 rows (where a=2). The main effect for factor B (or C) contrasts the 4 rows where b=1 (or c=1) with the 4 rows where b=2 (or c=2). The interactions (analogous to how df are determined) are just the products of the main effect coefficients. So, for the AB interaction, multiple the A column with the B column, and the AB column results.

Condition	GM	A	B	C	AB	AC	BC	ABC
$a_1b_1c_1$	1	-1	-1	-1	1	1	1	-1
$a_1b_1c_2$	1	-1	-1	1	1	-1	-1	1
$a_1b_2c_1$	1	-1	1	-1	-1	1	-1	1
$a_1b_2c_2$	1	-1	1	1	-1	-1	1	-1
$a_2b_1c_1$	1	1	-1	-1	-1	-1	1	1
$a_2b_1c_2$	1	1	-1	1	-1	1	-1	-1
$a_2b_2c_1$	1	1	1	-1	1	-1	-1	-1
$a_2b_2c_2$	1	1	1	1	1	1	1	1

That's the table for the full factorial. Next, we think up whichever conditions we wish to run, and for any fractional factorial, make a new table, keeping the rows associated with the proposed design, and delete the other rows. So using the 2nd example (above), in which we're planning to run the conditions $a_1b_1c_2$, $a_1b_2c_1$, $a_2b_1c_1$, and $a_2b_2c_2$, thus we keep those rows and delete the others to obtain this "fractional" table:

Condition	GM	A	B	C	AB	AC	BC	ABC
$a_1b_1c_2$	1	-1	-1	1	1	-1	-1	1
$a_1b_2c_1$	1	-1	1	-1	-1	1	-1	1
$a_2b_1c_1$	1	1	-1	-1	-1	-1	1	1
$a_2b_2c_2$	1	1	1	1	1	1	1	1

From this fractional table, we can identify the confounds by looking for equal columns. For example, the pattern of -1's and 1's for columns A and BC are identical, thus A and BC are aliased, as are B and AC, as well as C and AB.

Let's look at another, bigger example. We'll add 1 more factor such that if we were to run the full design, we'd run a 2^4 factorial. We'll use the rule of defining the simplest effect (GM) as being confounded with the most complicated effect (ABCD). Using GM = ABCD, we multiple all terms by ABCD, delete the terms that appear more than once, and the confounding is what's left. In the 1st row, A = BCD, then B = ACD, C = ABD, etc.

Source in 2^4	confounds in fractional factorial
A	$A(ABCD) = A^2BCD = BCD$
B	$B(ABCD) = AB^2CD = ACD$
C	$ABC^2D = ABD$
D	$ABCD^2 = ABC$
AB	$A^2B^2CD = CD$
AC	$A^2BC^2D = BD$
AD	$A^2BCD^2 = BC$
BC	$AB^2C^2D = AD$
BD	$AB^2CD^2 = AC$
CD	$ABC^2D^2 = AB$
ABC	$A^2B^2C^2D = D$
ABD	$A^2B^2CD^2 = C$
ACD	$A^2BC^2D^2 = B$
BCD	$AB^2C^2D^2 = A$
ABCD	GM

These patterns indicate that we cannot untangle the main effects A, B, C, and D, from BCD, ACD, ABD, ABC (respectively). If we assume the 3-way interactions away, then we could say that we're testing the main effects (but be sure that assumption is reasonable). In addition to the main effects being confounded with the 3-ways, notice that there is aliasing of the 2-ways: AB = CD, AC = BD, AD = BC.

Fractional factorials provide efficiencies when many factors are to be studied simultaneously yet the large sample size that would be required is unachievable or daunting. Yet nothing in life comes free: that efficiency brings the cost of the confounding. Clearly if some higher-order interactions are expected, and those are often of central focus in theory-based studies, we need to think hard before using this design. (It's interesting that these designs are frequently used in conjoint analysis, due to the massive numbers of factors that may vary in, say, a new product design, yet here too, the confounding implies that the results are not unambiguously interpretable.)

QUASI-EXPERIMENTS

In the experiments described in this chapter, study participants are randomly assigned to conditions, then in the study we do something to them (the manipulation or independent variable) and then we measure some result (the dependent variable). Random assignment is critical if we wish to have confidence in the internal validity of our study, to make progress in examining "if-then" sorts of questions. Prior to the study's manipulation or intervention, the groups start out essentially at the same baseline, thus any group differences at the end of the study may be attributed to the experimental factors.

If subjects cannot be assigned across experimental conditions, the resulting study is called a "quasi-experimental design" (or sometimes, a "natural" experiment). Giving up random assignment is huge and problematic, but sometimes it seems unavoidable, and when it occurs, we simply do our best to eliminate alternative explanations. Here are 2 examples.

- Many cities and states have tried installing electronic signs that read how fast a car is going on the highway as it approaches the sign. If the question is, "Does the presence of the radar sign make people drive slower?" it would be difficult to run a "true" experiment—we couldn't sample a group of drivers and randomly assign some to drive on that highway past that point, and others not, or for some to be speedy drivers and others not. However, we could install the sign and measure the average speed, then remove the sign and measure the average speed, and do so, back and forth a few times, to see if the speeds are negatively correlated with the presence of the radar signs.

- Or, say the Hershey's company developed an animated talking Hershey's kiss, as a spokes-character like the talking M&M's. The company might wonder whether the new characters boost sales. In national broadcasting, Hershey's cannot randomly assign some consumers in a market to see the ad and others to not see the ad. Let's say Hershey's airs a series of TV ads with the new character, and sales in the next 2-3 weeks show significant lift. That's great, but was the lift attributable to the ads? What if the 2-3 weeks during which sales increased were those weeks right before Valentine's Day or Halloween? That would be a seasonality effect we'd need to tease out before we were confident in drawing conclusions about the effectiveness of the talking Hershey kiss.

Data resulting from quasi-experiments are often analyzed with techniques such as time series analyses, rather than ANOVA. For more information, see Campbell's and Stanley's classic (1963), as well as the others listed below. Marketing research texts also address the issues faced with quasi-experiments (cf., Iacobucci and Churchill, 2015).

SUMMARY

This chapter introduces a number of experimental designs that are different from factorial designs. There's no question that simple (e.g., 2-way, indeed 2×2) factorials dominate the literature, however it's always good to have options. The alternative designs discussed in this chapter include: randomized block designs (in which a "blocking" factor is used to control for some extraneous variation, often a characteristic of the study participants), Latin Squares (in which 2 blocking factors are used, in a particular lattice structure), nested designs (in which 1 factor is contained in another, beyond the usual nesting of subjects in an experimental treatment factor), split-plot designs (in which 1 experimental manipulation is applied to whole plots and another factor is applied to divisions of them into "split-plots"), and fractional factorials (in which a subset of conditions from a full factorial are run).

This chapter is focused on so-called "true" experiments in which a factor is manipulated and a dependent variable is measured. With random assignment of subjects to conditions, statements of causality, if-then statements, may be achievable.

REFERENCES

Classics (reviewers would take these as high credibility):

1. Box, George E. P., J. Stuart Hunter, and William G. Hunter (2005), *Statistics for Experimenters: Design, Innovation, and Discovery* 2nd ed., New York: Wiley.
2. Box, George E. P., William G. Hunter and J. Stuart Hunter (1978), *Statistics for Experimenters: An Introduction to Design, Data Analysis and Model Building*, NY: Wiley.

3. Cochran, William G. and Gertrude M. Cox (1957), *Experimental Design*, (2nd ed.), New York: Wiley.
4. Cox, D. R. (1958), *Planning of Experiments*, NY: Wiley.
5. Hicks, Charles R. (1982), *Fundamental Concepts in the Design of Experiments* 3rd ed., New York: CBS College Publishing.
6. John, Peter M. W. (1971), *Statistical Design and Analysis of Experiments*, New York: Macmillan.
7. Kirk, Roger E. (1982), *Experimental Design: Procedures for the Behavioral Sciences* (2nd ed.), Belmont, CA: Brooks/Cole (pp.778-805).
8. Snedecor, George W. and William G. Cochran (1980) *Statistical Methods* 7th ed., Ames, IA: The Iowa State University Press.
9. Winer, B.J., Donald R. Brown, and Kenneth M. Michels (1991), *Statistical Principles in Experimental Design*, (3rd ed.), New York: McGraw-Hill.

Also very good:
1. Berger, Paul D. and Robert W. Maurer (2002), *Experimental Design: With Applications in Management, Engineering, and the Sciences*, Belmont, CA: Wadsworth.
2. Brown, Steven R. and Lawrence E. Melamed (1990), *Experimental Design and Analysis*, Newbury Park, CA: Sage.
3. and Aronson, Elliot, Phoebe C. Ellsworth, J. Merrill Carlsmith, and Marti Hope Gonzales (1990), *Methods of Research in Social Psychology*, New York: McGraw-Hill, especially Ch.1 "An Introduction to Experiments" (pp.1-39); Ch.3 "Ethical Issues" (pp.83-113).

More references on research designs, specifically experimentation compared with field studies and quasi-experimentation, internal and external validity issues:
1. Campbell, Donald T., and Julian C. Stanley (1963), *Experimental and Quasi-Experimental Designs for Research*, Chicago: Rand McNally.
2. Cook, Thomas D. and Donald T. Campbell (1979) *Quasi-Experimentation: Design & Analysis Issues for Field Settings*, Boston, MA: Houghton Mifflin.
3. Greenberg, Jerald and Robert Folger (1988), *Controversial Issues in Social Research Methods*, New York: Springer-Verlag, especially: Ch.4 "The Laboratory Experiment vs. Field Research" (pp.61-71); Ch.5 "Experiments versus Quasi-experiments" (pp.79-93).
4. Rosenthal, Robert, and Ralph L. Rosnow (1991) *Essentials of Behavioral Research: Methods and Data Analysis* 2nd ed., Boston, MA: McGraw-Hill.
5. Spector, Paul E. (1981), *Research Designs*, Newbury Park, CA: Sage (pp. 7-32, 39-54, 78-80).

Special topics:
1. Davison, Mark L. and Anu R. Sharma (1990), "Parametric Statistics and Levels of Measurement: Factorial Designs and Multiple Regression," *Psychological Bulletin*, 107 (3) 397-400.
2. Dey, Aloke (2010), *Incomplete Block Designs*, World Scientific Publishing Co.
3. Gardner, David M. and Russell W. Belk (1980), *A Basic Bibliography on Experimental Design in Marketing*, Bibliography Series No.37, Chicago: AMA.
4. Heck, Ronald H. and Scott L. Thomas (2000), *An Introduction to Multilevel Modeling Techniques*, Mahwah, NJ: Erlbaum.

5. Hinkelmann, Klaus and Oscar Kempthorne (1994), *Design and Analysis of Experiments, Volume 1: Introduction to Experimental Design*, NY: Wiley.
6. Jaccard, James (1998), *Interaction Effects in Factorial Analysis of Variance*, Thousand Oaks, CA: Sage.
7. John, J. A. (1987), *Cyclic Designs*, London: Chapman & Hall.
8. John, Peter W. M. (1980), *Incomplete Block Designs: Lecture Notes in Statistics*, Marcel Dekker.
9. Lawson, John (2010), *Design and Analysis of Experiments with SAS*, Chapman & Hall / CRC Texts in Statistical Science.
10. Maxwell, Scott E. and Harold D. Delaney (1990) *Designing Experiments and Analyzing Data: A Model Comparison Perspective*, Belmont, CA: Wadsworth.
11. Moen, Ronald, Thomas Nolan, and Lloyd Provost (2012), *Quality Improvement Through Planned Experimentation* 3rd ed., McGraw-Hill.
12. Pedhazur, Elazar J. and Liora Pedhazur Schmelkin (1991), *Measurement, Design, and Analysis: An Integrated Approach*, Hillsdale, NJ: Erlbaum.
13. Raudenbush, Stephen W. and Anthony S. Bryk (2002), *Hierarchical Linear Models: Applications and Data Analysis Methods*, 2nd ed., Newbury Park, CA: Sage.
14. Tabachnick, Barbara G. and Linda S. Fidell (2001), *Computer-Assisted Research Design and Analysis*, Needham Heights, MA: Allyn & Bacon.
15. Toubia, Olivier and John R. Hauser (2007), "On Managerially Efficient Experimental Designs," *Marketing Science*, 26 (6), 851-858.

Within-Subjects Designs:
1. Girden, Ellen R. (1992), *ANOVA: Repeated Measures*, Newbury Park, CA: Sage.
2. Greenwald, Anthony G. (1976), "Within-Subjects Designs: To Use or Not To Use?," *Psychological Bulletin*, 83 (2), 314-320.

Random vs. Fixed Factors and Designs:
1. Jackson, Sally and Dale E. Brashers. (1994), *Random Factors in ANOVA*, Thousand Oaks, CA: Sage.
2. And my review of that book in the *Journal of Marketing Research* 32 (May), 238-239.

Managerial articles:
1. Almquist, Eric and Gordon Wyner (2001), "Boost Your Marketing ROI with Experimental Design," *Harvard Business Review*, 79 (9), 135-141.
2. Anderson, Eric T. and Duncan Simester (2011), "A Step-by-Step Guide to Smart Business Experiments," *Harvard Business Review*, 89 (3), 98-105.

CHAPTER 8

WITHIN-SUBJECTS (OR REPEATED MEASURES) ANOVA

Questions to guide your learning:
Q_1: What is a "within-subjects" or "repeated measures" design?
Q_2: How is such a design analyzed using ANOVA?

In basic statistics, we usually begin with 1-sample t-tests and proceed to 2-sample t-tests, from which we launch into ANOVA and F-tests. The 2-samples (or the "a" samples in a 1-way ANOVA, etc.) are said to be "independent" samples. The study participants in 1 group have nothing to do with each other, and indeed should not be influenced by each other. In contrast, a different kind of design is when we have 2-samples that are "matched." The observations in group 1 are somehow linked to those in group 2. The idea is that such a design might be more sensitive (powerful) in detecting effects of the experimental treatment because the variability in the one group is as close to the variability in the other group as is possible.

Perhaps the most famous form of a matched sample design is a study on twins. One twin is randomly assigned to group 1, and the other twin to group 2. Doing so over numerous twins means we've controlled certain biological factors, and sometimes environmental issues as well. Here's what the twins' matched samples would look like:

sample 1	sample 2
twin 1a	twin 1b
twin 2a	twin 2b
twin 3a	twin 3b …

The matching doesn't have to be biological. In marketing research studies, it's fairly common to match stores on their sales volume. If we are studying the impact of some manipulation like different package designs, or price points, or end-of-aisle displays on a dependent variable like sales, we know that larger stores, just due to their greater traffic, will sell more. We wouldn't want to put 1 price point in small stores and another price point in the large stores because then the price manipulation would be confounded with store size. So when we draw a random sample of stores from our chain (to be able to generalize back to all our stores), we first note their sales, pick some point, probably a median, under which we'd say, "that's a small store," over which we'd put the store in the "big store" group. That design would look like this:

sample 1	sample 2
big store 1	big store 2
big store 3	big store 4
small store 1	small store 2
small store 3	small store 4
small store 5	small store 6 …

The null hypothesis for 2-groups (matched or independent) is $H_0\colon \mu_1 = \mu_2$. When the samples are matched, the t-test has to be modified slightly. For matched t-tests, the data point is not so much x_{ij} for group i and twin or store j, as the difference score between the twins or stores in each pair, $d_j = d_{1j} - d_{2j}$. That difference score is how we translate the notion of matching into an explicit comparison.

Here's another (weird) example. Say Reebok is testing the springiness of new rubber soles for their sneakers. We could run an experiment in which we gave a sample of consumers new sneakers but with the old kind of rubber soles, and another sample of consumers new sneakers with the new rubber soles, we could ask them to run and play in these sneakers, and then send them into Reebok in 3 months at which point the company would measure how much the soles have worn down, comparing the old rubber to the new. That would be fine.

A matched sample approach would be to give every consumer a new pair of sneakers wherein the left shoe had the old kind of sole, and the right shoe had the new rubber. Each study participant yields 2 data points—the wear of each sole. The stress that one pair of shoes endures is totally matched within each subject ("within-subjects"), and we can use a smaller sample (essentially half as many participants).

In the computation for a matched sample t-tests, notice we translate the scores (X's) immediately to difference scores (d's):

Left sole	Right sole	Difference
X_{11}	X_{12}	$d_1 = X_{11} - X_{12}$
X_{21}	X_{22}	$d_2 = X_{21} - X_{22}$
	…	
X_{n1}	X_{n2}	$d_n = X_{n1} - X_{n2}$

The d's are our new variable. We compute a mean and standard deviation on it, where "n" is the number of pairs of shoes:

$$d = \frac{1}{n}\sum_{i=1}^{n} d_i \qquad s_d^2 = \frac{1}{n-1}\sum_{i=1}^{n}\left(d_i - \bar{d}\right)^2$$

We compute a t-statistic on (n-1) df: $t = \frac{(\bar{d}-0)}{s_d/\sqrt{n}}$. This t-statistic tests the null hypothesis $H_0\colon D = 0$; i.e., that the old and new soles will wear the same vs. $H_A\colon D \neq 0$ (one or the other type of sole will be tougher). If our t-value exceeds the critical value on n-1 df, we reject the null, it's significant, yada yada.

In the next section, we'll see how the matched sample design generalizes and how the matched t-test generalizes into an ANOVA counterpart. It's important to keep in mind the advantages of these designs, so let's get a preview. We are going to see how these matched, or "within-subjects" or "repeated measures" designs can be more efficient than a comparable between-subject designs (e.g., in requiring fewer subjects), and more sensitive (i.e., we can detect treatment differences more easily) because we'll be comparing a subject's behavior to his/her own behavior under different conditions, not to someone else's under different conditions. These designs can be especially useful in studying behavior change such as learning, memory or forgetting, attitude change, etc., where the difference scores are essentially d = post-test – pre-test.

Diagrammatic Representation of Several Simple Within-Subjects Designs

Let's draw some representations of between-subjects designs and within-subjects designs for visual comparison. Up to this chapter, we've concentrated on "between-subjects designs" or the "completely randomized design" in which we draw a random sample of subjects, and randomly assign each subject to only 1 treatment condition. There would be 4 sub-samples of subjects, S1-S4 across a 2×2 factorial:

	b=1	b=2
a=1	S1	S2
a=2	S3	S4

By comparison, in a totally "within-subjects design" we'd still a draw random sample of subjects, but now each subject yields data under all the treatment conditions (or, in a moment we'll see, under a subset of treatment conditions). In this design, 1 subject is run in all conditions:

	b=1	b=2
a=1	S1	S1
a=2	S1	S1

Notice that a subject is crossed with factors A and B. S1 is not nested in 1 and only 1 condition, the way we had assumed when we had written error terms like S(AB). Rather, S1 is completely crossed with the experimental factors. Thus, we'll use the notation (A×B×S).

The simplest design would be a 1-factor within-subject design. When the factor takes on only 2 levels, it is the matched t-test. The design would be denoted (A×S), representing 1 treatment factor, A, and each subject serves in all "a" treatments:

a=1	S1
a=2	S1

In the fuller 2×2 factorial, the design would be called (A×B×S) to represent 2 treatment factors, A and B, and each subject is run in all "a×b" treatment combinations:

	b=1	b=2
a=1	S1	S1
a=2	S1	S1

There can also be the situation where we have a design that has both a between-subjects factor and a within-subjects factor. In the design B×(A×S), there are 2 treatment factors, A and B. The notation says that A is crossed with S, that is, each subject is run in all "a" treatment conditions, so A is the within-subjects factor. The notation puts the factor B outside the parentheses—it is crossed with A but S is not crossed with it; i.e., a subject sees 1 and only 1 level of the between-subjects factor B. It would look like this:

	b=1	b=2
a=1	S1	S2
a=2	S1	S2

The obverse of that design would have factor A as the between-subjects factor, and B as the within-subjects factor. The design would be called A×(B×S):

	b=1	b=2
a=1	S1	S1
a=2	S2	S2

These last 2 designs, B×(A×S) and A×(B×S), are sometimes called "mixed" designs, because there is a between-subjects factor (B in B×(A×S), and A in A×(B×S)) and a within-subjects factor (A in B×(A×S), and B in A×(B×S)) in each design. (FYI, note that these 2 studies may be considered to be split-plot designs, where a subject in a_i (for A×(B×S)) is a whole-plot, and application of all b_j's to that subject is effectively deriving split-plots.)

Within-Subjects—Advantages over Between-Subjects Designs

Let's look at some of the reasons that within-subjects designs are attractive. First is the issue of efficiency: the number of subjects can be fewer, while maintaining comparable power. For example, say we'll run a 2-factor study with n = 5 subjects per cell, and and a=b=2. In the CRD, a fully between-subjects design, with 5 subjects in each of the 4 cells, we'd need a total sample size of 20. In a fully within-subjects design, we'd only need the 5 subjects because they'd provide data in each of the 4 conditions. In the mixed designs, we'd need an intermediate number of subjects:

CRD A×B	between-subjects	a×b×n	= 2×2×5 = 20
A×(B×S)	mixed	a×n	= 2×5 = 10
B×(A×S)	mixed	b×n	= 2×5 = 10
(A×B×S)	within-subjects	n	= 5

While the savings in sample size can be hugely important and practical, statistically it is even more impressive that within-subjects designs are typically more powerful or sensitive. We're comparing the data point from a subject in 1 condition to the data point from that same subject in another condition. Thus, each study participant serves as his or her own control group. With these designs, we are reducing the error term by effectively removing a source of error variability, the individual differences:

- In between-subjects design, when we compare $\bar{x}_{a1}$ to $\bar{x}_{a2}$, the difference between these means reflects treatment differences (a_1 vs a_2) as well as subject differences. Even with random assignment, the groups are likely to be at least a little different even at the start of the experiment, at least to the extent that the people in 1 group are different from the people in the other group.
- By comparison, in within-subjects design, the differences $\bar{x}_{a1}$ to $\bar{x}_{a2}$ will reflect primarily treatment differences of the factor (a_1 vs a_2), because the group of subjects run in a_1 is exactly the same as (not just "statistically the same as") the group of subjects run in a_2.

We'll also see evidence for this notion of greater sensitivity in the computation of SS shortly.

A third strength of within-subjects designs is that they can be the only, or best, way to study certain phenomena, anything that shows the effects of experience, such as learning and practice. For example, we might estimate the form of a learning curve over some number of trials (or some number of exposures to some task, or advertisement).

A fourth advantage is that the experiment might proceed more quickly. There will be some initial instructions that introduce the study participant to the task for the first condition, but later treatment conditions will probably not require lengthy instructions.

Within-Subjects—Disadvantages vs. Between-Subjects Designs

If within-subjects designs only had advantages over between-subjects designs, that's all we would run. But within-subjects designs have a few issues of their own.

For example, it's true that if we run 1 subject through all "a" conditions, we can expect more consistency (less random error variability) across treatments than if each of some number of subjects is run in 1 and only 1 of the "a" conditions. However, a subject in a within-subjects design is not exactly the same from condition to condition.

A subject's performance might get worse over time. The subject might get bored, tired, or careless over time, especially if the study participants are run in many conditions and each condition is lengthy or taxing. Possible solutions to reduce these detrimental effects would be to increase the subjects' motivation (e.g., increase their rewards for "good" performances), or by giving subjects the opportunity to rest, giving breaks between conditions if that is possible and if doing so is not too disruptive to the flow of the experiment.

Alternatively, a subject's performance might improve over time. The subject might get more practiced and knowledgeable over time, so that performance in later treatments will be better than performance in earlier sessions, even if the population treatment means should be equal; i.e., even if there is no treatment effect, there might be an effect of time and practice. Possible solutions to this issue of enhanced performance run in one of two directions. First, if learning per se is not of interest, we could have subjects practice for several trials until their behavior (as measured by our dependent variable) is up to some criterion level (out to some asymptote above which further learning is expected to be minimal). Second, if learning is of interest, we might control for practice effects by counterbalancing treatments presented across subjects. Thus, subject 1 would be run through a_1, a_2, a_3 and subject 2 run through a_2, a_3, a_1 etc., so the treatment is not confounded with time, order, or practice. For a=3, there are $3!=3\times2\times1=6$ possible orders for counter-balancing the different treatments:

$\{a_1\ a_2\ a_3\}$, or $\{a_1\ a_3\ a_2\}$, or $\{a_2\ a_1\ a_3\}$, or $\{a_2\ a_3\ a_1\}$, or $\{a_3\ a_1\ a_2\}$, or $\{a_3\ a_2\ a_1\}$

More general than effects of practice would be "context" effects. We can't wipe a subject's memory clean and refresh him/her between experimental sessions, so there might be residual "carry over" from earlier conditions to later ones. Not all forms of carry over or contamination can be resolved by counterbalancing. Say some condition a_2 is atypical (it's more difficult or more aversive, etc.), then that experience may affect all remaining sessions. For example, say we're asking consumers to rate various attributes of movies. Say the a=3 conditions are the kinds of movie clips they'll see. In condition a_1 = they see animated films like *Bambi*, a_2 = chainsaw slasher movies, a_3 = scifi movies. If subjects rate a_1 then a_2, they might rate movies in a_2 as more gory than usual just due to the contrast with the movies in a_1. If rating a_2 movies affected all remaining sessions, a_2 would have to go last; we can't overcome the effect with counterbalancing.

If we don't expect differential carry-over effects, we could run a complete within-subjects design. Or, we could run a mixed design where the within-subjects factor is the one we're expecting to show smaller effects (so we need the greater sensitivity).[13]

The last issue regarding within-subjects (or repeated measures) designs is that the statistical analysis can be more complicated. And for the first time in this book, we have to worry a bit about model assumptions. Let's look at each.

---

[13] Anthony G. Greenwald (1976), "Within-Subjects Designs: To Use or Not To Use?," *Psychological Bulletin*, 83 (2), 314-320, argues that between-subjects designs are not without their own context effects. It's a good article—good arguments and easy to read.

Within-Subjects Design: Notation and Analysis

We'll begin with the simplest of within-subjects designs—a single experimental factor A, and each subject, 1, 2, …, n, is exposed to, and provides data in, all "a" conditions. This design is referred to as (A×S), in that there is a complete crossing of all levels of factor A and all subjects. To the left we see that factor A, here with a=4 levels, has subject 1 in all levels, subject 2 in all levels, etc. To the right, we see the data, x_{ij}, for i=factor A (i = 1, 2, …, a), and j for subject (j = 1, 2, …, n).

(A×S)	a_1	a_2	a_3	a_4
	s_1	s_1	s_1	s_1
	s_2	s_2	s_2	s_2
	…			

yields	a_1	a_2	a_3	a_4
s_1	x_{11}	x_{21}	x_{31}	x_{41}
s_2	x_{12}	…		
s_3	x_{13}			
s_4	…			

With such data, we can compute treatment means, as always: $\bar{x}_i = \sum_{j=1}^{n} x_{ij}$. What's new is that we also now have more than 1 score per subject, so we can compute subject means as well, averaging over all of the conditions they were in: $\bar{x}_j = \sum_{i=1}^{a} x_{ij}$.

Next, let's look at the within-subjects ANOVA model. Usually, when we have a design with 2 factors, they are factors A and B. We don't speak of subjects as a factor, and they are nested in the AB cells, per S(AB). For the standard 2-factor factorial design, we know how the variability may be apportioned over effects in the model: $SS_T = SS_A + SS_B + SS_{AB} + SS_{S(AB)}$. In the within-subjects case, we don't have a factor B; our second factor is really "S." In addition, we don't have an $SS_{S(AB)}$ term (or $SS_{S(A)}$ for just 1 factor), because subjects aren't nested in a factor S(A) (or S(AB)), instead, subjects are crossed with a factor A×S (or A×B×S). So if we don't have an error term like $SS_{S(A)}$ (or $SS_{S(AB)}$), we have to derive a new error term.

In the figure below, we see on the left a standard 2-factor table, in which any (i,j) cell has n subjects. To the right is a 2-way table but it's a cross-classification of the 1 factor A with subjects, and any (i,j) cell is just a single data point—the subject j in that i[th] level of factor A. In the right table, we see that we do not have replications in our cells for the within-subjects design.

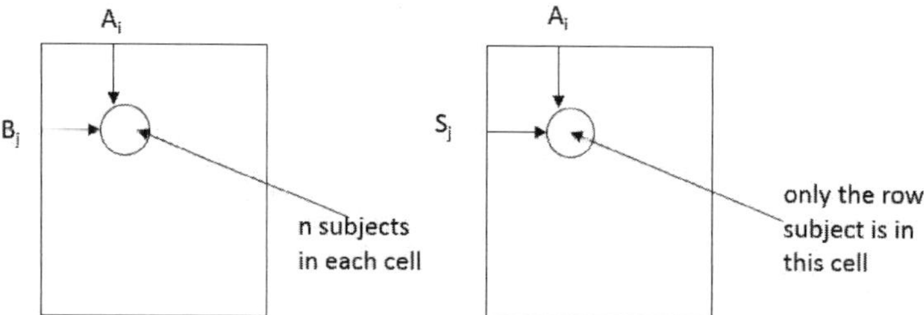

Recall from Chapter 3, that when we had only 1 observation in each cell, we had to use the interaction SS as the error SS, and that's essentially what we'll do for the within-subjects design. With replications in each cell, in a 2-factor design, we have the basic ANOVA table structure to the left. Without replications, in a 1-factor within-subject design, we have the ANOVA table structure to the right, with "a" levels of factor A, and n subjects, but no multiple reps or multiple observations in each A×S cell.

In A×B CRD effect	df		In (A×S) within-subjects design effect	df
A	a-1		A	a-1
B	b-1	vs.	S	n-1
A×B	(a-1)(b-1)		A×S	(a-1)(n-1)
error	ab(n-1)		error	an(#reps in cell -1) = an(1-1) = 0
total	abn-1		total	an-1

The ANOVA table for the 1-factor within-subjects design derives straight from the design and df:

source	df	SS	MS	F
A	a-1	$\sum\sum(\bar{x}_i - \bar{x})^2$	$\frac{SS_A}{a-1}$	$\frac{MS_A}{MS_{AS}}$
S	n-1	$\sum\sum(\bar{x}_j - \bar{x})^2$	$\frac{SS_S}{n-1}$	
A×S=error	(a-1)(n-1)	$\sum\sum(\bar{x}_{ij} - \bar{x}_i - \bar{x}_j + \bar{x})^2$	$\frac{SS_{AS}}{(a-1)(n-1)}$	
total	an-1	$\sum\sum(x_{ij} - \bar{x})^2$		

If we began to write a 1-factor within-subjects model, we might express the data point as: $x_{ij} = \mu + \alpha_i + \pi_j + (\alpha\pi)_{ij} + \epsilon_{ij}$. We're using π for "population." We cannot distinguish between the terms $(\alpha\pi)_{ij}$ and ϵ_{ij} because they're confounded (they have the same subscripts). This identity is just another illustration of the notion of 0 df for the error term. In such a model, we assume the "interaction" between A and S is not of interest (because it's confounded with error and we need an error term, and we can't test A×S against any error term), so $(\alpha\pi)_{ij}$ becomes ϵ_{ij}.

Still, from the sums of squares, we can see more precisely what is meant when within-subjects designs are characterized as being sensitive or more powerful than between-subjects designs. In a CRD ("completely randomized design," what we've been calling between-subjects), we know that the error SS is whatever is left after we subtract all the effects from the total SS:

$$\begin{aligned} SS_{error} &= SS_{total} - SS_{alleffects} \\ &= SS_{total} - SS_A = SS_{S(A)} \end{aligned}$$

In within-subjects designs, the logic is the same. For a 1-factor design, note that we subtract off more an additional SS term:

$$SS_{error} = SS_{total} - SS_A - SS_S$$

SS are always $\geq$0, (and almost always >0), so the SS_{error} for the within-subjects design will be $\leq SS_{error}$ for the comparable between-subjects design. We essentially start with the error SS of the CRD and then we subtract off even more—the SS of subject variability. We're subtracting off the individual differences.

2-Factors in a Within-Subjects Design

Let's make the design more complicated, now with 2 experimental factors, A and B, and S for subjects who we'll say are crossed with B and nested in A, thus factor A is the between-subjects factor (i = 1, 2, …, a) and factor B is the within-subjects factor (j= 1, 2, …, b) in the design: A×(B×S), with subjects as k = 1, 2, …, n.

	b=1	b=2
a=1	S1	S1
a=2	S2	S2

The ANOVA table for this design follows;

source	df	SS definitions	MS	F
A	(a-1)	$\sum\sum\sum(\bar{x}_i - \bar{x})^2$	$\frac{SS}{df}$	$\frac{MS_A}{MS_{S(A)}}$
S(A)	a(n-1)	$\sum\sum\sum(\bar{x}_{ik} - \bar{x}_i)^2$		–
B	(b-1)	$\sum\sum\sum(\bar{x}_j - \bar{x})^2$		$\frac{MS_B}{MS_{BS(A)}}$
A×B	(a-1)(b-1)	$\sum\sum\sum(\bar{x}_{ij} - \bar{x}_i - \bar{x}_j + \bar{x})^2$		$\frac{MS_{AB}}{MS_{BS(A)}}$
B×S(A)	a(b-1)(n-1)	$\sum\sum\sum(\bar{x}_{ijk} - \bar{x}_{ij} - \bar{x}_{ik} + \bar{x}_i)^2$		–
total	abn-1	$\sum\sum\sum(x_{ijk} - \bar{x})^2$		

Notice that in this analysis, there are 2 error terms, $MS_{S(A)}$ and $MS_{BS(A)}$. For the main effect of A, and any follow-up contrasts on A, we use $MS_{S(A)}$ as the error term in an F-test. For the main effect of B and the interaction, and any of their contrasts, we'd use $MS_{BS(A)}$ as the error term.

Deriving the EMS is complicated, but very intuitively the F-ratio is formed with the MS for the effect of interest in the numerator, as always, and the denominator is whatever MS we have that has that "effect" plus the S (subject) term. In general, the F-test is the ratio of the MS_{effect} to $MS_{effect-w-subject}$. Specifically for a 2-factor within-subjects design: $MS_A/MS_{S(A)}$ and $MS_B/MS_{BS(A)}$.

Generalize to 3 Factors: A, B, C, and Subjects

Let's go up to one more level of complication. Let's say we have 3 experimental factors, as well as subjects. We'll see 3 varieties of a 3-factor design, with 1, 2, or all 3 factors being within-subjects factors.

3-Factors: 1 Within-Subjects Factor, 2 Between-Subjects Factors

Let's say we have factors A and B as between-subjects factors, and factor C is a within-subjects (or repeated measures) factor. The design A×B×(C×S) would look like this:

c=1

	b=1	b=2
a=1	S1	S2
a=2	S3	S4

c=2

	b=1	b=2
a=1	S1	S2
a=2	S3	S4

The ANOVA table, sources of variation, definitions of SS, and forms of the F-tests follow:

	source	df	SS	MS	F
betw	A	(a-1)	$\sum\sum\sum\sum(\bar{x}_i - \bar{x})^2$	$\frac{SS}{df}$	$\frac{MS_A}{MS_{S(AB)}}$
betw	B	(b-1)	$\sum\sum\sum\sum(\bar{x}_j - \bar{x})^2$		$\frac{MS_B}{MS_{S(AB)}}$
betw	A×B	(a-1)(b-1)	$\sum\sum\sum\sum(\bar{x}_{ij} - \bar{x}_i - \bar{x}_j + \bar{x})^2$		$\frac{MS_{AB}}{MS_{S(AB)}}$
betw	S(AB)	ab(n-1)	$\sum\sum\sum\sum(\bar{x}_{ijk} - \bar{x}_{ij})^2$		–
w/in	C	(c-1)	$\sum\sum\sum\sum(\bar{x}_k - \bar{x})^2$		$\frac{MS_C}{MS_{CS(AB)}}$
w/in	A×C	(a-1)(c-1)	$\sum\sum\sum\sum(\bar{x}_{ik} - \bar{x}_i - \bar{x}_k + \bar{x})^2$		$\frac{MS_{AC}}{MS_{CS(AB)}}$
w/in	B×C	(b-1)(c-1)	$\sum\sum\sum\sum(\bar{x}_{jk} - \bar{x}_j - \bar{x}_k + \bar{x})^2$		$\frac{MS_{BC}}{MS_{CS(AB)}}$
w/in	A×B×C	(a-1)(b-1)(c-1)	$\sum\sum\sum\sum(\bar{x}_{ijk} - \bar{x}_{ij} - \bar{x}_{ik} - \bar{x}_{jk} + \bar{x}_i + \bar{x}_j + \bar{x}_k - \bar{x})^2$		$\frac{MS_{ABC}}{MS_{CS(AB)}}$
w/in	C×S(AB)	ab(c-1)(n-1)	$\sum\sum\sum\sum(x_{ijk\ell} - \bar{x}_{ij\ell} - \bar{x}_{ijk} + \bar{x}_{ij})^2$		–
	total	abcn-1	$\sum\sum\sum\sum(x_{ijk\ell} - \bar{x})^2$		

In this ANOVA table, we can detect the pattern. The between-subjects effects are tested against a between-subjects error term, and the within-subjects effects are tested against against a within-subjects error term.

3-Factors: 2 Within-Subjects Factor, 1 Between-Subjects Factors

In this example, let's look at 2 within-subjects factors (B and C), with A being a between-subjects factor: A×(B×C×S).

c=1

	b=1	b=2
a=1	S1	S1
a=2	S2	S2

c=2

	b=1	b=2
a=1	S1	S1
a=2	S2	S2

source	df	SS	F
A	(a-1)		$MS_A/MS_{S(A)}$
S(A)	a(n-1)	crunch	—
B	(b-1)	crunch	$MS_B/MS_{BS(A)}$
C	(c-1)	crunch	$MS_C/MS_{CS(A)}$
A×B	(a-1)(b-1)		$MS_{AB}/MS_{BS(A)}$
A×C	(a-1)(c-1)		$MS_{AC}/MS_{CS(A)}$
B×C	(b-1)(c-1)		$MS_{BC}/MS_{BCS(A)}$
A×B×C	(a-1)(b-1)(c-1)		$MS_{ABC}/MS_{BCS(A)}$
B×S(A)	a(b-1)(n-1)		—
C×S(A)	a(c-1)(n-1)		—
B×C×S(A)	a(b-1)(c-1)(n-1)		—
total	abcn-1		

Notice that we have 4 different error terms!

3-Factors: All 3 Being Within-Subjects Factors

As the last permutation, consider all 3 factors, A, B, C as within-subjects (repeated measures) factors: (A×B×C×S).

c=1	b=1	b=2
a=1	S1	S1
a=2	S1	S1

c=2	b=1	b=2
a=1	S1	S1
a=2	S1	S1

source	df	SS	F
A	(a-1)		MS_A/MS_{AS}
B	(b-1)	crunch	MS_B/MS_{BS}
C	(c-1)	crunch	MS_C/MS_{CS}
A×B	(a-1)(b-1)	crunch	MS_{AB}/MS_{ABS}
A×C	(a-1)(c-1)		MS_{AC}/MS_{ACS}
B×C	(b-1)(c-1)		MS_{BC}/MS_{BCS}
A×B×C	(a-1)(b-1)(c-1)		MS_{ABC}/MS_{ABCS}
S	(n-1)		—
A×S	(a-1)(n-1)		—
B×S	(b-1)(n-1)		—
C×S	(c-1)(n-1)		—
A×B×S	(a-1)(b-1)(n-1)		—
A×C×S	(a-1)(c-1)(n-1)		—
B×C×S	(b-1)(c-1)(n-1)		—
A×B×C×S	(a-1)(b-1)(c-1)(n-1)		—
total	abcn-1		

Now there are nearly as many error terms as there are effects terms!

In these ANOVA tables, the sources of variation, the df and sums of squares, all look similar to between-subjects designs, it's just that the factor "S" for the the within-subjects designs replaces the B, and of course the selections of error terms have changed. As a practical matter, we'll have the computer calculate the df, SS, and MS, and then it would be a good idea to compute the F-tests by hand to verify that the computer had used correct the error term as the denominator.

Within-Subjects Model Assumptions

The between-subjects ANOVA model carried with it certain assumptions. One assumption was the homogeneity of variance, that the variability within each cell was roughly (statistically) the same across groups, and that assumption is what allowed us to aggregate all the error SS across all cells in the design to obtain a MS_{error}. The within-subjects ANOVA carries a similar assumption, but things get more complicated quickly.

In a 1-factor within-subjects (repeated measures) design, the homogeneity of variance assumption gets translated as follows. Say a=4 in (A×S). With 4 levels, we might look at 3 difference scores. In the dataset below, to the left is a subjects-by-levels-of-factor-A listing of the data that each subject gave under each experimental condition. In the middle are the difference scores we could calculate (the computer will do this automatically for us). At the right is an example of what 1 person's data might look like, or even what the means in the "a" conditions might look like, and therefore, what the difference scores, the d variables might look like.

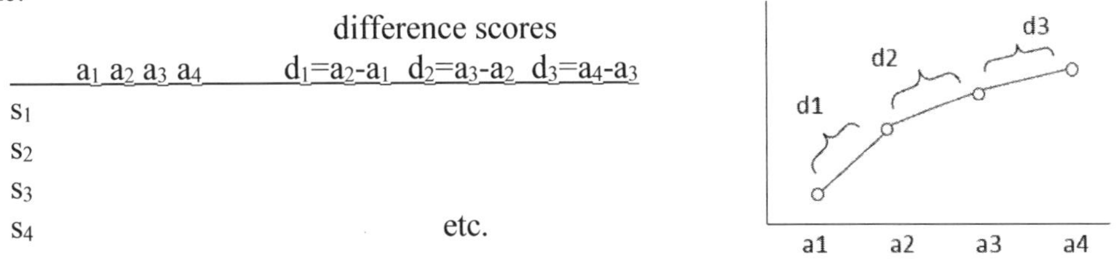

Given that the point of a within-subjects design is to compare explicitly a subject's data in 1 condition to the subject's data in another condition, the model is working directly off the difference score d's, not the raw data x's. Thus, where this model assumes homogeneity of variance, it's the homogeneity of the variances of these difference scores. Each difference score has a variance across the subjects in the sample or population: σ_{d1}^2, σ_{d2}^2, σ_{d3}^2. The homogeneity of σ^2 assumption states that in the population, $\sigma_{d1}^2 = \sigma_{d2}^2 = \sigma_{d3}^2$. In within-subjects designs, this assumption is known as the "homogeneity of treatment differences' variances" or h.o.t.d.v. (yes, really), or also "sphericity." (We'll say more about this assumption in Chapter 16.)

Also, in this example, some would say that more difference scores could be extracted. That seems unwise to me in a simple df sense; i.e., with 4 levels, 3 difference scores probably covers it. It is true that a different set of difference scores could be extracted. The differences we just calculated between adjacent time points are called "profile" difference scores. An alternative approach would be to take a difference score between each treatment a_1, a_2, a_3, with the last a_4 (or each a_2, a_3, a_4 with the first a_1). Hopefully the research context will be suggestive of which set of difference scores are likely to be most enlightening.

Satisfying the sphericity assumption naturally gets more challenging as the design gets bigger. With more levels of a factor, or more factors, there can be more difference scores, meaning there are more σ^2's to be equal, and an assumption of equality is less and less likely to hold, the more things there are to be equal. That is, assuming $\sigma_{d1}^2 = \sigma_{d2}^2 = \sigma_{d3}^2$ is one thing, but assuming: $\sigma_{d1}^2 = \sigma_{d2}^2 = \cdots = \sigma_{d\_many}^2$ gets increasingly restrictive.

What's worse is that while the CRD or between-subjects ANOVA models are very robust, meaning, only "extreme" violations of the assumption of homogeneity of variances are cause for concern, and only then if cell sizes are not equal, in within-subjects designs (or for the repeated measures variable in a mixed between- and within- design), the violation of the assumption can be less severe and it would still affect the F-test. The effect is one that makes the F-test too liberal. That is, we think we have $\alpha = .05$, but really we have >>.05. We'd make more Type I errors than we should, rejecting H_0's when we shouldn't, and "finding" effects that don't really exist.

There are 2 possible solutions to this problem. The first solution is called the "Geisser-Greenhouse correction" and it works by simply choosing critical F-values that are more conservative. The way this is done is by decreasing both the numerator and denominator df:

$$df_{numerator} = \frac{df_{effect}}{df_{repeated\ factors}} \qquad df_{denominator} = \frac{df_{error}}{df_{repeated\ factors}}$$

For example, in the design A×(B×S)

	b=1	b=2
a=1	S1	S1
a=2	S2	S2

we'd be using these mean squares terms: $MS_A = \frac{SS_A}{a-1}$, $MS_B = \frac{SS_B}{b-1}$, $MS_{S(A)} = \frac{SS_{S(A)}}{a(n-1)}$, $MS_{BS(A)} = \frac{SS_{BS(A)}}{a(b-1)(n-1)}$. A is a between-subjects factor, not a within-subjects (repeated measures) factor, so we leave its F-test alone. Its F-value is still: $= \frac{MS_A}{MS_{S(A)}}$, on (a-1) and a(n-1) df. But B is the repeated measures factor, so we apply the correction to the df as follows:

$$F = \frac{MS_B}{MS_{BS(A)}} \text{ on } \frac{b-1}{b-1} = 1 \text{ and } \frac{a(b-1)(n-1)}{b-1} = a(n-1) \text{ df.}$$

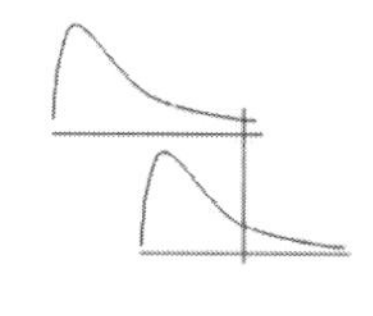

That adjustment to the df changes the critical value for the F-test on factor B. Say we have a = 3, b = 3, and n = 5. The usual F-test would have (b-1) = 2 and a(b-1)(n-1) = 24 df, and the critical value would be: $F_{2,24,.05} = 3.40$. The more conservative F-test with the adjusted df would have 1 and a(n-1) = 12 df, for a critical value of $F_{1,12,.05} = 4.75$.

This solution is an approximate solution. It has the great advantage that it is simple. Unfortunately, its disadvantage is that it tends to over-correct; thus, the new adjusted test is no longer too liberal, but perhaps too conservative (so it would be difficult to find any significant effects).

In Chapter 16, we shall see that a better solution will be to use a multivariate ANOVA, or MANOVA. In that model, the assumptions are not so restrictive, thus the likelihood of violating them is less.

SUMMARY

This chapter presented the logic and ANOVA for within-subjects, or repeated measures, designs. The design is an extension of a matched t-test. Any time we have obtained more than 1 data point from our subjects that we wish to model simultaneously, we have to consider whether we have a within-subjects, repeated measures situation (a sort of "multivariate" situation).

For example, we might wish to have a sample of subjects rate 2 brands on several attributes. We might pose the question as to whether the 2 brands differ statistically with respect to consumer perceptions on any of those attributes. If we obtained ratings about 1 brand from 1 sample, and ratings about the other brand from another sample, that would be a CRD or between-subjects design, and indeed, we run these all the time. But if we want an efficient, and sensitive (powerful) test in which we compare 1 subject's thoughts about 1 brand to his or her thoughts about another brand, we would do so using the logic of the differences, which make the comparisons explicit.

The terminology of "within-subjects" and "repeated measures" may be used interchangeably. The "repeated measures" phrase comes from the fact that in such a study, we do something (a=1) to subjects, then measure their responses, then we do something else to the same subjects (a=2) and measure their responses under the new condition. Thus, we are repeatedly measuring the study participants.

REFERENCE

Greenhouse, Samuel W. and Seymour Geisser (1959), "On Methods in the Analysis of Profile Data," *Psychometrika*, 24 (2), 95-112.

APPENDIX

This appendix works through a within-subjects (repeated-measures) example using the univariate ANOVA approach described in this chapter. (The multivariate (MANOVA) approach is superior and it is described in Chapter 16.) These data come from the article by David D. S. Poor (1973), "Analysis of Variance for Repeated Measures Designs: Two Approaches," *Psychological Bulletin*, 80 (3), 204-209.

```
data rept;
*This design is a 2×3×3. A is a between-subjects factor, and B and C are within subjects factors. Track one subject (e.g., #1) in the data listed below to verify the design, i.e., any subject is in 1 and only 1 level of A and sees all levels of B and C. Thus, this is an A×(B×C×S) design. X is the dependent variable, and S=subject#;
input a b c X s; datalines;
1 1 1 45 1
1 1 1 35 2
1 1 1 60 3
1 1 2 53 1
1 1 2 41 2
1 1 2 65 3
1 1 3 60 1
1 1 3 50 2
1 1 3 75 3
1 2 1 40 1
1 2 1 30 2
1 2 1 58 3
1 2 2 52 1
1 2 2 37 2
```

```
1 2 2 54 3
1 2 3 57 1
1 2 3 47 2
1 2 3 70 3
1 3 1 28 1
1 3 1 25 2
1 3 1 40 3
1 3 2 37 1
1 3 2 32 2
1 3 2 47 3
1 3 3 46 1
1 3 3 41 2
1 3 3 50 3
2 1 1 50 4
2 1 1 42 5
2 1 1 56 6
2 1 2 48 4
2 1 2 45 5
2 1 2 60 6
2 1 3 61 4
2 1 3 55 5
2 1 3 77 6
2 2 1 25 4
2 2 1 30 5
2 2 1 40 6
2 2 2 34 4
2 2 2 37 5
2 2 2 39 6
2 2 3 51 4
2 2 3 43 5
2 2 3 57 6
2 3 1 16 4
2 3 1 22 5
2 3 1 31 6
2 3 2 23 4
2 3 2 27 5
2 3 2 29 6
2 3 3 35 4
2 3 3 37 5
2 3 3 46 6
proc print; proc glm; class a b c s;
model x = A B C A*B A*C B*C A*B*C S(A) B*S(A) C*S(A) B*C*S(A) / SS3;
*There is no error term left in that model statement. Even B*C*S(A) is specified as an effect.
Doing so forces us to choose the right error terms;
test H = A            E = S(A);
test H = B            E = B*S(A);
test H = C            E = C*S(A);
```

```
test H = A*B        E = B*S(A);
test H = A*C        E = C*S(A);
test H = B*C        E = B*C*S(A);
test H = A*B*C      E = B*C*S(A); run;
```

Those commands result in the following statistical tests. Note that SAS very helpfully provides us feedback as to which MS term is being used as the error term in the F-test.

Tests of Hypotheses Using the Type III MS for S(A) as an Error Term

Source	DF	Type III SS	Mean Square	F Value	Pr > F
A	1	468.1666667	468.1666667	0.75	0.4348

Tests of Hypotheses Using the Type III MS for B*S(A) as an Error Term

Source	DF	Type III SS	Mean Square	F Value	Pr > F
B	2	3722.333333	1861.166667	63.39	<.0001
A*B	2	333.000000	166.500000	5.67	0.0293

Tests of Hypotheses Using the Type III MS for C*S(A) as an Error Term

Source	DF	Type III SS	Mean Square	F Value	Pr > F
C	2	2370.333333	1185.166667	89.82	<.0001
A*C	2	50.333333	25.166667	1.91	0.2102

Tests of Hypotheses Using the Type III MS for B*C*S(A) as an Error Term

Source	DF	Type III SS	Mean Square	F Value	Pr > F
B*C	4	10.66666667	2.66666667	0.34	0.8499
A*B*C	4	11.33333333	2.83333333	0.36	0.8357

HOMEWORK

This HW covers repeated measures.

Problem 1. Create the ANOVA table for the design: A×(B×C×D×S), with "source," df, and F.

CHAPTER 9

ANALYSIS OF COVARIANCE

Questions to guide your learning:
Q_1: What is ANCOVA?
Q_2: How are covariates useful?

This chapter looks at the analysis of covariance (ANCOVA). It's a fabulous technique, totally under-used, and it helps make our models super flexible. We'll see how those claims are true, and then we'll see how ANCOVA can be viewed as a mix of ANOVA and regression.

Think about the F-statistic in ANOVA:

$$F = \frac{MS_{effect}}{MS_{error}} \tag{1}$$

We spend a lot of time thinking about how to make the experimental manipulations clear and strong, so as to maximize the numerator, MS_{effect}, but notice that we can also obtain a larger F-value (increase the sensitivity and power of the test) if we can make the MS_{error} term smaller. One way we can do that is by using "covariates."

There are other ways, of course, to make the MS_{error} term smaller, so as to make the F-value larger: 1) When we speak of enhancing power (for larger F-tests) by making our sample size as large as possible, a large N feeds directly into making the MS_{error} term smaller (e.g., $MS_{S(AB)} = SS_{S(AB)}/df_{S(AB)}$, where $df_{S(AB)} = ab(n-1)$, and $N = abn$). 2) We can exercise direct control over error variability (the SS_{error} part of the MS_{error}) or "individual differences" or "heterogeneity" by: a) using a blocking design, and creating groups of subjects who are relatively homogeneous, pulling their variability out of the SS_{error}, or b) using a within-subjects (repeated measures) design. The option discussed in this chapter is based not on mechanical control in the design stage, but on statistical control in the modeling stage, using "covariates." A covariate is a variable we might include in our analyses to help remove extraneous variation from the dependent variable (to increase the precision of the analyses), or to remove bias due to pre-existing group differences.[14]

We'll see in a moment that ANCOVA is a mix of ANOVA and regression models, so it comes with a couple of conditions. An ANCOVA might be appropriate if:

1. There are 1 or more extraneous sources of variation expected to affect the dependent variable.
2. The extraneous variable(s), i.e., the covariate(s), can be measured on an interval- or ratio-level scale, because we'll need to compute correlations and means.
3. The covariate is linearly related to the dependent measure.

---

[14] Regarding terminology: "Analysis of Covariance" is like "Analysis of Variance" and is not to be confused with analysis of "Covariance Structures" (i.e., structural equations models, path models, Lisrel, etc.).

4. Direct experimental control (blocking or within subjects designs) is not possible or feasible (sometimes simply because the study has already been run and the data collected). Experimental solutions of blocking or within-subjects designs must be conducted prior to the data collection, whereas ANCOVA is a means of statistically controlling for variability even after we have the data (if we had measured some variable that can serve as a covariate).
5. The measurement of the covariate cannot reflect the experimental treatments (or else the covariate would in effect be another dependent variable). So, for example, we might measure the covariate prior to administering the treatment. Alternatively, we could measure the covariate after the experimental treatment, if we can assume that the treatment didn't affect the covariate. For example, in a memory skills experiment, we might use "G.P.A." as a proxy for intelligence. We could probably ask subjects for this information before or after the treatment or experimental manipulation since it shouldn't change as a function of the experimental treatments we're imposing on the study participants.

Let's look at the ANCOVA model in the context of an example. Recall from Chapter 5 (on contrasts), we studied an experiment in which we presented lists of 40 words to subjects to test which presentation format would yield the greatest successes in terms of recall of those 40 words. Factor A was the presentation style (on paper or a laptop/tablet, etc.) and the dependent variable was the measure of the number of words recalled. Even if we expected factor A to have differential treatment errors (some presentations will aid study participants' recall more than others), we might also suspect that subjects who are simply more intelligent will already have good memory strategies and be able to recall more words. Our interest is in factor A, and while "intelligence" may affect the dependent variable, it is a "nuisance variable." We don't care about it theoretically; we need to acknowledge it and control for it.

If the ANCOVA is a mix of ANOVA and regression, let's briefly review regression. Figure 1 shows a scatterplot depicting the relationship between 2 variables, X and V. (In this context, X is our dependent variable, and V will be our covariate.) The regression line assumes a linear relationship between the 2 variables, and posits a model with an intercept and a slope. The slope reflects the strength of the relationship between X and V, and the measure of association, r or r^2 (or R or R^2) is stronger to the extent that our knowledge of a person's score on V could help us predict the person's score on X.

Figure 1
Simple regression

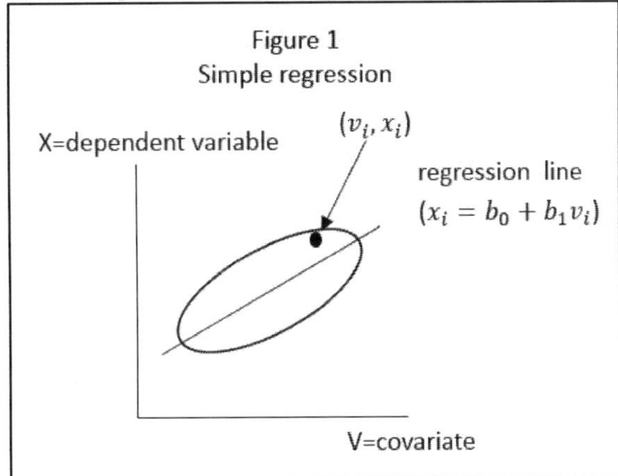

Within that larger scatterplot, in Figure 2, we're zooming in on a single person's data point, defined at (v_i, x_i). Figure 2 shows that in a regression model, there are 3 deviations of interest: $(x_i - \bar{x}) = (x_i - \hat{x}_i) + (\hat{x}_i - \bar{x})$. (2)

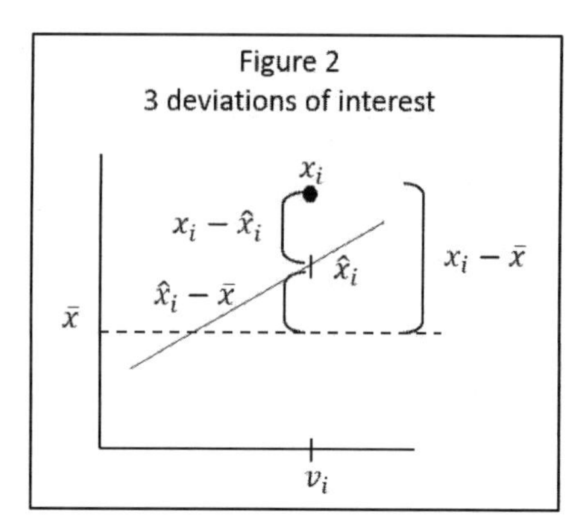

Figure 2
3 deviations of interest

If we sum and square these terms:

$$\sum(x_i - \bar{x})^2 = \sum(x_i - \hat{x}_i)^2 + \sum(\hat{x}_i - \bar{x})^2 \quad (3)$$

then the left hand side is "total variability in X" and the last term on the right hand side is the variability due to the linear regression (the difference from predicting the flat line corresponding to the mean on X to a line with a slope of / or \); i.e., is there a relationship between the covariate V and the dependent variable X.

The first term on right hand side is the variability left in X after removing the effect of the linear regression. This term is usually called $SS_{residuals}$ or SS_{error}. In the ANCOVA application, we'll call it $SS_{x(adj)}$; that is, SS in X after adjusting for the relationship with the covariate V.

In an ANCOVA, it's as if we ran the regression between the covariate V and the dependent variable X, and computed $(x_i - \hat{x}_i)$ for each subject, and then did an ANOVA on those $(x_i - \hat{x}_i)$ scores. That is, we'll be running an ANOVA on the "adjusted" scores. In ANOVA, the simplest definition of error SS may be stated as:

$$SS_{error} = SS_{total} - SS_{between} \quad (4)$$

The regression SS_{error} is analogous. In the ANCOVA context, let's refer to SS_{error} as $SS_{x(adj)}$, and recognize that SS_{total} for the dependent variable X is SS_x, and finally, switch out "between" SS (which is more of an ANOVA term) to a more regression term of $SS_{(regr)}$:

$$SS_{x(adj)} = SS_x - SS_{(due\ to\ lin\ regr)} \quad (5)$$

$$= SS_x - \frac{[\sum(x_i - \bar{x})(v_i - \bar{v})]^2}{\sum(v_i - \bar{v})^2} = SS_x - \frac{SS_{xv}}{SS_v}$$

The variance on the dependent variable X is the usual:

$$var(x) = \frac{SS_x}{df_x} = \frac{\sum(x_i - \bar{x})^2}{n-1} \quad (6)$$

The variance on the error or adjusted scores is similar:

$$var(x_{adjusted-scores}) = \frac{SS_{x(adj)}}{df_{x(adj)}} = \frac{SS_{x(adj)}}{n-1-1} \quad (7)$$

In comparing equations (6) and (7), the numerator has changed from SS_x to $SS_{x(adj)}$ (which we hope is smaller), and in the denominator we have lost 1 df for having estimated the slope parameter b. The loss of any error df is a bummer, but in ANCOVA, we hope that the reduction in $SS_{x(adj)}$ (from SS_x) is greatly worth it, more than compensating for the loss of a single df.

Structurally, the ANOVA and ANCOVA look similar. For a simple one-way, between-subjects design, we'd have a dataset that looked something like this:

		1	2	3	…	a	Factor A
subjects	1						
	2						
	…						
	n						

The notation is the standard:

$i = 1....a$ (factor A)

$j = 1....n$ (subject in group i)

And now each subject gives 2 data points: their score on the dependent variable x_{ij}
and a score on the covariate v_{ij}

In the ANOVA we model: $x_{ij} = \mu + \alpha_i + \epsilon_{ij}$ (8)

$\uparrow$ α_i is the effect of treatment i

In the ANCOVA, we will model: $x_{ij} = \mu + \alpha_i + \beta(v_{ij} - \bar{v}) + \epsilon_{ij}$ (9)

the regression/slope coefficient, β $\uparrow$

Alternatively, we could think of this analysis as using an ANOVA model on the "adjusted" dependent measures: $x_{ij(adj)} = x_{ij} - \beta(v_{ij} - \bar{v}) = \mu + \alpha_i + \epsilon_{ij}$ (10)

Let's see an example

In this (admittedly odd) example, each "subject" is a municipality in a rural area in Finland (from Wildt and Ahtola, 1978, p.33). Factor A takes on 1 of 3 levels (a=3) defined by the type of liquor license granted to the municipality.

X = the dependent measure, defined as the number of traffic accidents during some time period after the license had been granted.

V = the covariate defined as the number of traffic accidents during a time period of the same length prior to when the liquor license had been granted.

We are interested in the effect of the license (factor A) on the traffic accidents (X). We are not particularly interested in the number of traffic accidents prior to the granting of licenses, but we recognize that the municipalities probably differ (populations, road conditions, etc.), and we want to control for that prior state.

As a brief aside, an alternative modeling approach would be to analyze these data as a repeated measures design, and look at the difference score (X-V). The researchers had chosen not to analyze the data in this manner because they had already expected a decrease in the trend of number of accidents due to lighter traffic (due, in turn, to higher gasoline prices).

Here are the data:

Factor A = 1, 2, 3 = type of license
V = pre-license # accidents; covariate
X = post-license # accidents; dependent variable

ID	A	V	X
1	1	190	177
2	1	261	225
3	1	194	167
4	1	217	176
5	2	252	226
6	2	228	196
7	2	240	198
8	2	246	206
9	3	206	226
10	3	239	229
11	3	217	215
12	3	177	188

In a plot of X by V, a linear relationship look vaguely reasonable, both within groups 1-3 and across them:

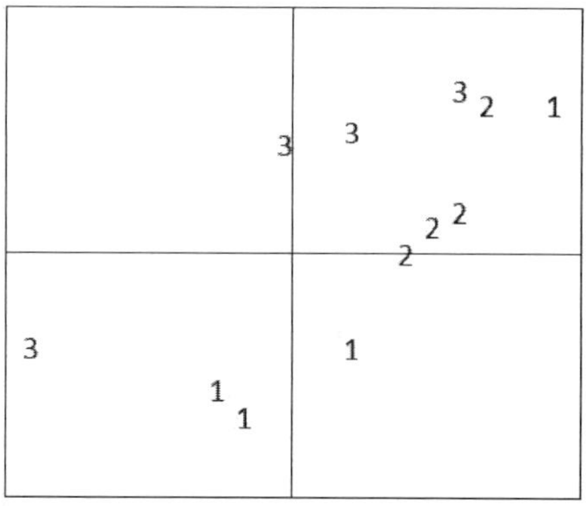

In the analyses that follow, we've run two models to make comparisons. In both the ANOVA and ANCOVA models, the dependent variable is X, and the single experimental factor is A. In SAS, both models would specify: class A;

1. For an ANOVA, we specify the model as usual: model X = A;
2. For an ANCOVA, we specify the model: model X = A V; This model statement says there are 2 predictors of X, both A and V. SAS knows that "A" is a factor (because it was included in the "class" statement) and "V" is a covariate predictor and not a factor (because it had not been listed in the "class" statement).

The 2 ANOVA tables follow.

ANOVA

Source		df	SS		MS		F	Pr>F
A		2	1696.167	← 4	848.083		2.08	0.181
error	1 →	9	3670.750		407.861	← 2		
total		11	5366.917					

ANCOVA

Source		df	SS		MS		F	Pr>F
A		2	2401.757	← 4	1200.879		12.83	0.0032
V		1	2922.071		2922.071	3 →	31.22	0.0005
error	1 →	8	748.679		93.585	← 2		
total		11	5366.917					

Look at the arrows and numbers in the 2 tables. There are 4 things to note comparing the ANOVA and ANCOVA model results:

1) Notice that we did indeed lose 1 error df, since we had to estimate the slope coefficient in the linear regression between X and V in the ANCOVA. (That price to be paid is similar to losing df when using blocking factors.)
2) The loss of 1 df is more than compensated for in these data since the MS_{error} is reduced from 407.861 to 93.585. So we were successful in removing some extraneous variability. As a result of the reduction in MS_{error}, notice that the F-statistic jumps from $F_{2,9} = 2.08$ (p=.181) to $F_{2,8} = 12.83$ (p=.003).
3) F = 31.22 is a test of $H_0\colon \beta = 0$. Since we reject H_0, we conclude the covariate has a significant relationship with the dependent variable X, and so, is necessary or helpful in the model. (We know SS must be ≥ 0, usually >0, so even if the covariate itself had not been significant, it is very likely that the SS_{error} will still have been reduced.)
4) Also note that the treatment SS increased after adjusting for the covariate (from 1696.167 to 2401.757). That doesn't have to happen, it just depends on the group differences on V and on r_{xv}. That is, the basic goal of the inclusion of a covariate is to reduce the error variability. The covariate can increase or decrease the treatment effects (the numerator SS) depending on the group differences on the covariate and on the correlation between the covariate and the dependent variable.

A final comment regarding ANOVA vs. ANCOVA... By default, SAS produces "Type I" sums of squares and "Type III" sums of squares. (The typical difference between these types of SS is defined precisely in Chapter 10.) When running an ANCOVA, the difference is as follows: The Type III SS are the SS having adjusted for the covariate—that is, these are the proper SS to report for the ANCOVA. What's handy is that the Type I SS are the SS we'd have obtained had we run an ANOVA without the covariate, so it's easy to compare the 2 models. Anyway, be sure to report the results that are based on the Type III SS.

SUMMARY

This chapter presents the analysis of covariance (ANCOVA). The model is shown to be a mix of regression and ANOVA. ANCOVAs are very useful for reducing SS_{error} so as to enhance the F-test for the effects in the model that are of greater theoretical interest than the covariate(s) which we are using as statistical controls.

REFERENCES

1. Edwards, Allen L. (1979), *Multiple Regression and the Analysis of Variance and Covariance*, NY: Freeman.
2. Maxwell, Scott E., Harold D. Delaney, and Charles A. Dill (1984), "Another Look at ANCOVA Versus Blocking," *Psychological Bulletin*, 95 (1), 136-147.
3. Wildt, Albert R., and Olli T. Ahtola (1978), *Analysis of Covariance*, Beverly Hills, CA: Sage.

CHAPTER 10

UNBALANCED DESIGNS (UNEQUAL n's)

Questions to guide your learning:
Q_1: What constitutes an unbalanced design?
Q_2: What problems arise with unbalanced data?
Q_3: What is the solution to the problem?
Q_4: What are "Types of SS"? And why will we vow to use only Type III?

Balanced: all $n_{ij} = n$
Unbalanced: cell sizes vary.

In this chapter, we take a look at what happens when the sample sizes in each cell in the experimental design aren't precisely equal. Let's start with a tiny example that is balanced.

The table below depicts a 2×2 design, with only 2 observations per cell (to keep things simple). To get a rough feel for the data, look at the cell means (the $\bar{x}_{ij}$'s: 8, 5, 8, 5). They indicate that there is no A×B interaction; and no main effect for factor A, but the mean for b=1 is larger than the b=2 mean, so there may be a B main effect.

	b_1	b_2	
a_1	$x_{111} = 7$ $x_{112} = 9$	$x_{121} = 5$ $x_{122} = 5$	←raw data (2 per cell)
	$\bar{x}_{11} = \mathbf{8}$ $n_{11} = 2$	$\bar{x}_{12} = \mathbf{5}$ $n_{12} = 2$	←cell mean ←cell size
a_2	$x_{211} = 8$ $x_{212} = 8$	$x_{221} = 4$ $x_{222} = 6$	
	$\bar{x}_{21} = \mathbf{8}$ $n_{21} = 2$	$\bar{x}_{22} = \mathbf{5}$ $n_{22} = 2$	

In this little data set, $n_{ij} = 2$, so N = a×b×n = 2×2×2 = 8. Given that the n's in each cell are identical, this is a "balanced" design (i.e., that's what we've been assuming for most of the book). When the n's are off even slightly from perfectly equal, they are said to be unbalanced. Unbalanced data affect (at least) two things: first, how the main effect means are calculated, and second, the nature of the hypotheses we test. Let's examine each issue first in a balanced data set and then in an unbalanced data set.

How Unbalancedness Affects the Means

The first issue is the computation of the means. The means for the main effects can be computed in either of 2 ways:[15]

---

[15] Main effect means are also called marginal means because they are found from, and often represented in, the margins of the data table, thus each row for factor A would have a mean aggregating over all levels of factor B, and each column for factor B would have a mean aggregating over all levels of A.

1) As means of cell means. We'll call these μ's or $\hat{\mu}$'s:

$$\hat{\mu}_{a1} = \frac{\hat{\mu}_{11} + \hat{\mu}_{12}}{b} = \frac{\bar{x}_{11} + \bar{x}_{12}}{b} = \left(\frac{1}{2}\right)(8+5) = 6.5$$

$$\hat{\mu}_{a2} = \frac{\hat{\mu}_{21} + \hat{\mu}_{22}}{b} = \frac{\bar{x}_{21} + \bar{x}_{22}}{b} = \left(\frac{1}{2}\right)(8+5) = 6.5$$

2) Or equivalently, as means of the raw data. We'll call these μ^*'s or $\hat{\mu}^*$'s:

$$\hat{\mu}^*_{a1} = \frac{\sum_{k=1}^{n} x_{11k} + \sum_{k=1}^{n} x_{12k}}{n_{11} + n_{12}} = \frac{(7+9)+(5+5)}{2+2}$$
$$= \frac{1}{4}(7+9+5+5) = \frac{26}{4} = 6.5$$

$$\hat{\mu}^*_{a2} = \frac{\sum_{k=1}^{n} x_{21k} + \sum_{k=1}^{n} x_{22k}}{n_{21} + n_{22}} = \frac{(8+8)+(4+6)}{2+2}$$
$$= \frac{1}{4}(8+8+4+6) = 6.5$$

This approach of calculating marginal means (μ^*) from the raw data can also be shown as a function of weighting each cell mean with its cell size:

$$\hat{\mu}^*_{a1} = \frac{n_{11}}{n_{a1}}(\hat{\mu}_{11}) + \frac{n_{12}}{n_{a1}}(\hat{\mu}_{12}) = \frac{2}{4}(8) + \frac{2}{4}(5) = 6.5$$

$$\hat{\mu}^*_{a2} = \frac{n_{21}}{n_{a2}}(\hat{\mu}_{21}) + \frac{n_{22}}{n_{a2}}(\hat{\mu}_{22}) = \frac{2}{4}(8) + \frac{2}{4}(5) = 6.5$$

Similarly we can compute the factor B main effect means either way:

1) As means of cell means:

$$\hat{\mu}_{b1} = \frac{\hat{\mu}_{11} + \hat{\mu}_{21}}{a} = \left(\frac{1}{2}\right)(8+8) = 8$$
$$\hat{\mu}_{b2} = \frac{\hat{\mu}_{12} + \hat{\mu}_{22}}{a} = \left(\frac{1}{2}\right)(5+5) = 5$$

2) Or equivalently, as means of the raw data:

$$\hat{\mu}^*_{b1} = \frac{1}{4}(7+9+8+8) = 8$$
$$\hat{\mu}^*_{b2} = \frac{1}{4}(5+5+4+6) = 5$$

→So far, **lesson #1** is that for balanced data, the
μ's, derived as means of cell means, and the
μ^*'s, derived as means of raw data (or the weighted means of cell means)
are the same. That is, for balanced data, $\mu = \mu^*$.

Now, the question is, what happens with unbalanced data? Here is our previous data set but with 1 observation removed from the (1,2) cell and 1 observation deleted from the (2,1) cell. These particular observations were removed so that we wouldn't be disturbing the cell means. Thus the cell means (or the means we'd use to look at the interaction term) are identical to those in the previous, balanced example, so it looks like there is no interaction.

	b_1	b_2	
a_1	$x_{111} = 7$ $x_{112} = 9$	$x_{121} = 5$	←raw data (2 per cell)
	$\bar{x}_{11} = 8$ $n_{11} = 2$	$\bar{x}_{12} = 5$ $n_{12} = 1$	←cell mean ←cell size
a_2	$x_{211} = 8$	$x_{221} = 4$ $x_{222} = 6$	
	$\bar{x}_{21} = 8$ $n_{21} = 1$	$\bar{x}_{22} = 5$ $n_{22} = 2$	

The issue regarding the effect of unbalanced data on computing the marginal means is that if we use μ or μ^*, we obtain different estimates. Beginning with factor A, we see the marginal means computed first as:

1) As means of cell means for factor A:

$$\hat{\mu}_{a1} = \frac{\hat{\mu}_{11} + \hat{\mu}_{12}}{b} = \left(\frac{1}{2}\right)(8 + 5) = 6.5$$

$$\hat{\mu}_{a2} = \frac{\hat{\mu}_{21} + \hat{\mu}_{22}}{b} = \left(\frac{1}{2}\right)(8 + 5) = 6.5$$

2) NOT equivalently, as means of the raw data:

$$\hat{\mu}^*_{a1} = \frac{1}{3}(7 + 9 + 5) = 7$$

$$\hat{\mu}^*_{a2} = \frac{1}{3}(8 + 4 + 6) = 6$$

Or using the idea of weighting the means by their relative cell sizes:

$$\hat{\mu}^*_{a1} = \frac{n_{11}}{n_{a1}}(\hat{\mu}_{11}) + \frac{n_{12}}{n_{a1}}(\hat{\mu}_{12}) = \frac{2}{3}(8) + \frac{1}{3}(5) = 7$$

$$= \frac{1}{3}((2 \times 8) + 5) = 7$$

$$\hat{\mu}^*_{a2} = \frac{n_{21}}{n_{a2}}(\hat{\mu}_{21}) + \frac{n_{22}}{n_{a2}}(\hat{\mu}_{22}) = \frac{1}{3}(8) + \frac{2}{3}(5) = 6$$

$$= \frac{1}{3}(8 + (2 \times 5)) = 6$$

Note that the means for factor A are now different. If we calculate the means the first way, then there is no main effect for A. If we calculate the means the second way, there might be a main effect for A.

Analogously, let's do the calculations for factor B:

1) As means of cell means:

$$\hat{\mu}_{b1} = \frac{\hat{\mu}_{11} + \hat{\mu}_{21}}{a} = \left(\frac{1}{2}\right)(8+8) = 8$$
$$\hat{\mu}_{b2} = \frac{\hat{\mu}_{12} + \hat{\mu}_{22}}{a} = \left(\frac{1}{2}\right)(5+5) = 5$$

2) And means of the raw data:

$$\hat{\mu}^*_{b1} = \frac{1}{3}(7+9+8) = 8$$
$$\hat{\mu}^*_{b2} = \frac{1}{3}(5+4+6) = 5$$

In general, the means are not equal, $\mu \neq \mu^*$ (these are equal by coincidence).

→So, **lesson #2**: in general, as it happened for factor A, for unbalanced data, $\mu \neq \mu^*$. That ambiguity is not good. For something as simple as a main effect, like for factor A in this example, we don't know whether we have a main effect or not. (Ohmigosh I hope there is a solution! There is. Hold on.)

How Unbalancedness Affects the Hypotheses

The second way (that we'll discuss) that unbalanced data raises a concern is that it affects the hypotheses we test, or the hypotheses we think we're testing. Let's focus on testing the main effect for factor A (but obviously the logic and math are analogous for factor B). We have the 2-way ANOVA model: $x_{ijk} = \mu + \alpha_i + \beta_j + (\alpha\beta)_{ij} + \epsilon_{ijk}$, and the null hypothesis is that there is no main effect for factor A: H_0: $\alpha_1 = \alpha_2 = 0$.

Let's use the model parameters to look at the structure of each data point $x_{ijk} = \mu + \alpha_i + \beta_j$ (to keep things simple, we'll forget $(\alpha\beta)_{ij} + \epsilon_{ijk}$ for the moment). In this table, for the balanced data ($n_{ij} = 2$ for all i,j), we have written every data point in terms of the grand mean and the 2 main effect terms:

	b_1	b_2
a_1	$7 = \mu + \alpha_1 + \beta_1$ $9 = \mu + \alpha_1 + \beta_1$	$5 = \mu + \alpha_1 + \beta_2$ $5 = \mu + \alpha_1 + \beta_2$
a_2	$8 = \mu + \alpha_2 + \beta_1$ $8 = \mu + \alpha_2 + \beta_1$	$4 = \mu + \alpha_2 + \beta_2$ $6 = \mu + \alpha_2 + \beta_2$

To test H_0: $\alpha_1 = \alpha_2 = 0$ (or $\alpha_1 - \alpha_2 = 0$), we'd compare $\bar{x}_{a1}$ to $\bar{x}_{a2}$ as follows.

$$\bar{x}_{a1} - \bar{x}_{a2} =$$
$$= \left(\frac{1}{4}\right)[(\mu + \alpha_1 + \beta_1) + (\mu + \alpha_1 + \beta_1) + (\mu + \alpha_1 + \beta_2) + (\mu + \alpha_1 + \beta_2)]$$
$$- \left(\frac{1}{4}\right)[(\mu + \alpha_2 + \beta_1) + (\mu + \alpha_2 + \beta_1) + (\mu + \alpha_2 + \beta_2) + (\mu + \alpha_2 + \beta_2)]$$
$$= \left(\frac{1}{4}\right)(4\mu + 4\alpha_1 + 2\beta_1 + 2\beta_2)$$
$$- \left(\frac{1}{4}\right)(4\mu + 4\alpha_2 + 2\beta_1 + 2\beta_2)$$
$$= \mu + \alpha_1 + (.5)\beta_1 + (.5)\beta_2 - \mu - \alpha_2 - (.5)\beta_1 - (.5)\beta_2$$
$$= \alpha_1 - \alpha_2$$

That's exactly what we want. In comparing $\bar{x}_{a1}$ and $\bar{x}_{a2}$, the μ's cancel, the β's cancel, and we're left with the main effect terms, the α's. We have isolated precisely the effects we're interested in comparing—the group effects for a=1 and a=2.

Lesson #3: in balanced data, by comparing the means, we're testing the correct hypothesis.

Let's test the same null hypothesis, $H_0\text{: } \alpha_1 = \alpha_2 = 0$ in the unbalanced data set:

	b_1	b_2
a_1	$7 = \mu + \alpha_1 + \beta_1$ $9 = \mu + \alpha_1 + \beta_1$	$5 = \mu + \alpha_1 + \beta_2$
a_2	$8 = \mu + \alpha_2 + \beta_1$	$4 = \mu + \alpha_2 + \beta_2$ $6 = \mu + \alpha_2 + \beta_2$

If we seek to test the main effect for factor A, obviously logic dictates that we compare a mean that reflects what's going on for a=1 to a mean that reflects what's going on in conditions where a=2. The problem is that when we compare $\bar{x}_{a1}$ to $\bar{x}_{a2}$ the results differ depending on whether we use $(\mu_{a1} - \mu_{a2})$ or $(\mu^*_{a1} - \mu^*_{a2})$.

Using $(\mu_{a1} - \mu_{a2})$, the means of the cell means, we get the same result as before (the one we want):

$$\begin{aligned}
\mu_{a1} - \mu_{a2} &= \\
&= \left(\tfrac{1}{2}\right)\left[\left(\tfrac{1}{2}\right)[(\mu + \alpha_1 + \beta_1) + (\mu + \alpha_1 + \beta_1)] + (\mu + \alpha_1 + \beta_2)\right] \\
&\quad - \left(\tfrac{1}{2}\right)\left[(\mu + \alpha_2 + \beta_1) + \left(\tfrac{1}{2}\right)[(\mu + \alpha_2 + \beta_2) + (\mu + \alpha_2 + \beta_2)]\right] \\
&= \left(\tfrac{1}{2}\right)[(\mu + \alpha_1 + \beta_1) + (\mu + \alpha_1 + \beta_2)] \\
&\quad - \left(\tfrac{1}{2}\right)[(\mu + \alpha_2 + \beta_1) + (\mu + \alpha_2 + \beta_2)] \\
&= \mu + \alpha_1 + (.5)\beta_1 + (.5)\beta_2 - \mu - \alpha_2 - (.5)\beta_1 - (.5)\beta_2 \\
&= \alpha_1 - \alpha_2
\end{aligned}$$

However, when we use the raw data, we get a different result.

$$\begin{aligned}
(\mu^*_{a1} - \mu^*_{a2}) &= \left(\tfrac{1}{3}\right)[(\mu + \alpha_1 + \beta_1) + (\mu + \alpha_1 + \beta_1) + (\mu + \alpha_1 + \beta_2)] \\
&\quad - \left(\tfrac{1}{3}\right)[(\mu + \alpha_2 + \beta_1) + (\mu + \alpha_2 + \beta_2) + (\mu + \alpha_2 + \beta_2)] \\
&= \left(\tfrac{1}{3}\right)(3\mu + 3\alpha_1 + 2\beta_1 + \beta_2) - \left(\tfrac{1}{3}\right)(3\mu + 3\alpha_2 + \beta_1 + 2\beta_2) \\
&= \mu + \alpha_1 + \left(\tfrac{2}{3}\right)\beta_1 + \left(\tfrac{1}{3}\right)\beta_2 - \mu - \alpha_2 - \left(\tfrac{1}{3}\right)\beta_1 - \left(\tfrac{2}{3}\right)\beta_2 \\
&= \alpha_1 - \alpha_2 + \left(\tfrac{1}{3}\right)(\beta_1 - \beta_2)
\end{aligned}$$

What's that! We wanted to test (and we think we're testing) $H_0\text{: } \alpha_1 - \alpha_2 = 0$, but we'd actually be testing $H_0\text{: } \alpha_1 - \alpha_2 + \left(\tfrac{1}{3}\right)(\beta_1 - \beta_2) = 0$. The difference is picking up on some effect of B in addition to A, or we would say that the test for the effect of factor A is biased by B (or confounded in part by B). In all our statistical tests, we need to estimate an effect having controlled for the other effects in the model. Here we need to adjust the means of A to remove the contamination of factor B. We want the effects to be "orthogonal," i.e., the tests of 1 factor are independent of other effects, or the effects are "adjusted" for the others, and said still differently, we seek to estimate the effects after other effects are partialled out.

So, **Lesson #4**: in unbalanced data, hypothesis tests may be biased, or contaminated by other effects in the model. The effect to be tested is not purely isolated so the test reflects a bit of a confound. In this example, even if there were no main effect for factor A, that is, $\alpha_1 - \alpha_2 = 0$, the "main effect" test for factor A might be deemed significant if this term were sufficiently big: $\left(\frac{1}{3}\right)(\beta_1 - \beta_2)$.

The Solution

The goal is to estimate the effects we wish to test purely, isolating it from confounds or biases imposed by other effects. The question is how to do so, how to "partial out" the other effects. We know it can be done, because we saw that using the cell means (rather than the raw data) yielded the correct hypothesis tests. So let's see, in general, the logic of partialling out effects.

The first column in the following table lists all the possible models we could fit for a 2-way factorial—we could include each main effect or not, and the interaction or not. (In reality, we'd always include both main effects and the interaction, all whether they're significant or not. We're just looking at these options to understand the partialling issue.) The next column shows of course the SS that would be estimated. The final column shows what the error term would be when we start with the variance in the variable, X, and extract (or partial out) the effects we're estimating by having included them in the model statement.

For example, in the first line, we have the usual—we're estimating both main effects and the interaction, so the error SS is what's left after estimating the grand mean, the main effects (A and B), and the interaction. In the 2nd line, we've left off the A main effect, so the error SS is what's left in the variance of X after we've estimated the grand mean, the main effect for B, and the interaction.

Models we could fit	SSeffects we'd get	SS_{error}
1. model X = a b a*b;	SS_A SS_B SS_{AB}	$Ss_e(\mu,\alpha, \beta,(\alpha\beta))$
2. model X = b a*b;	SS_B SS_{AB}	$SS_e(\mu, \beta,(\alpha\beta))$
3. model X = a a*b;	SS_A SS_{AB}	$SS_e(\mu,\alpha, (\alpha\beta))$
4. model X = a b;	SS_A SS_B	$SS_e(\mu,\alpha, \beta)$
5. model X = b;	SS_B	$SS_e(\mu, \beta)$
6. model X = a;	SS_A	$SS_e(\mu,\alpha)$
7. model X = ;	(just grand mean)	$SS_e(\mu)$

Model #1 is referred to as the full model, in which we estimate all effects. Models #2-7 are said to be restricted or reduced models. If we compare models, we can see how much error is reduced by adding terms to models. This logic is like saying, we'll run a regression to predict variable X as a function of variables V_1 and V_2, and then run another regression predicting X using V_1, V_2, and V_3. The difference in R^2 between these 2 models will reflect whether the 3rd variable is significant and helps us to understand X better (that's what is reflected when the β for V_3 has a significant t-test).

So for example, if we compare the error SS for models 1 and 2, we obtain the SSre, or the amount the SS error has been reduced:

$$SSe(\mu,\beta,(\alpha\beta)) - SSe(\mu,\alpha,\beta,(\alpha\beta)) = SSre(\alpha|\mu,\beta,(\alpha\beta)).$$

That reduction term (of the effect of α, conditional upon the other effects being in the model) has to be ≥0 because SSe(μ,β,(αβ)) has to be ≥ SSe(μ,α,β,(αβ)). If there is no main effect for factor A, and its SS are literally 0, then the SSe in models 1 and 2 would be equal, and this reduction SS would be 0. Usually though, SS_A will be >0, even if it's not significant, so the error SS when we exclude the A main effect will be larger (because we've lumped the SS_A into the SS_{error}) than the error SS when we include the A main effect in the model, that is, when it gets estimated and partialled out (in this case, out of the error variability).

A comparison of models 1 and 2 is not the only way we could test for the main effect of A; we could also compare models 4 and 5, or models 6 and 7. In theory, the results should all converge. Each pair of models has a different set of terms in it, but in each pair, the way they differ is only in whether the A main effect is included in the model or not. In general, there are numerous pairwise comparisons:

a.	SSre(α|μ,β,((αβ))	= SSe(μ,β,((αβ))	- SSe(μ,α,β,((αβ))
b.	SSre(β|μ,α,((αβ))	= SSe(μ,α,((αβ))	- SSe(μ,α,β,((αβ))
c.	SSre((αβ)|μ,α,β)	= SSe(μ,α,β)	- SSe(μ,α,β,((αβ))
d.	SSre(α|μ,β)	= SSe(μ,β)	- SSe(μ,α,β)
e.	SSre(β|μ,α)	= SSe(μ,α)	- SSe(μ,α,β)
f.	SSre(α|μ)	= SSe(μ)	- SSe(μ,α)
g.	SSre(β|μ)	= SSe(μ)	- SSe(μ,β)

In balanced designs, effects are orthogonal, that is, the set of other effects included in the model do not matter, so, for example, the tests for the main effect for A (tests a, d, and f) will all be equal. Similarly the tests b = e = g. For unbalanced designs, these equalities (a=d=f, b=e=g) do not hold.

Thus, **lesson #5**: for balanced data, the model comparisons will all converge on the same result. For unbalanced data, the model comparisons will differ. The model comparisons are translated into "Types of SS," which we see next.

The 4 Types of SS

In this final section of this chapter, we see the definition of 4 "types" of sums of squares. To date, different kinds of SS meant SS_A, SS_B, etc. Here, the idea is that for each of those effects, A, B, A×B, there are different ways to compute their respective sums of squares.

SAS Type I SS

SAS computes 4 types of sums of squares. All 4 sets of SS are equal when the design is balanced (exactly equal cell sizes), but when the design is unbalanced (even slightly), the different types of SS yield different values, ultimately because they test different sets of H_0's. Type I SS are called the "sequential sum of squares" (in SAS), and they reflect the incremental improvement (i.e., decrease in error SS) as each effect is added to the model.[16] These SS are 1 of 2 defaults that are printed by SAS. The Type I SS estimate the following effects:

Source	SS	
A	SSre(α | μ)	SS for A ignoring B and the interaction
B	SSre(β|μ,α)	SS for B ignoring A×B, adjusting for A
A×B	SSre((αβ)|μ,α, β)	SS for A×B adjusting for A and B effects

[16] Additional terminology is found in the appendix to this chapter.

One positive property of Type I SS (and likely a reason why it has been a, or the, default in many statistical computing packages) is that it is the only SS for which the SS add up to the total SS; that is, $SS_A + SS_B + SS_{AB} + SS_{error} = SS_{total}$. However, there are negative properties of Type I SS as well. The SS_A tests the null H_0: $\mu^*_{a1} = \mu^*_{a2}$ (recall these were the means of raw data and they varied with unbalanced data). That's not good.

Furthermore, the approach to the model fitting is such that each subsequent term is adjusted for all terms previously fit in the model. That sounds neutral until we consider the implication: this is the only SS of the 4 types in which the hypotheses that are tested depend on the order of the factors as they are specified in the model statement. So for example, if we state the model as X = a b a*b, we will obtain SSre(α|μ), SSre(β|μ,α), and SSre((αβ)|μ,α,β). The first piece considers the effect of factor A conditional upon only the grand mean being in the model (α|μ). This is referred to as factor A being "unadjusted," or factor A ignoring the effect of B and A×B. The second term will consider the effect of factor B conditional upon the grand mean and the main effect for A being in the model (β|μ,α). This effect is referred to as factor B having been adjusted for (or eliminating the effect of) factor A. The highest order interaction in an ANOVA is always ok, and this example shows why: the interaction term will be estimated conditioning on the grand mean and both main effects being in the model ((αβ)|μ,α,β), and those are all the effects there are, so the interaction term SS is what exists when partialling out all the other effects.

Alternatively, if we instead state the model as X = b a a*b, we will obtain SSre(β|μ), SSre(α|μ,β), and SSre((αβ)|μ,β,α). This approach to estimation yields a main effect for B that is unadjusted because it ignores factor A and the interaction (β|μ), and a main effect for A that is adjusted for B but not the interaction (α|μ,β). The overall issue is that if the order in which effects are specified matters, we need a theoretical reason for an *a priori* ordering, which we rarely have.

Type II SS

Type II SS are often referred to as "EAD," as in "Each (main effect is) ADjusted for the other." The idea underlying the Type II SS is that the main effect for factor A is estimated having adjusted for the grand mean and also the other main effect(s) B. Analogously, B is adjusted for the grand mean and A. Both still ignore the interaction A×B.

The order of specification of the effects in a model statement in a statistical computing package does not matter. The ANOVA "sources" of variation of A, B, and A×B will reflect SSre(α|μ,β) and SSre(β|μ,α), thus both main effects have indeed been adjusted for each other (yet both also do not partial out A×B), and the interaction term adjusts for every lower order term, so it is fine, SSre((αβ)|μ,α,β).

If we could assume the interaction is zero, the hypotheses tested would be the same as in the next method and this method (Type II) would be more powerful. However, given that neither the SS_A nor SS_B corrects for SS_{AB}, some suggest that the use of Type II SS is inappropriate if the interaction term cannot be assumed to be zero (which is most of the time).

Type III SS

Type III SS are printed by default in SAS. Type III SS are called the "partial SS" (by SAS) and are said to be the most appropriate sums of squares to be used for most research settings (Searle, 1987, pp.86-92). The model does the best job, of all the types of SS, of adjusting for or partialling out other effects in the model such that the effect we wish to test is isolated as cleanly as possible with no confound or bias, and the SS estimates are the closest to what the true values would be had the data been balanced; this "complete linear model analysis, involves estimation of independent effects of each factor adjusted for all others include in the model" (Overall and Spiegel, 1969, p.315; Overall, Lee and Hornick, 1981).

Type III SS estimate each effect we wish to test, by adjusting for all the other effects in the model.

Source	SS	adjusting out all other terms
A	SSre(α|μ,β,(αβ))	SS for A adjusted for B and A×B
B	SSre(β|μ,α,(αβ))	SS for B adjusted for A and A×B
A×B	SSre((αβ)|μ,α,β)	SS for A×B adjusting for A and B

All along, for Type I and Type II SS and here for Type III SS, the interaction SS has been the same, and it estimates the SS_{AB} adjusting for the grand mean and both main effects (all "lower order" or simpler effects). However, the main effect tests using Type I or Type II SS were not partialled out as cleanly. Now with Type III SS, the estimation of SS_A (or SS_B) has all the other effects statistically controlled for. Thus, when we examine SS_A, it tests the null $H_0\colon \mu_{a1} = \mu_{a2} = \cdots = \mu_{aa}$, or $H_0\colon \alpha_1 = \alpha_2 = \cdots = \alpha_a = 0$, and when we examine SS_B, it tests the appropriate null $H_0\colon \mu_{b1} = \mu_{b2} = \cdots = \mu_{bb}$, or $H_0\colon \beta_1 = \beta_2 = \cdots = \beta_b = 0$. These hypotheses are the most analogous to the H_0's tested in balanced designs. As a result, if we have unbalanced designs, and we usually do, given that the precise definition of balanced requires that all cell sizes n_{ij} be precisely equal to each other, then the Type III SS are the ones to use, test, and report.

Type IV SS

With that endorsement of Type III SS, one might wonder why there is a Type IV. The only caveat from the norm that Type III SS should always be used is that Type III SS are valid only if all $n_{ij} > 0$; that is, if there are no missing cells (and indeed it is rarely the case that we have any cell in an experimental design that is entirely empty).

If there are no missing cells, that is, there is no $n_{ij} = 0$, then the Type III SS and Type IV SS are identical.[17] If we have a missing cell, the statistical computing package will warn us that the "Type IV SS solution may not be unique" which sounds rather scary but it's pretty easy to understand. Imagine a 2×2 factorial, with some sample in each of 3 conditions, but a cell size of 0 in one:

$n_{11} > 0$	$n_{12} > 0$
$n_{21} > 0$	$n_{22} = 0$

Now imagine plotting the 3 cell means and trying to determine whether there is an A×B interaction. We won't be able to know—do the lines converge, diverge, or cross over? We cannot answer because there are not 2 lines—there will be 1 line and 1 point. Obviously it's a good idea to avoid experimental designs for which one or more conditions will have zero observations.[18]

Our conclusion is easy. In any situation, always look at the Type III sums of squares. If the data are balanced, then Type I SS = Type II SS = Type III SS = Type IV SS, so it doesn't

---

[17] Given the equality between Type III and IV SS, when there are no missing cells, SAS will compute Type III (and does so by default), because the Type IV SS are more time consuming to compute (and that time can translate into $).

[18] For more on the analysis of designs that contain missing cells, see Searle (1987, pp.132-168).

matter. If the data are unbalanced, then Type III SS tests the proper hypotheses in which each effect has been isolated, partialling out or statistically controlling for all the other effects in the model.

A Bigger Example

How about an example? These hypothetical data come from Overall, Spiegel, and Cohen (1975, p.185). On the left is the balanced data set, and on the right, the unbalanced. The cell means in the unbalanced data set resemble those in the balanced data set. The cell sample sizes are different, with the balanced data being replicated in the unbalanced data.

Balanced data:

	b_1	b_2	b_3
a_1	3	6	9
	6	4	12
	9	7	14
a_2	5	6	3
	6	5	5
	8	7	8
a_3	10	11	12
	12	13	8
	9	15	10

Unbalanced data:

	b_1	b_2	b_3
a_1	3	6	9
	6	4	12
	9	7	14
	3		
	6		
	9		
a_2	5	6	3
	6	5	5
	8	7	8
		6	
		5	
		7	
a_3	10	11	12
	12	13	8
	9	15	10
		11	12
		13	8
		15	10

The cell means follow in the 3×3 table, and they're the same for the balanced and unbalanced data. The main effect means (the row and column marginal means) follow for the balanced data (those means are slightly different in the unbalanced data). (Within cell variability will also differ between the balanced and unbalanced data sets.)

	b_1	b_2	b_3	$\bar{x}_i$'s
a_1	6.000	5.667	11.667	7.778
a_2	6.333	6.000	5.333	5.889
a_3	10.333	13.000	10.000	11.111
$\bar{x}_j$'s	10.333	13.000	10.000	

Let's run an ANOVA on these 2 data sets. First we'll see that the Types of SS are the same in the balanced data, but they vary in the unbalanced data.

ANOVA tables for the balanced data

For balanced data, note that: 1) all Types of SS are equal to each other, and 2) the SS add up to the total SS: $SS_A + SS_B + SS_{AB} + SS_{error} = SS_{total}$.

Source	df	Type I SS	F	p	Type II SS	F	p
A	2	125.852	15.04	.000	125.852	15.04	.000
B	2	9.407	1.12	.347	9.407	1.12	.347
A×B	4	76.593	4.58	.010	76.593	4.58	.010
error	18	75.333			75.333		
Total	26	287.185			287.185		

Source	df	Type III SS	F	p	Type IV SS	F	p
A	2	125.852	15.04	.000	125.852	15.04	.000
B	2	9.407	1.12	.347	9.407	1.12	.347
A×B	4	76.593	4.58	.010	76.593	4.58	.010
error	18	75.333			75.333		
Total	26	287.185			287.185		

ANOVA tables for the unbalanced data

For unbalanced data, the 4 Types of SS are not the same, and they don't necessarily add up to the total SS.

Source	df	Type I SS	F	p	Type II SS	F	p
A	2	210.919	28.42	.000	194.321	26.18	.000
B	2	11.988	1.62	.216	11.988	1.62	.216
A×B	4	95.196	6.41	.001	95.196	6.41	.001
error	30	111.333			111.333		
Total	38	429.436			429.436		

Source	df	Type III SS	F	p	Type IV SS	F	p
A	2	171.292	23.08	.000	171.292	23.08	.000
B	2	11.292	1.52	.235	11.292	1.52	.235
A×B	4	95.196	6.41	.001	95.196	6.41	.001
error	30	111.333			111.333		
Total	38	429.436			429.436		

For the results on the unbalanced data, note:

1) The Type III = Type IV SS here because there are no completely empty cells, for which $n_{ij} = 0$.
2) Types I, II, III SS differ.
3) SS_{AB} is the same for all 4 types of SS. This property holds for the highest order interaction in a factorial ANOVA.
4) The summing property holds only for Type I SS (i.e., $SS_A + SS_B + SS_{AB} + SS_{error} = SS_{total}$). That seems like a good thing, but it's less important than partialling out the other effects in the model, which is what Type III does.

One more important thing to note is that in this particular data set, the different SS yield different numeric values, yet they all converge in the same qualitative assessment, the same conclusion as to which effects are significant or not. The p-values for the main effect of A and the A×B interaction are all <0.05, so these effects are consistently significant across all 4 types of SS. That convergence doesn't always happen. For unbalanced data, we're to look at Type III SS, so if an effect is significant by Type III and not by Type I, or II, or IV, then yay. However, if an effect is not significant by Type III and it is via Type I, or II, or IV, then tough noogies, we still must conclude, per the Type III diagnostic, that the effect is not significant.

SAS prints the Type I and Type III SS by default. Be sure to look at and report only the Type III results. To obtain the different SS estimates (if you're curious), use the SS option on the model statement: model x = a b a*b / SS1 SS2 SS3 SS4;

The bottom line couldn't be clearer. *For unbalanced designs, use Type III SS*. In fact, if the data set is balanced, Type III SS will still be correct (all 4 types of SS will be equal), so always use Type III SS and ignore all others.[19]

→TYPE III SS!

→TYPE III SS!

→TYPE III SS!

SUMMARY

This chapter demonstrated effects of unbalanced data on the computation of means and SS and the hypotheses they test. There are other effects of unbalanced data as well, generally hurting both power and robustness (Milligan, Wong, and Thompson, 1987). Thus it is a good idea to aim for cell sizes n_{ij}'s to be as close to equal as possible. Examine and report F-tests and p-values based on the use of Type III SS.

REFERENCES

1. Appelbaum, Mark I. and Elliot M. Cramer (1974), "Some Problems in the Nonorthogonal Analysis of Variance," *Psychological Bulletin*, 81 (6), 335-343.
2. Carlson, James E., and Neil H. Timm (1974), "Analysis of Nonorthogonal Fixed-Effects Designs," *Psychological Bulletin,* 81 (9), 563-570.
3. Herr, David G. (1986), "On the History of ANOVA in Unbalanced, Factorial Designs: The First 30 Years," *The American Statistician,* 40 (4), 265-270.
4. Herr, David G. and Jacquelyn Gaebelein (1978) "Nonorthogonal Two-Way Analysis of Variance," *Psychological Bulletin*, 85 (1), 207-216.
5. Hocking, R. R. and F. M. Speed (1975), "A Full Rank Analysis of Some Linear Model Problems," *Journal of the American Statistical Association*, 70 (351), 706-712.
6. Horst, Paul and Allen L. Edwards (1982) "Analysis of Nonorthogonal Designs: The 2^k Factorial Experiment," *Psychological Bulletin*, 91 (1), 190-192.

---

[19] To get SAS to print only Type III SS, so the misleading Type I SS won't be printed by default, add SS3 alone as an option on the model statement: model x = a b a*b / SS3;

7. Hosking, James D. and Robert M. Hamer (1979), "Nonorthogonal Analysis of Variance Programs: An Evaluation," *Journal of Educational Statistics*, 4 (2), 161-188.
8. Iacobucci, Dawn (1995). "The Analysis of Variance for Unbalanced Data," in David W. Stewart and Naufel J. Vilcassim (Eds.), *Marketing Theory and Applications, 6*, Chicago: AMA, 337-343.
9. Iacobucci, Dawn (1994). "Analysis of Experimental Data," in Richard Bagozzi (Ed.), *Principles of Marketing Research*, Cambridge, MA: Blackwell, 224-278.
10. Little, Roderick J. A. and Donald B. Rubin (1987), *Statistical Analysis with Missing Data*, NY: Wiley.
11. Milligan, Glenn W., Danny S. Wong, and Paul A. Thompson (1987), "Robustness Properties of Nonorthogonal Analysis of Variance," *Psychological Bulletin*, 101 (3), 464-470.
12. Overall, John E. and Douglas K. Spiegel (1969), "Concerning Least Squares Analysis of Experimental Data," *Psychological Bulletin*, 72 (5), 311-322.
13. Overall, John E., Dennis M. Lee and Chris W. Hornick (1981), "Comparison of Two Strategies for Analysis of Variance in Nonorthogonal Designs," *Psychological Bulletin*, 90 (2), 367-375.
14. Overall, John E., Douglas K. Spiegel and Jacob Cohen (1975), "Equivalence of Orthogonal and Nonorthogonal Analysis of Variance," *Psychological Bulletin*, 82 (2), 182-186.
15. Perreault, William D. and William R. Darden (1975), "Unequal Cell Sizes in Marketing Experiments," *Journal of Marketing Research,* 12 (3), 333-342.
16. Schendel, U. (1989), *Sparse Matrices: Numerical Aspects with Applications for Scientists and Engineers*, NY: Wiley.
17. Searle, Shayle R. (1987), *Linear Models for Unbalanced Data*, NY: Wiley.
18. Speed, F. M., R. R. Hocking, and O. P. Hackney (1978), "Methods of Analysis of Linear Models with Unbalanced Data," *Journal of the American Statistical Association*," 73 (361), 105-112.
19. Yates, Frances (1934), "The Analysis of Multiple Classification with Unequal Numbers in the Different Classes," *Journal of the American Statistical Association*, 29, 51-66.

Also related, on sampling:
1. Kish, Leslie (1965), *Survey Sampling*, NY: Wiley.
2. Thompson, Steven K. (1992), *Sampling*, NY: Wiley.

APPENDIX: TERMINOLOGY

The concepts and issues that arise with unbalanced data are not tremendously complicated, but the terminology used in the literature is dreadful, not just because it shows little consistency, but also because some labels sound like "X" whereas others sound like "not X." At this point, the clearest labels for the 4 Types of SS are "Type I" through "Type IV," just as SAS produces them and depicts them in all of its online documentation; at least it is a common base to which we all might refer. This appendix mentions some of the other labels for the types of SS, in case a mapping to the literature is necessary.

Type I SS

Type I SS have also been called "method 3" or the "step-down" method (Overall and Spiegel, 1969, p.316). SAS calls Type I SS the "sequential sum of squares," as a reminder that the order in which the effects are listed in the model statement is important.

For example, if, in SAS, we specify the model, "X = a b a*b," we will obtain SSre(α|μ), SSre(β|μ,α), and SSre((αβ)|μ,α,β). The SS_A will not have been "adjusted" for the effects of B or A×B. This order has been called HRC, standing for "hierarchical—rows then columns" (Herr, 1986, p.265).

Instead, if we fit the model with B listed before A, "X = b a a*b," then we will obtain SSre(β|μ), SSre(α|μ,β), and SSre((αβ)|μ,β,α). Now, SS_B will not have been adjusted for A or A×B, but SS_A will have been adjusted for B, but not A×B. This order has been called HCR, "hierarchical—columns then rows" (Herr 1986, p.265). For more on these labels, see Carlson and Timm (1974, pp.508-509), and Herr and Gaebelein (1978, p.709).

Type I SS are also referred to as the SS based on the "weighted averages of cell means" or the computation of "marginal means via weighted means," as in, we're weighting each cell mean as a function of its cell size (Carlson and Timm, 1974; Searle 1987, pp.92-94). These SS compare the μ^*'s, which we saw were derived from the raw data or in terms of the weighted means of cell means (with the weights being based on the relative cell sizes), and which did not test the hypotheses we wished to test.

Type I SS are 1 of the 2 defaults (with Type III) produced in SAS. If desired, they can be obtained in SPSS through the modeling dialog box, and they are (used to be?) the default in R.

Type II SS

Type II SS are sometimes called EAD, the point being that "Each" main effect is "ADjusted" for the other main effect, but neither is adjusted for the interaction (Herr, 1986). Type II SS are also called "method 2" (Overall and Spiegel, 1969). Given that the SS for the main effects A and B are not adjusted for A×B, these SS are not typically recommended because, "In most real experimental situations, it is unrealistic to believe that true interaction effects are totally absent" (Overall, Lee, and Hornick, 1981, p.369).

Type III SS

Type III SS are the ones we want. The SS for each effect (e.g., A) is pure, having partialled out the other effects (e.g., B, A×B, and μ). These SS provide the "general linear model analysis" that is closest to the "true values" (as if the data were actually balanced; Overall, Lee, and Hornick, 1981). These SS refer to those that examine the μ's which tested the hypotheses we want to test, via the "unweighted means of the cell means" (Carlson and Timm, 1974). Type III SS are also variously called "method 1" (Overall and Spiegel, 1969), the "unique" or "regression" SS in SPSS, and a "complete least squares analysis" in SAS. Yates (1934) had called this approach to the calculation the "weighted squares of means analysis" (cf., Speed, Hocking, and Hackney, 1978, pp.107-108).

Given that some techniques for computing SS yield results wherein 1 or more effects might not have been partialled out of 1 or more other effects, the analysis of unbalanced data is often called a nonorthogonal ANOVA, implying the results are somewhat correlated, that there is a multicollinearity of sorts present in the effects being estimated (recall the B effects biasing the A hypothesis when using μ^*'s). The Type III SS restores orthogonal estimation to the extent possible when cell sizes are not all equal (Overall, Spiegel, and Cohen, 1975). Type

III yields estimates of the independent contribution of an effect (A, B, or A×B) over and above the other effects, or having adjusted for the other effects in the model (Overall and Spiegel, 1969).

In SAS, Type I SS and Type III SS are produced by default. In SPSS, Type III is the default in its ANOVA and MANOVA procedures (but double-check via the menu dialog boxes; look for "unique" or "regression" but not "sequential").

Type IV SS

Type IV SS don't have any nicknames. They are like Type III but they are used if 1 or more cell size in a design =0. If all $n_{ij} > 0$, then the Type III and Type IV SS solutions will be identical.

CHAPTER 11
SAS

Questions to guide your learning:
Q_1: What are some basic procedures to run in SAS?
Q_2: How do I use SAS to run an ANOVA, ANCOVA, and MANOVA?

In this chapter, we'll look at some SAS procedures for data analysis—simple ones to get started, and the main proc, or procedure, in SAS for ANOVA: proc glm, for the "general linear model."[20] Some people type data and commands into a Word file, and then cut and paste it into SAS, and others like to work interactively right in the SAS screen environments. Different strokes. For output, it's a good idea to copy and paste from SAS into Word in part to edit out the many excessive pages and extraneous graphs that SAS likes to insert.

SAS is syntax-based (though JMP is a menu version of SAS if you prefer that mode of processing).[21] Each command in SAS ends in a semi-colon, a ";" symbol. A statement that begins with an asterisk "*" is dismissed by SAS as a comment. Any line of code can contain more than 1 command, and SAS is not picky about spacing between words. While Word will wrap around onto a next line any line that gets long, I try to insert hard returns so that most of my SAS command lines are not wider than my screen.[22]

Let's first begin with several data issues: 1) how to enter data and import datasets, 2) how to combine variables and do variable transformations, 3) how to use proc append and merge statements, 4) how to use arrays, 5) how to generate plots to look at basic data properties.

Data Entry and Importing Data Sets

In the "editor" screen in SAS, we could type (or cut and paste from a Word file):

```
data mine; input
 x y  z; datalines;
 1 2  3
 4 5  6
 7 8  9
10 11 12
;
proc print data=mine; var z;
run;
```

---

[20] SAS has plentiful online documentation. Begin here: http://support.sas.com/documentation/ and click on "SAS Procedures by name and product." Click on "g" for GLM.

[21] SAS has generously made available a version of their software with many of their procedures. Go to: www.sas.com/en_us/software/university-edition.html or begin at www.sas.com, got to "Products & Solutions" and then "SAS University Edition" (or Google "SAS free") to download a free copy of SAS!

[22] I guess I don't trust SAS not to lop off the right hand side ending of my commands. I'm sure my fear is residual from when SAS required each line be no longer than 80 characters, even that a holdover from card punching. If you have to ask, enjoy your youth.

Here's what that tiny SAS program says. First, the "data" statement says, "read into a dataset I'd like to call 'mine' the variables listed in the input statement." The input statement tells SAS to read in 3 variables, call them X, Y, and Z. A "free format" is assumed in that input statement because we didn't specify any format or tell SAS which columns the data are in, etc. With free formatted input statements, missing data must be entered as "." in the data file. It's better to specify a formatted input statement, and we'll see how to do so in a moment.

The "datalines" statement tells SAS that the very next thing are the data. If we wish to create any new variables (such as transformations of the variables we're reading in), these statements must follow the input statement and precede the datalines statement. For example:

```
input x y z;
q = (y + z)/2;
if x >  52 then q4=z; else q4=0;
datalines;
```

Finally, in the small SAS program example, a semi-colon cues SAS that the data are finished, and proc print will print all the observations for all the variables. In this example, we specified "var z" so SAS will print all observations only for variable Z (if we left "var z" off, then all variables would be printed). Also note that the print procedure specifies the dataset called "mine" and that's helpful because sometimes we'll have more than 1 dataset going on at the same time. The program ends with the command to execute all the commands, and that is "run;".

There are many alternative forms of the input statement. If we have entered data into a text file, we can specify what columns contain which variables. Here are some examples:

- input x 1-2 y 4-5 z 7-8; *x is in columns 1-2, y in 4-5, etc. The data must be right-justified, so if x is the single digit 8, in the data file, it must read: "space 8";
- input (x y z) (2.0 2*3.1); *this is old Fortran formatting where 2.0 indicates the value takes 2 spaces with 0 decimals, and 2*3 indicates that for the next 2 variables, they take 3 spaces with 1 decimal. Again, the data must be right-justified;
- input (x1-x3) (3*10.3); *variable names may be referred to in sequence, as this x1-x3 example indicates;
- input x y $ z; *the $ denotes that variable y is alphabetical, not numerical;
- input x y z / q; *the slash indicates that there are 2 lines of data per subject, with x, y, and z on line 1, and q on line 2;

The data don't have to exist in the same file as the SAS commands. In fact, typically our datasets are large enough that that would be clumsy. Data might exist in an Excel file, and such a file can be directly imported into SAS.

Alternatively, say we have copied data in a Word file (if we copy data from Excel into Word, highlight the data in Word and translate the "table" into text, using space delimiters and not tabs). Call that data file "project1."

Next, in SAS, we'll use the following commands:

```
libname myproject 'c:\myHD\mydata\project1';
data mine; input x y; datalines;
1 2
3 4
5 6
; run;
```

The “libname” command sets a “library” name, essentially a folder, called “myproject” which is located on the c drive, at myHD (my hard drive), in a folder called “mydata” in a subfolder called “project1.”

The dataset “mine” is a temporary dataset, meaning if we close SAS, the dataset is gone. We can save it, obviously in a Word file, say, but we can also create a “SAS dataset.” To do so, say:

```
data myproject.bigdata; set mine;
```

With that command, we’ve defined a new dataset called “bigdata” that is associated with the library called “myproject” and the new file can be found in the “c:\myHD…” location that defined the library. The “set” command says we’re creating that new dataset drawing from the current dataSET called “mine.”

Given that the dataset is now permanent, we can quit SAS and the data will still live in the project1 subfolder. The next time we enter SAS and wish to access and use that dataset, we say:

```
libname myproject 'c:\myHD\mydata\project1';
data new; set myproject.bigdata;
```

Here we are reading the permanent dataset called bigdata into a temporary dataset called new. It’s a good idea to use a temporary dataset that way in case we do anything stupid (no, never), the old dataset won’t be overwritten and destroyed.

Variable Transformations

If we wish to transform any of our variables, the commands must be inserted after the “input” statement and before the “datalines” statement. Here are common transformations:

- scale=(q1+q2+q3)/3; *creates an average of variables q1, q2, q3 called scale;
- reverseq4=(8-q4); *if we wish to reverse code a variable measured on a 7-point scale;
- if q5 = ‘ ‘ then q5 = . ; if q6 = ’99’ then q6= . ; *change missing data from space or 99 to SAS’s preferred notation of a period;
- if q7 >30 then dummy7=1; else dummy7=0; *create a dummy variable;
- if q8 <= 4.56 then mdn=0; else mdn=1; *create a median split variable at mdn=4.56;
- if q9=1 or q9=2 or q9=3 then categ9=1; else categ9=2; *recode a categorical variable;
- q10= q1**2; *square; q11=q2**(.5); *square root; q3inv=q3**(-1); *reciprocal;

Putting Two Data Sets Together—Proc Append (Top to Bottom)

Proc append stacks one dataset on top of another, effectively increasing sample sizes on variables common across the datasets.

```
data one; input
ID a  b  c  d; datalines;
01 1  2  3  4
02 5  6  7  8
03 9 10 11 12
data two; input
ID  a  b  e; datalines;
04 13 14 15
05 16 17 18
06 19 20 21
07 22 23 24
proc append base=one data=two; run;
```

That proc append command yields the following combined dataset:

ID	a	b	c	d	e
01	1	2	3	4	.
02	5	6	7	8	.
03	9	10	11	12	.
04	13	14	.	.	15
05	16	17	.	.	18
06	19	20	.	.	21
07	22	23	.	.	24

If we wanted to add more datasets, keep the “base” the same and say: proc append base=one data=three; etc.

Putting Two Data Sets Together—Merge (Side by Side)

The merge statement is not a proc and it works the other way—rather than placing one dataset on top of another, it places two datasets next to each other, side by side.

```
data one; input
ID  a  b  c  d; datalines;
01  1  2  3  4
02  5  6  7  8
03  9 10 11 12
data two; input
ID  a  b  e; datalines;
04 13 14 15
05 16 17 18
06 19 20 21
07 22 23 24
```

If we merge the 2 datasets, the variables “ID,” “a,” and “b” in data=two will write over the variables with the same names in data=one. So first, let’s rename them.

```
data two; set two; idd=id; aa=a; bb=b; delete id a b;
```

Now we can merge the 2 datasets.

```
data new; merge one two; proc print;
```

That print statement would show us the newly merged dataset:

ID	a	b	c	d	IDD	aa	bb	e
01	1	2	3	4	04	13	14	15
02	5	6	7	8	05	16	17	18
03	9	10	11	12	06	19	20	21
.	.	.	.	.	07	22	23	24

If we had not renamed the variables before merging, we would have gotten:

ID	a	b	c	d	e
04	13	14	3	4	15
05	16	17	7	8	18
06	19	20	11	12	21
07	22	23	.	.	24

Regarding merging, a frequently encountered scenario is when additional measures have been obtained on the same subjects that had been in the first sample. If we're trying to combine variables a, b, c, and d from one dataset with variables e and f from another dataset, and they're all different variables measured on the same people, it is important to "sort" the data before merging the datasets. Doing so involves only a few more lines of code.

```
data one; input
ID a  b  c  d; datalines;
01  1  2  3  4
02  5  6  7  8
03  9 10 11 12
data two; input
ID   e  f; datalines;
01  13 14
02  15 16
03  17 18
proc sort data=one; by id; proc sort data=two; by id;
data new; merge one two; by id;
proc print data=new; run;
```

That would give us the following dataset.

```
ID a  b  c  d  c  f
01 1  2  3  4 13 14
02 5  6  7  8 15 16
03 9 10 11 12 17 18
```

Arrays

Sometimes we wish to perform the same operation on a number of variables. Writing the commands to do so variable by variable is not as efficient as the more succinct use of an "array."

```
data mine;
array sunshine{3} x v w; *the array is named sunshine (array of sunshine, get it?),
SAS is told it has 3 variable elements, and the variable names are listed;
do i=1 to 3;
if sunshine{i} = . then sunshine{i}=0; *or sunshine{i}=(-1)*sunshine{i} etc.;
end;
```

If the array is large in size, instead of counting its elements, we can say "array sunshine{*}" and then list the numerous variables to be included in the array.

The Plot Thickens

Plots are always a great way to get a feel for our data. SAS has quite intricate graphing capabilities, but "proc plot" is a handy, easy procedure.

```
proc plot; plot y*x; run;
*plots y on the vertical axis and x on the horizontal axis;
```

Next let's look at some of SAS's procedures for data analysis: proc freq, univariate, means, corr, reg, and of course, proc glm. Let's then look at an issue we need to deal with for within-subjects data. Then we'll end the chapter with several procs that are not related to ANOVA but are very useful, such as how to obtain coefficient alpha for multi-item scales.

Several Useful SAS Procs: Freq, Univariate, Means, Corr, Reg

proc freq data=my; tables q1; run; *1-way tables to check histogram-like frequencies;
proc freq; tables q1 q2 q3; run; *1-way tables on several variables;
proc freq; tables q1*q2; run; *cross-tabs;

proc univariate plot data=ex; var q1 q2 q3; run; *basic descriptive statistics, useful plots;

proc means; var q1-q5; run; *very easy, just list all the variables for which means and standard deviations are sought;

proc corr data=ex; var x y z; run; *results in a full correlation matrix, like below, left;
proc corr data=ex; var y z; with x; run; *results in the correlations between the "var" and "with" sets, like that below, right;

	x	y	z
x	1		
y	r_{yx}	1	
z	r_{zx}	r_{zy}	1

	x
y	r_{yx}
z	r_{zx}

Proc GLM

The big proc for an ANOVA user is proc glm. SAS has a proc ANOVA but it assumes perfectly balanced data, which is rare. In SAS, an ANOVA dataset for an A×B factorial might look like the following. For example, the third data point, 5.4, is X_{113}.

```
data ex1; input a 1-2 b 3-4 subjid 5-7 score 8-12;
datalines;
 1 1   1   1.3
 1 1   2  12.0
 1 1   3   5.4
 1 2   4   7.6
 1 2   5   7.0
 1 2   6  12.1
 2 1   7  11.9
 2 1   8   4.2
 2 1   9   5.7
 2 2  10   7.5
 2 2  11  14.3
 2 2  12  -3.2
 3 1  13   1.0
 3 1  14   1.1
 3 1  15  12.1
 3 2  16   4.2
 3 2  17   5.7
 3 2  18   5.8
proc print data = ex1;
```

```
proc glm data = ex1;
  class a b;                    *here we list our factors;
  model score = a b a*b / SS3;  *score is the dependent variable, a and b are main effects;
  means a b a*b;                *and a*b is the interaction, and note specification of SS3;
  contrast 'factor a' a 1 -.5 -.5; run;
```

To run simple effects, recall from Chapter 5 that the contrast statement is a bit more complicated. In a 2×2 factorial, here are the 4 varieties.

```
contrast 'a @ b1' a 1 -1 a*b 1 0 -1 0; *simple effect of A at level b=1;
contrast 'a @ b2' a 1 -1 a*b 0 1 0 -1; *simple effect of A at level b=2;

contrast 'b @ a1' b 1 -1 a*b 1 -1 0 0; *simple effect of B at level a=1;
contrast 'b @ a2' b 1 -1 a*b 0 0 1 -1; *simple effect of B at level a=2;
```

To run an **ANCOVA**, we list our factors in the class statement, and any predictors in the model statement that are not in the class statement are covariates. For example:

```
proc glm; class a b;
model depvar = a b a*b v1 v2;
means a b a*b;   run;
```

To run a **MANOVA**, we obviously will include more than one dependent variable, and otherwise, it's easy in that only one more statement is necessary.

```
data try; input a b v1 v2; datalines;
      1  1  30  35
      1  2  25  33
      1  1  21  41
      1  2  15  21
      1  1  27  32
      1  2  25  34
      1  1  35  34
      1  2  31  36
      1  1  20  37
      1  2  14  21
      1  1  23  38
      1  2  19  25
      2  1  20  15
      2  2  26  25
      2  1  11   3
      2  2  18  19
      2  1   7   2
      2  2  11   8
      2  1  21  15
      2  2  24  22
      2  1  15  11
      2  2  17  13
      2  1  13  12
      2  2  20  15
```

```
proc glm; class a b;
model v1 v2 = a b a*b; *note that 2 dependent variables are listed;
manova h=a   / printh printe summary; *printh and printe will print the H and E matrices;
manova h=b   / printh printe summary; *summary will print the ANOVAs for v1 and v2;
manova h=a*b  / printh printe summary;
means a b a*b; run;
```

The manova keyword tells SAS to model variables v1 and v2 simultaneously, i.e., in a MANOVA.

If variables v1 and v2 had been **repeated** measures, e.g., the same variable measured at times 1 and 2, we would run the proc glm as a MANOVA approach to analyzing repeated measures or within-subjects data. The tweak is in the use of the "repeated" statement below, specifying a super-variable called "v" that has 2 levels. (Chapter 16 provides more information, including keywords like "profile" that may be added to the "repeated" statement.)

```
proc glm; class a b; model v1 v2 = a b a*b;
repeated v 2;
means a b a*b; run;
```

Here are some other SAS procs, rather randomly selected!

Regression

```
proc reg; model y = v1 v2 v3 / stb; run; *this proc runs a regression, and the option stb provides standardized beta coefficients in addition to the raw b-weights provided by default;
```

Coefficient Alpha for Reliability on Multi-Item Scales

```
proc corr data=ex alpha; var q1 q2 q3; run; *just insert the "alpha" keyword;
```

For multi-factor:

```
Proc corr alpha; var q1-q3;
Proc corr alpha; var q4-q6; run;
```

Proc Logistic

```
proc logistic;
class a b;
model ycateg = a b q1 q2 / expb; run; *this syntax looks much like glm, but the dependent variable is categorical. The option "expb" translates the otherwise natural log results back into the scale of the raw data;
```

Proc IML—Interactive Matrix Language

```
proc iml; *anything you want to do with matrices, you can do in SAS's proc iml;
x={ 0 1 0 1,
    0 0 1 1,
    0 0 0 0,
    0 0 1 0}; *define a matrix, delineating rows by commas;
xpx=(x`)*x;  *that computes X'X; print xpx; quit; run;
```

Proc Mixed

Proc mixed is used on hierarchical linear models. Say we have students, named by an "id" and each student is nested in 1 and only 1 school.

```
proc mixed data=hier  info method = reml covtest;
class school id;
model y = x z  / solution chisq;
random int / subject=school type=vc;
random int / subject=id(school) type=vc; run;
```

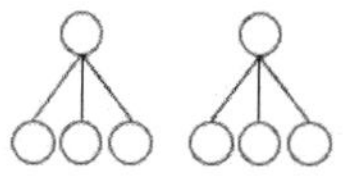

SUMMARY

This chapter introduces basic syntax and procedures in SAS. The major SAS proc for ANOVA is proc glm.

REFERENCES

1. Cody, Ron and Jeffrey K. Smith (2005) *Applied Statistics and the SAS Programming Language*, 5th ed., Pearson.
2. Littell, Ramon, Walter Stroup, and Rudolf Freund (2002), *SAS for Linear Models*, 4th ed., Cary, NC: SAS Institute.

And for alternatives:

3. Gaur, Ajai S. and Sanjaya S. Gaur (2009), *Statistical Methods for Practice and Research: A Guide to Data Analysis Using SPSS*, Sage. Also Google "SPSS documentation" and it will provide links (currently hosted at IBM). For the basics in ANOVA, see Appendix A.
4. For XLSTAT, see Appendix B to this chapter.

→For more SAS help at the online reference and documentation, start here: http://support.sas.com/documentation/

The following screen opens:

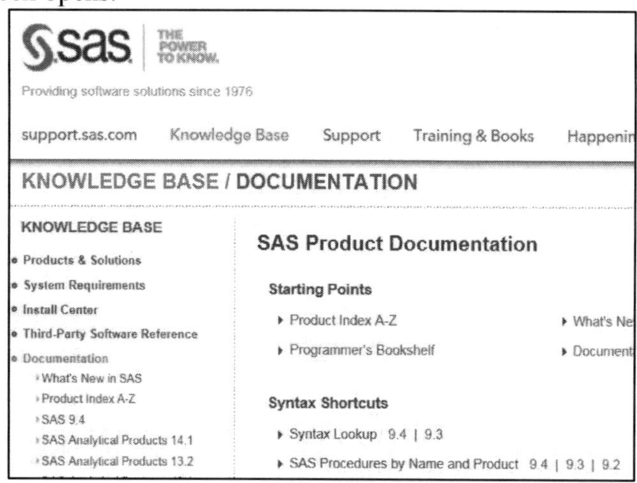

In the last line, click on "SAS Procedures by Name and Product" 9.4 (or a more recent version, if one is available). Then click on "SAS Procedures by Name" for an alphabetic listing. For proc glm, go to "g," for proc reg, go to "r," etc.

APPENDIX A: SPSS

SPSS (like XLSTAT next) is menu-driven (as opposed to SAS's syntax). The choices are fairly straightforward.

1. To run an ANOVA, click on the tabs: Analyze, General Linear Model, Univariate.
   a. Move your dependent variable over to the "dependent variable" box.
   b. Move your factors to the "fixed factors" box (unless you have a "random" factor, other than the study participants, or "subjects").
   c. For the "Model" you want the defaults (full factorial and Type III SS).
   d. You might want to plot the data.
   e. Under "Options" there are 2 things:
      i. In the top half of the dialog box, click on "display means."
      ii. Among the choices in the lower half of the dialog box, you can get effect sizes printed (unfortunately they are η^2 not ω^2).
   f. When you're ready, click "OK"
2. To run an ANCOVA, after steps 1.a. and 1.b. above, move your covariate variable over to the "covariate(s)" box. Then proceed to steps. 1.c through 1.f.
3. To run a MANOVA, click on the tabs: Analyze, General Linear Model, Multivariate. Now there will be more than 1 dependent variable to move to the "dependent variable" box.

SPSS output begins with commands in syntax form (which you can click on, edit, and rerun, if you want to tweak something). For example, for ANOVA:

```
UNIANOVA x BY a
  /METHOD=SSTYPE(3)
  /INTERCEPT=INCLUDE
  /CONTRAST (a)=SPECIAL(1 0 -1)
  /POSTHOC=a(BTUKEY SCHEFFE)
  /PLOT=PROFILE(a)
  /EMMEANS=TABLES(a)
  /PRINT=ETASQ
  /CRITERIA=ALPHA(.05)
  /DESIGN=a.
```

The ANOVA table is produced:

Dependent Variable:x

Source	Type III Sum of Squares	df	Mean Square	F	Sig.	Partial Eta Squared
Corrected Model	1696.167[a]	2	848.083	2.079	.181	.316
Intercept	491670.083	1	491670.083	1205.484	.000	.993
a	1696.167	2	848.083	2.079	.181	.316
Error	3670.750	9	407.861			
Total	497037.000	12				
Corrected Total	5366.917	11				

The means are printed:

Dependent Variable:x

a	Mean	Std. Error	95% Confidence Interval	
			Lower Bound	Upper Bound
1.00	186.250	10.098	163.407	209.093
2.00	206.500	10.098	183.657	229.343
3.00	214.500	10.098	191.657	237.343

For a MANOVA, the 4 test statistics are provided:

Effect		Value	F	Hypothesis df	Error df	Sig.
Intercept	Pillai's Trace	.993	535.842[a]	2.000	8.000	.000
	Wilks' Lambda	.007	535.842[a]	2.000	8.000	.000
	Hotelling's Trace	133.961	535.842[a]	2.000	8.000	.000
	Roy's Largest Root	133.961	535.842[a]	2.000	8.000	.000
a	Pillai's Trace	.990	4.408	4.000	18.000	.012
	Wilks' Lambda	.168	5.766[a]	4.000	16.000	.005
	Hotelling's Trace	4.023	7.040	4.000	14.000	.003
	Roy's Largest Root	3.774	16.982[b]	2.000	9.000	.001

APPENDIX B: XLSTAT

There are some add-ins and programs complementary to Excel that can provide simple analyses. The programs can be handy for users familiar with Excel. One such program is called XLSTAT.

Consider the following example. There are 2 columns in the Excel spreadsheet dataset.

group	x
1	4
1	5
1	6
1	5
2	5
2	6
2	6
2	7
3	9
3	10
3	11
3	10

To access ANOVA, go to the "Modeling data" tab and click on "ANOVA." In the dialog box, specify the dependent variable by highlighting it for the "Y / Dependent variables: Quantitative" and specify the independent variable(s) by highlighting the group (1, 2, 3) membership. Under the "Outputs" tab, click on "Type III SS" and then click ok.

The output is simple: the ANOVA table and a plot of means.

Analysis of variance:

Source	DF	Sum of squares	Mean squares	F	Pr > F
Model	2	56.000	28.000	42.000	< 0.0001
Error	9	6.000	0.667		
Corrected Total	11	62.000			

Computed against model Y=Mean(Y)

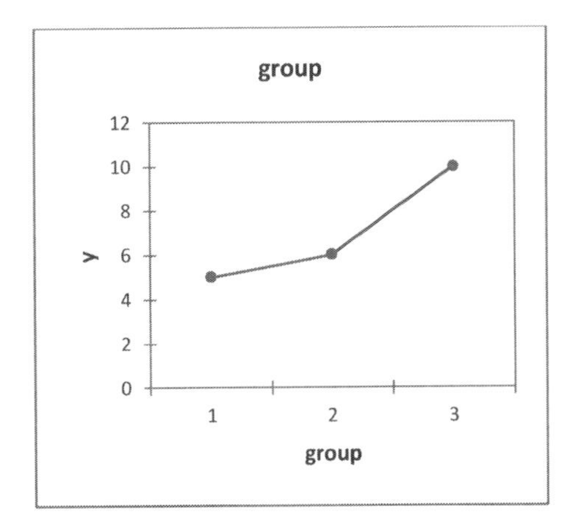

To run an ANOVA and include interaction terms among 2 or more factors, click on the "options" tab, click on interactions, and then in that pop-up window, click on all the effects desired to be included in the model (usually all of them).

For help, go to www.xlstat.com, click on "support" and then "tutorials."

CHAPTER 12

MATRIX ALGEBRA

Questions to guide your learning:
Q_1: What is a matrix? Why are matrices helpful?
Q_2: What are some easy things to do with matrices?
Q_3: What are more challenging things to do with matrices?

In this chapter, we go through the gory details of matrix algebra. The equations used in basic stats classes are pretty straightforward; a mean is simple, a variance is simple. Mode is best of all. Each equation is fairly manageable. Well that's about to change. The reason some people have trouble with multivariate statistics, like multivariate ANOVA, into which we're transitioning, is that the equations get more complicated. They're not more complicated just because statisticians are sadistic, though that's probably also true. They're more complicated by definition because we're balancing more variables—*multi*variate, right?

So one result of looking at more than one variable at a time is that wherever we had a single mean, we'd now have "p" means. Looking at more than one variable at a time means we can see how the variables are inter-related. (I'd say "correlated," but correlated is only one way that we can look at the variables' associations.) Now imagine trying to communicate results for just p=3 variables. We'd have 3 means and standard deviations, and we'd have to also somehow capture how variables 1 and 2 relate, 1 and 3, and 2 and 3. And imagine how much more there would be with even more variables than just 3.

> Matrix language takes some practice, but helps in multivariate models.

So one thing matrices will do for us is store information in an organized manner, which facilitates their presentation and communication. In addition, if we create matrices, we can follow rules of "matrix algebra" to derive many interesting results.

Matrix algebra is a short-hand way of referring to many things—variables, respondents—in a way that we can say "**A**" when we mean "100 people's answers to 15 questions." Impressive. Also, when we write out equations for the various kinds of modeling we'll do, the matrix version of the equation is more succinct and elegant.

For example, in multiple regression, if we want to estimate the beta weights, the matrix equation turns out to be: $\widehat{\boldsymbol{\beta}} = (\boldsymbol{X}'\boldsymbol{X})^{-1}(\boldsymbol{X}'\boldsymbol{Y})$. If we were to write the same information out as a series of equations, it would be a real mess. So, the good news is that matrices are efficient for us, and they yield more information (as we shall see). Unfortunately, they're a lot like learning a foreign language. Some of the stuff will come easy—you'll see that adding two matrices, for example, is a piece of cake. Other things will be a little harder, like matrix multiplication, and still other things are complete killers (e.g., eigenvalues and eigenvectors). But just like with foreign languages, the more you practice, the better it'll get.

In any data collection task and resulting data set, we'll probably have many measures on each subject (i.e., p is big), but for most statistical models, each analysis will probably be done on a subset of the variables. Some variables we'll want to run through a regression, and others we'll want to factor analyze etc., so just because we have 100 variables in our data set, for any given analysis, the working matrix will effectively have fewer variables. It's also important for most statistical models to have n be much larger than p (n>>p). So, for example, if we collect data on 200 study participants with 100 items on a questionnaire, and wish to submit 15 of those items to a "factor analysis," p will be 15 for the purposes of the factor analysis. Rarely would our substantive research questions require that we analyze all of the variables (e.g., all 100) simultaneously, and furthermore, we'd be asking for trouble when our number of variables (p) approaches our number of subjects (n).

In this chapter, we'll cover the following matrix algebra concepts:

- Basic matrix terms
    - What is a matrix, matrix order in terms of #rows and #columns, square matrices, symmetric matrices, matrix transposes, vectors and scalars, diagonal matrices, scalar matrices, identity and null matrices, equality of matrices
- Simple matrix manipulations
    - Matrix addition and subtraction, scalar multiplication, matrix multiplication, and traces
- Useful matrix steps for data analysis
    - Sums and means, the data matrix, the deviation matrix, sums of squares and cross-products matrices (SSCPs), covariance matrices, and correlation matrices
- Killer matrix algebra stuff
    - Linear transformations, determinants, the matrix inverse, orthogonal and orthonormal matrices, the rank of a matrix, eigenvalues and eigenvectors, Kronecker products, singular value decomposition, and generalized inverses

What is a Matrix?

A matrix is a table of numbers, sort of like a spreadsheet. It has rows and columns. It is standard (for both "models" and statistical computing packages) to use rows to represent "respondents" (or "subjects," "observations," "entities," etc.), and columns to represent the "variables" (or "rating scales," "measures," "attributes," "scores," etc.) measured on those respondents. We'll call the data matrix, "**X**," which has "n" rows (subjects) and "p" columns (variables).

variable: j=1,2,...,p

Data matrix "**X**": $$\begin{bmatrix} x_{11} & x_{12} & \cdots & x_{1p} \\ x_{21} & x_{22} & \cdots & x_{2p} \\ \vdots & & x_{ij} & \\ x_{n1} & x_{n2} & \cdots & x_{np} \end{bmatrix}$$

subject: i=1,2,...n

x_{ij} is value of subject i on variable j

Subjects are often individuals—consumers, students in an experiment, etc., but they can be other entities—states, schools, stores, test markets, companies, etc. The first subscript on a matrix element (the "i") indicates what row we're referring to, and the second subscript (the "j") tells us what column we're in. This is standard too—always think: rows first, then columns.

The **order** of a matrix refers to its size (or its denomination!). It is the number of rows by the number of columns (r by c), e.g., n×p, 100×15, etc.

Matrices are variously denoted: **A** or ***A*** or $_r\mathbf{A}_c$. The matrix is comprised of elements: **A** = { a_{ij} }, and as always, i = row, j = column.

Let's look at two examples. The matrix on the left has the matrix elements labeled with matrix algebra notation, and the one on the right might be a real data set—filled with some numbers.

a 2×3 example

$$\boldsymbol{A} = \begin{bmatrix} a_{11} & a_{12} & a_{13} \\ a_{21} & a_{22} & a_{23} \end{bmatrix}$$

a 2×3 numerical example:

$$\boldsymbol{A} = \begin{bmatrix} 1 & 2 & 3 \\ 4 & 5 & 0 \end{bmatrix}$$

Square Matrices

Most data matrices are not square. Given that we normally want n to be much larger than p, most data matrices are actually rectangular (tall, not wide), and not square. And you know, to call a rectangular matrix "square" would be, well, stretching it!

Square matrices are just as they sound—matrices that look square because the #rows=#columns (r=c); e.g., this [ ] or this [ ] but not this [ ] or [ ].

Symmetric Matrices

Among square matrices, a special set are called **symmetric**. The definition is that a matrix **B** is symmetric if: 1) it is square, and 2) $b_{ij} = b_{ji}$. For example, this matrix is symmetric:

$$\boldsymbol{B} = \begin{bmatrix} b_{11} & b_{12} & b_{13} \\ b_{21} & b_{22} & b_{23} \\ b_{31} & b_{32} & b_{33} \end{bmatrix} = \begin{bmatrix} b_{11} & b_{12} & b_{13} \\ \underline{b_{12}} & b_{22} & b_{23} \\ \underline{b_{13}} & \underline{b_{23}} & b_{33} \end{bmatrix} = \begin{bmatrix} b_{11} & b_{12} & b_{13} \\ & b_{22} & b_{23} \\ sym. & & b_{33} \end{bmatrix}$$

Note in the left-most matrix, all the elements are labeled with their standard subscripts (row then column). In the middle matrix, the elements on the lower triangle below the diagonal are underlined to point out that they're the same as the elements on the upper triangle. In the symmetric **B** matrix, $b_{12} = b_{21}$, $b_{13} = b_{31}$, $b_{23} = b_{32}$.

Think about folding over the matrix along the diagonal (upper left to lower right), or putting a mirror on the diagonal. The values in the upper triangle will reflect those on the lower triangle (or vice versa). For short-hand, too, note the right-most matrix just has the diagonal and one of the triangles entered, and then it just says, "sym," as in, "It's symmetric, why are you looking down here?" Here is a numeric example of a symmetric matrix:

$$\begin{bmatrix} 1 & 4 & 5 \\ 4 & 2 & 6 \\ 5 & 6 & 3 \end{bmatrix}$$

Later, we'll use correlation matrices (and covariance matrices). They are both symmetric. Here is a correlation matrix with p=4:

$$\boldsymbol{R} = \begin{matrix} \\ var1 \\ var2 \\ var3 \\ var4 \end{matrix} \begin{matrix} \begin{matrix} var1 & var2 & var3 & var4 \end{matrix} \\ \begin{bmatrix} 1.000 & r_{12} & r_{13} & \boldsymbol{r_{14}} \\ r_{21} & 1.000 & r_{23} & r_{24} \\ r_{31} & r_{32} & 1.000 & r_{34} \\ \boldsymbol{r_{41}} & r_{42} & r_{43} & 1.000 \end{bmatrix} \end{matrix}$$

→ $r_{14} = r_{41}$

The correlation between variables 1 and 4 = correlation between variables 4 and 1.

The Transpose of a matrix

In a **transpose**, we are turning a matrix on its side. If $\boldsymbol{A} = \begin{bmatrix} 1 & 2 & 3 \\ 4 & 5 & 6 \end{bmatrix}$ then its transpose, $\boldsymbol{A}' = \begin{bmatrix} 1 & 4 \\ 2 & 5 \\ 3 & 6 \end{bmatrix}$. The 1st row of **A** becomes 1st column of **A'**, the 2nd row of **A** becomes 2nd column of **A'** (or the 1st column of **A** becomes the 1st row of **A'**, etc.), and the notation changes from **A** = $\{a_{ij}\}$ to **A'** = $\{a_{ji}\}$.

So the definition of symmetry earlier could also be given as, matrix **B** is symmetric if: 1) it is square and 2) **B=B'**. For example, this matrix is symmetric because if we turn it on its side, it is the same as the original matrix:

B=B': $\begin{bmatrix} 1 & 4 & 5 \\ 4 & 2 & 6 \\ 5 & 6 & 3 \end{bmatrix}$

If we transpose a matrix twice, we get back the original matrix (like a matrix toggle). That is: (**A'**)' = **A**. For example,

matrix: **A** **A'** **(A')'**

$[\quad]$ $[\quad\quad]$ $[\quad]$

Vectors and Scalars

More special names are used when r=1 or c=1 or r=c=1:

It's a "**column vector**" if c=1 (i.e., matrix is r×1): $\begin{bmatrix} 1 \\ 2 \end{bmatrix}$

It's a "**row vector**" if r=1 (i.e., matrix is 1×c): [1 2]

It's a "**scalar**" if r=c=1 (i.e., matrix is 1×1—in the real world, we'd call this a "number," but matrix analysts make up the rules here...): [2]

More Special Matrices

In a ***Variance-Covariance Matrix*** (or "covariance matrix" for short), we have the variances for each of the p variables along the diagonal $\hat{\sigma}_i^2$, and the covariances between the (i,j) pairs of variables in the off-diagonal cells:

$$\boldsymbol{\Sigma} = \begin{array}{c} \\ 1 \\ 2 \\ \\ i \\ j \\ \\ p \end{array} \begin{array}{c} \begin{array}{ccccccc} 1 & 2 & & i & j & & p \end{array} \\ \begin{bmatrix} \hat{\sigma}_1^2 & & & & & & \\ \hat{\sigma}_{21} & \hat{\sigma}_2^2 & & & & & \\ & & & & & & \\ \hat{\sigma}_{i1} & \hat{\sigma}_{i2} & & \hat{\sigma}_i^2 & & & \\ \hat{\sigma}_{j1} & \hat{\sigma}_{j2} & & \hat{\sigma}_{ji} & \hat{\sigma}_j^2 & & \\ & & & & & & \\ \hat{\sigma}_{p1} & \hat{\sigma}_{p2} & & \hat{\sigma}_{pi} & \hat{\sigma}_{pj} & & \hat{\sigma}_p^2 \end{bmatrix} \end{array}$$

Recall that a covariance measures the linear relationship between two variables, like the correlation coefficient. If "j" goes up as "i" goes up, the covariance will be positive. If "j" goes down as "i" goes up, the covariance will be negative. If there is no linear relationship between "i" and "j," the covariance will be near zero. A covariance is different from a correlation coefficient in that a correlation is normed to range from -1 to +1, but the actual value of a covariance depends on the standard deviations of variables "i" and "j". Mathematically, the relationship between a correlation, r_{ij} and a covariance, $\hat{\sigma}_{ij}$ is: $r_{ij} = \frac{\hat{\sigma}_{ij}}{s_i s_j}$.

We will use covariance matrices frequently.

Diagonal, Scalar, and Identity Matrices

A **diagonal** matrix is defined as: 1) the matrix is square and 2) all the off-diagonal entries = 0, that is, d_{ij}=0 for i and j, $i \neq j$. Note they are also symmetric. Here is an example:

$$\begin{bmatrix} 1 & 0 & 0 \\ 0 & 2 & 0 \\ 0 & 0 & 3 \end{bmatrix}$$

A **scalar matrix** is a matrix that is diagonal and all the entries along the diagonal are equal. For example: $\begin{bmatrix} 2 & 0 & 0 \\ 0 & 2 & 0 \\ 0 & 0 & 2 \end{bmatrix}$

The **identity matrix** is a scalar matrix and the value on the diagonal is one. This is a **VIM**—very important matrix! It is the matrix equivalent of the number one. (We'll see a "null matrix" next—the matrix equivalent of the number zero.)

$\mathbf{I}_3$ =identity matrix 3×3: $\begin{bmatrix} 1 & 0 & 0 \\ 0 & 1 & 0 \\ 0 & 0 & 1 \end{bmatrix}$ $\mathbf{I}_4$ = identity matrix 4×4: $\begin{bmatrix} 1 & 0 & 0 & 0 \\ 0 & 1 & 0 & 0 \\ 0 & 0 & 1 & 0 \\ 0 & 0 & 0 & 1 \end{bmatrix}$

Null Matrix and Equality of Matrices

A **null matrix** can be square or rectangular, but the important thing is that all elements=0. For example, $\mathbf{0}_{2\times3} = \begin{bmatrix} 0 & 0 & 0 \\ 0 & 0 & 0 \end{bmatrix}$.

Regarding the **equality of matrices, A** = **B** if they are of the same order and $a_{ij} = b_{ij}$ for all i and j. The comparison is done elementwise. First note that matrices **A** and **B** below are of the same order: the number of rows in **A** = the number of rows in **B**, and the number of columns in **A** = the number of columns in **B**. Next, **A**=**B** if: $a_{11}=b_{11}$ (the lines) and $a_{12}=b_{12}$ (the bolds) and similarly $a_{21}=b_{21}$ and $a_{22}=b_{22}$:

$$\boldsymbol{A} = \begin{bmatrix} a_{11} & \boldsymbol{a_{12}} \\ a_{21} & a_{22} \end{bmatrix}$$

$$\boldsymbol{B} = \begin{bmatrix} b_{11} & \boldsymbol{b_{12}} \\ b_{21} & b_{22} \end{bmatrix}$$

As a smaller example, consider: **A**=[4 1] and **B**=[4 1], **A**=**B**.

Alternatively, we can define the equality of matrices using the null matrix:

A = **B** if **A-B** = 0, where **A**, **B**, and **A-B** are all (r×c).

Matrix Addition

Now we're getting ready for action. Adding and subtracting 2 matrices require that they be the same order, and the operations are done elementwise. That is,

A+B=C $c_{ij} = a_{ij} + b_{ij}$ elementwise addition

A, B, C all have same number of rows and the same number of columns.
If **A** and **B** have different numbers of rows and/or columns, we can't add them (they're "not conformable"):

$$A = \begin{bmatrix} & \end{bmatrix}, \quad B = [\quad]$$

Here is a numerical example of adding **C=A+B**:

$$A = \begin{bmatrix} 1 & 2 & 3 \\ 4 & 5 & 6 \end{bmatrix}$$

$$B = \begin{bmatrix} 2 & 2 & 3 \\ 1 & 4 & 0 \end{bmatrix}$$

$$C = A + B = \begin{bmatrix} 1+2 & 2+2 & 3+3 \\ 4+1 & 5+4 & 6+0 \end{bmatrix} = \begin{bmatrix} 3 & 4 & 6 \\ 5 & 9 & 6 \end{bmatrix}$$

Matrix Subtraction is done the same way; elementwise.

E= [2 1 3]
F= [1 1 5]
D=**E**-**F**= [2-1 1-1 3-5] = [1 0 -2]

For addition and subtraction, the associative property holds:

(**A**+**B**)+**C** = **A**+(**B**+**C**) = **A**+**B**+**C**

and sequence is not important: **A**+**B**-**C** = **B**-**C**+**A**.

Scalar Multiplication

In scalar multiplication, we simply multiply each element in a vector or matrix by a single number (the scalar). Here are 2 examples:

If $A = \begin{bmatrix} 1 & 2 & 3 \\ 4 & 5 & 6 \end{bmatrix}$ then $\left(\frac{1}{2}\right)A = \begin{bmatrix} .5 & 1 & 1.5 \\ 2 & 2.5 & 3 \end{bmatrix}$

If $D = [1 \quad 0 \quad -2]$ then $(-1)D = [-1 \quad 0 \quad 2]$.

Matrix Multiplication

To multiply 2 matrices, such as **X**=**YZ,** there is first a test of conformability. If **Y** has r rows and c columns, then **Z** must have c rows and say q columns. Then the resulting matrix, **X**, will be r by q. The conformability test compares the "inner" subscripts of **Y** and **Z** (the 2 c's) and the result (**X**) will be of the order specified by the "outer" subscripts of **Y** and **Z** (the r and q):

$${}_r\mathbf{X}_q \quad = \quad {}_r\mathbf{Y}_c \; {}_c\mathbf{Z}_q$$

X is r×q   **Y** is r×c   **Z** is c×q

The multiplication manipulation is defined as follows (we'll see an example in a moment). In general (c is the number of columns of the first/left matrix):

$$x_{ij} = \sum_{k=1}^{c} \left(Y_{ik} Z_{kj}\right)$$

In matrix multiplication, sequence is important: **X=YZ**. Sometimes the reversed sequence doesn't even exist. In the example, **Y** and **Z** are conformable for **X=YZ**, but they wouldn't conform to create **ZY** (unless q=r, and even if it did, **ZY** is not likely to = **YZ**).

Here is an example of matrix multiplication. First we'll check the "inner subscripts." They're both 3, so the matrices conform.

$$\boldsymbol{Y} = \begin{bmatrix} 1 & 2 & 3 \\ 2 & 3 & 4 \end{bmatrix} \qquad \boldsymbol{Z} = \begin{bmatrix} 1 & 1 & 3 & 3 \\ 0 & 2 & 4 & 1 \\ 0 & 1 & 0 & 1 \end{bmatrix}$$

Y is 2×3   **Z** is 3×4

\ they conform /

X=YZ, **X** will be of what order? 2×4 (see the outside dimensions)

Proceeding with the example. Here is the resulting **X**:

$$\underset{\mathbf{Y}_{(2\times3)}}{\begin{bmatrix} 1 & 2 & 3 \\ 2 & 3 & 4 \end{bmatrix}} \underset{\mathbf{Z}_{(3\times4)}}{\begin{bmatrix} 1 & 1 & 3 & 3 \\ 0 & 2 & 4 & 1 \\ 0 & 1 & 0 & 1 \end{bmatrix}} = \underset{\mathbf{X}_{(2\times4)}}{\begin{bmatrix} 1 & 8 & 11 & 8 \\ 2 & 12 & 18 & 13 \end{bmatrix}}$$

And here the elements in **X** are illustrated:

$x_{11} = \sum_{k=1}^{3}(Y_{1k}Z_{k1}) = (1 \times 1) + (2 \times 0) + (3 \times 0) = 1$

$$\begin{bmatrix} \boxed{1 \quad 2 \quad 3} \\ 2 \quad 3 \quad 4 \end{bmatrix} \begin{bmatrix} \boxed{1} & 1 & 3 & 3 \\ \boxed{0} & 2 & 4 & 1 \\ \boxed{0} & 1 & 0 & 1 \end{bmatrix} = \begin{bmatrix} \boxed{1} & 8 & 11 & 8 \\ 2 & 12 & 18 & 13 \end{bmatrix}$$

$x_{12} = (1 \times 1) + (2 \times 2) + (3 \times 1) = 8$

$$\begin{bmatrix} \boxed{1 \quad 2 \quad 3} \\ 2 \quad 3 \quad 4 \end{bmatrix} \begin{bmatrix} 1 & \boxed{1} & 3 & 3 \\ 0 & \boxed{2} & 4 & 1 \\ 0 & \boxed{1} & 0 & 1 \end{bmatrix} = \begin{bmatrix} 1 & \boxed{8} & 11 & 8 \\ 2 & 12 & 18 & 13 \end{bmatrix}$$

$x_{13} = (1 \times 3) + (2 \times 4) + (3 \times 0) = 11$

$$\begin{bmatrix} \boxed{1 \quad 2 \quad 3} \\ 2 \quad 3 \quad 4 \end{bmatrix} \begin{bmatrix} 1 & 1 & \boxed{3} & 3 \\ 0 & 2 & \boxed{4} & 1 \\ 0 & 1 & \boxed{0} & 1 \end{bmatrix} = \begin{bmatrix} 1 & 8 & \boxed{11} & 8 \\ 2 & 12 & 18 & 13 \end{bmatrix}$$

$x_{14} = (1 \times 3) + (2 \times 1) + (3 \times 1) = 8$

$$\begin{bmatrix} \boxed{1 \quad 2 \quad 3} \\ 2 \quad 3 \quad 4 \end{bmatrix} \begin{bmatrix} 1 & 1 & 3 & \boxed{3} \\ 0 & 2 & 4 & \boxed{1} \\ 0 & 1 & 0 & \boxed{1} \end{bmatrix} = \begin{bmatrix} 1 & 8 & 11 & \boxed{8} \\ 2 & 12 & 18 & 13 \end{bmatrix}$$

$x_{21} = (2 \times 1) + (3 \times 0) + (4 \times 0) = 2$

$$\begin{bmatrix} 1 & 2 & 3 \\ 2 & 3 & 4 \end{bmatrix} \begin{bmatrix} 1 & 1 & 3 & 3 \\ 0 & 2 & 4 & 1 \\ 0 & 1 & 0 & 1 \end{bmatrix} = \begin{bmatrix} 1 & 8 & 11 & 8 \\ 2 & 12 & 18 & 13 \end{bmatrix}$$

$x_{22} = (2 \times 1) + (3 \times 2) + (4 \times 1) = 12$

$$\begin{bmatrix} 1 & 2 & 3 \\ 2 & 3 & 4 \end{bmatrix} \begin{bmatrix} 1 & 1 & 3 & 3 \\ 0 & 2 & 4 & 1 \\ 0 & 1 & 0 & 1 \end{bmatrix} = \begin{bmatrix} 1 & 8 & 11 & 8 \\ 2 & 12 & 18 & 13 \end{bmatrix}$$

$x_{23} = (2 \times 3) + (3 \times 4) + (4 \times 0) = 18$

$$\begin{bmatrix} 1 & 2 & 3 \\ 2 & 3 & 4 \end{bmatrix} \begin{bmatrix} 1 & 1 & 3 & 3 \\ 0 & 2 & 4 & 1 \\ 0 & 1 & 0 & 1 \end{bmatrix} = \begin{bmatrix} 1 & 8 & 11 & 8 \\ 2 & 12 & 18 & 13 \end{bmatrix}$$

$x_{24} = (2 \times 3) + (3 \times 1) + (4 \times 1) = 13$

$$\begin{bmatrix} 1 & 2 & 3 \\ 2 & 3 & 4 \end{bmatrix} \begin{bmatrix} 1 & 1 & 3 & 3 \\ 0 & 2 & 4 & 1 \\ 0 & 1 & 0 & 1 \end{bmatrix} = \begin{bmatrix} 1 & 8 & 11 & 8 \\ 2 & 12 & 18 & 13 \end{bmatrix}$$

Again (it's important), in general, even if **AB** and **BA** are both conformable, they won't typically be equal. That is, if we calculate both **C**=**AB** and **D**=**BA**, usually **C**≠**D**. Here is an example to illustrate the typical inequality:

e.g., with **A** = 2×2 = $\begin{bmatrix} 1 & 2 \\ 2 & 3 \end{bmatrix}$ and **B**= 2×2 = $\begin{bmatrix} 0 & 1 \\ 0 & 1 \end{bmatrix}$

C = **AB** = $\begin{bmatrix} 1 & 2 \\ 2 & 3 \end{bmatrix} \begin{bmatrix} 0 & 1 \\ 0 & 1 \end{bmatrix} = \begin{bmatrix} 0 & 3 \\ 0 & 5 \end{bmatrix}$

D = **BA** = $\begin{bmatrix} 0 & 1 \\ 0 & 1 \end{bmatrix} \begin{bmatrix} 1 & 2 \\ 2 & 3 \end{bmatrix} = \begin{bmatrix} 2 & 3 \\ 2 & 3 \end{bmatrix}$

→Note that indeed, **C**≠**D**.

It's also important to understand what happens with the multiplication of a vector and itself. For example, say we have the vector **V**:

$$\mathbf{V} = \begin{bmatrix} 1 \\ 2 \\ 3 \end{bmatrix}$$

V'V= 1×3 3×1 1×1

$$\begin{bmatrix} 1 & 2 & 3 \end{bmatrix} \begin{bmatrix} 1 \\ 2 \\ 3 \end{bmatrix} = [14]$$ = its sum of squares

$= (1\times1) + (2\times2) + (3\times3) = 1^2 + 2^2 + 3^2$

VV' = 3×1 1×3 3×3

$$\begin{bmatrix} 1 \\ 2 \\ 3 \end{bmatrix} \begin{bmatrix} 1 & 2 & 3 \end{bmatrix} = \begin{bmatrix} 1 & 2 & 3 \\ 2 & 4 & 6 \\ 3 & 6 & 9 \end{bmatrix}$$

The matrix algebra equivalent of division comes later.

Trace of a Matrix

The **trace** of a matrix, trace(**B**) = tr(**B**) is the sum of its diagonal elements:

e.g., tr = 1+4+9: $\begin{bmatrix} \mathbf{1} & 2 & 3 \\ 2 & \mathbf{4} & 6 \\ 3 & 6 & \mathbf{9} \end{bmatrix}$

And tr(**V'V**) = tr (**VV'**). In the previous example, tr(**V'V**) = 14, tr(**VV'**) = 1+4+9=14.

More special matrix multiplications

If we want to add up the rows or columns of a data matrix, here's how we would proceed. Say $\boldsymbol{X} = \begin{bmatrix} 1 & 0 & 1 \\ 3 & 2 & 4 \\ 6 & 2 & 4 \end{bmatrix}$, we'll define $\mathbf{1}_3 = \begin{bmatrix} 1 \\ 1 \\ 1 \end{bmatrix}$. Then multiply them together.

1'X = 1×3 3×3 1×3

$$\begin{bmatrix} 1 & 1 & 1 \end{bmatrix} \begin{bmatrix} 1 & 0 & 1 \\ 3 & 2 & 4 \\ 6 & 2 & 4 \end{bmatrix} = \begin{bmatrix} 10 & 4 & 9 \end{bmatrix}$$ = column sums of **X**

These totals are important to finding means (in a second).

X1 = 3×3 3×1 3×1

$$\begin{bmatrix} 1 & 0 & 1 \\ 3 & 2 & 4 \\ 6 & 2 & 4 \end{bmatrix} \begin{bmatrix} 1 \\ 1 \\ 1 \end{bmatrix} = \begin{bmatrix} 2 \\ 9 \\ 12 \end{bmatrix}$$

= row sums of **X**

Mathematical Properties

The associative property holds: **A(BC)** = **(AB)C** = **ABC.**

Sequence is important. **ABC** is not necessarily = **ACB** or **CAB** etc.

The distributive property holds: **(A+B)C** = **AC** + **BC** (sequence is maintained), or **D(E+F)** = **DE** + **DF.**

If we multiple a matrix by **I** (the identity matrix), we get the same matrix: **FI** = **F** (just like if we multiply a number by 1, we get the same number back). For example:

$$\underset{\mathbf{F}}{\begin{bmatrix} 2 & 3 \\ 1 & 4 \end{bmatrix}} \underset{\mathbf{I}}{\begin{bmatrix} 1 & 0 \\ 0 & 1 \end{bmatrix}} = \underset{\mathbf{F}}{\begin{bmatrix} 2 & 3 \\ 1 & 4 \end{bmatrix}}$$

Useful Things: Data matrix and things to do with it…

We're ready to look at a data matrix. Say we have 5 subjects, and 2 variables, X_1 and X_2:

$$\mathbf{X} \text{ is } 5\times 2 \rightarrow \begin{bmatrix} 7 & 8 \\ 8 & 10 \\ 0 & 1 \\ 2 & 0 \\ 1 & 1 \end{bmatrix}$$

1'X = 1×5 5×2 1×2

$$\begin{bmatrix} 1 & 1 & 1 & 1 & 1 \end{bmatrix} \begin{bmatrix} 7 & 8 \\ 8 & 10 \\ 0 & 1 \\ 2 & 0 \\ 1 & 1 \end{bmatrix} = \begin{bmatrix} 18 & 20 \end{bmatrix} \quad \text{column sums}$$

$$\frac{1}{n}(\mathbf{1}'\boldsymbol{X}) = \frac{1}{5}[18 \quad 20] = \begin{bmatrix} \frac{18}{5} & \frac{20}{5} \end{bmatrix} = [3.6 \quad 4.0] \quad \text{means}$$

/ \ / \
#subjects scalar means: var1 var2

1'X is the matrix way of saying: $\sum_{i=1}^{n} x_i$, so, $\frac{1}{n}(\mathbf{1}'\boldsymbol{X})$ is the matrix way of saying: $\frac{1}{n}\sum_{i=1}^{n} x_i$.
Next we'll repeat the mean for each variable for each subject:

$$1\left[\frac{1}{n}(\mathbf{1}'\boldsymbol{X})\right] = \begin{bmatrix} 1 \\ 1 \\ 1 \\ 1 \\ 1 \end{bmatrix} [3.6 \quad 4.0] = \begin{bmatrix} 3.6 & 4.0 \\ 3.6 & 4.0 \\ 3.6 & 4.0 \\ 3.6 & 4.0 \\ 3.6 & 4.0 \end{bmatrix}$$

To get a "deviation" matrix:

$$\boldsymbol{X}_d = \boldsymbol{X} - \left\{1\left[\frac{1}{n}(\mathbf{1}'\boldsymbol{X})\right]\right\}$$

$$\boldsymbol{X}_d = \begin{bmatrix} 7 & 8 \\ 8 & 10 \\ 0 & 1 \\ 2 & 0 \\ 1 & 1 \end{bmatrix} - \begin{bmatrix} 3.6 & 4.0 \\ 3.6 & 4.0 \\ 3.6 & 4.0 \\ 3.6 & 4.0 \\ 3.6 & 4.0 \end{bmatrix} = \begin{bmatrix} 3.4 & 4.0 \\ 4.4 & 6.0 \\ -3.6 & -3.0 \\ -1.6 & -4.0 \\ -2.6 & -3.0 \end{bmatrix}$$

$\boldsymbol{X}$ $-\left\{1\left[\frac{1}{n}(\mathbf{1}'\boldsymbol{X})\right]\right\}$ call this $\mathbf{X}_d$ ↑

Matrix, "$\mathbf{X}_d$," contains deviations of each person's data from the column (variable) mean.

Now some magic happens!

X$_d$'**X**$_d$ =

p×n	n×p	p×p
variables × subjects	subjects × variables	variables × variables

$$\begin{bmatrix} & X_d' & \end{bmatrix} \begin{bmatrix} \\ X_d \\ \\ \end{bmatrix} = \begin{bmatrix} 53.2 & 65.0 \\ 65.0 & 86.0 \end{bmatrix} \leftarrow \underline{\text{SSCP}}$$

"SSCP" stands for Sums of Squares and Cross-Products:

- sums of squares: the numerator of a variance, $\sum(x_i - \bar{x})^2$
- cross products: the numerator of a correlation coefficient, $\sum(x_i - \bar{x})(y_i - \bar{y})$

The covariance matrix then is just 1 step from the SSCP matrix:

$$\mathbf{S} = (1/(n\text{-}1))\ \mathbf{X}_d'\mathbf{X}_d$$

$$= \frac{1}{(5-1)}\begin{bmatrix} 53.2 & 65.0 \\ 65.0 & 86.0 \end{bmatrix} = \begin{bmatrix} 13.30 & 16.25 \\ 16.25 & 21.50 \end{bmatrix}$$

$$= \begin{bmatrix} s_{x1}^2 & cov(x_1, x_2) \\ cov(x_1, x_2) & s_{x2}^2 \end{bmatrix}$$

We've reviewed what a "covariance" is earlier, but for a convenient reminder: A covariance is like a correlation in measuring the (positive or negative) linear relationship between two variables. Correlations are bounded between -1 to +1, but the range on covariances depends on the two variables' standard deviations. Specifically the relationship is: $r_{ij} = \frac{\sigma_{ij}}{s_i s_j}$. The main diagonal of a covariance matrix contains the variances of the variables, σ_i^2.

More magic: First, we'll put the variances of **S** into a diagonal matrix:

$$\boldsymbol{D}_S = \text{diag}(\mathbf{S}) = \begin{bmatrix} 13.30 & 0 \\ 0 & 21.50 \end{bmatrix}$$

To get the standard deviations in the diagonal, we'll take the square root (or raise the variance to the ½ power):

$$\boldsymbol{D}_S^{1/2} = \begin{bmatrix} \sqrt{13.30} & 0 \\ 0 & \sqrt{21.50} \end{bmatrix} = \begin{bmatrix} 3.647 & 0 \\ 0 & 4.637 \end{bmatrix} = \begin{bmatrix} s_{x1} & 0 \\ 0 & s_{x2} \end{bmatrix}$$

Take the reciprocals of the diagonals, now we have 1/(standard deviations):

$$\boldsymbol{D}_S^{-1/2} = \begin{bmatrix} \frac{1}{3.647} & 0 \\ 0 & \frac{1}{4.637} \end{bmatrix} = \begin{bmatrix} .274 & 0 \\ 0 & .216 \end{bmatrix} = \begin{bmatrix} \frac{1}{s_{x1}} & 0 \\ 0 & \frac{1}{s_{x2}} \end{bmatrix}$$

So what is $\boldsymbol{D}_S^{-1/2}\, \boldsymbol{S}\, \boldsymbol{D}_S^{-1/2} = ?$

$$\qquad \boldsymbol{D}_S^{-1/2} \qquad\qquad \mathbf{S} \qquad\qquad \boldsymbol{D}_S^{-1/2}$$

$$= \begin{bmatrix} \frac{1}{s_{x1}} & 0 \\ 0 & \frac{1}{s_{x2}} \end{bmatrix} \begin{bmatrix} s_{x1}^2 & cov(x_1, x_2) \\ cov(x_1, x_2) & s_{x2}^2 \end{bmatrix} \begin{bmatrix} \frac{1}{s_{x1}} & 0 \\ 0 & \frac{1}{s_{x2}} \end{bmatrix}$$

$$= \begin{bmatrix} s_{x1} & \frac{cov(x_1,x_2)}{s_{x1}} \\ \frac{cov(x_1,x_2)}{s_{x2}} & s_{x2} \end{bmatrix} \begin{bmatrix} \frac{1}{s_{x1}} & 0 \\ 0 & \frac{1}{s_{x2}} \end{bmatrix}$$

$$= \begin{bmatrix} 1 & \frac{cov(x_1,x_2)}{s_{x1}s_{x2}} \\ \frac{cov(x_1,x_2)}{s_{x1}s_{x2}} & 1 \end{bmatrix}$$

$$= \begin{bmatrix} 1 & r_{x1,x2} \\ r_{x2,x1} & 1 \end{bmatrix} = [\boldsymbol{R}] \qquad = \text{correlation matrix!}$$

$$\boldsymbol{R} = \boldsymbol{D}_s^{-1/2}\, \boldsymbol{S}\, \boldsymbol{D}_s^{-1/2}$$

R = correlation matrix, ***S*** = the covariance matrix
They're both symmetric ($r_{ij}=r_{ji}$ = correlation between variables i and j) and $r_{ii}=1$.

Advanced Matrix Algebra

These next matrix algebra concepts are a little more intense. We'll begin with linear combinations, then determinants, then inverses, the matrix equivalent to division.

Linear combinations/linear transformations

We can create a new variable "y" as a linear combination of $x_1, x_2, ..., x_p$.

$$\begin{bmatrix} y_1 \\ y_2 \\ \dots \\ y_n \end{bmatrix} = \begin{bmatrix} x_{11} & x_{12} & \cdots & x_{1p} \\ x_{21} & x_{22} & & x_{2p} \\ & \dots & & \dots \\ x_{n1} & x_{n2} & & x_{np} \end{bmatrix} \begin{bmatrix} w_1 \\ w_2 \\ \dots \\ w_p \end{bmatrix}$$

We start with our data matrix **X** which has n subjects and p variables, put weights in the vector **w**, and obtain the new vector of observations for each subject on the newly defined variable in the vector **y**:

$$\underset{n\times 1}{\mathbf{y}} = \underset{n\times p}{\mathbf{X}}\ \underset{p\times 1}{\mathbf{w}}$$

$$y_i = w_1 x_{i1} + w_2 x_{i2} + \cdots + w_p x_{ip}$$

As an example of a linear combination, we'll define a new variable "y" as the mean of x_1 and x_2 (and ignore variables 3-p):

$$\begin{bmatrix} y_1 \\ y_2 \\ \dots \\ y_n \end{bmatrix} = \begin{bmatrix} x_{11} & x_{12} & \cdots & x_{1p} \\ x_{21} & x_{22} & & x_{2p} \\ & \dots & & \dots \\ x_{n1} & x_{n2} & & x_{np} \end{bmatrix} \begin{bmatrix} .5 \\ .5 \\ 0 \\ \dots \\ 0 \end{bmatrix} = \begin{bmatrix} (.5)x_{11} + (.5)x_{12} + \cdots + (0)x_{1p} \\ (.5)x_{21} + (.5)x_{22} + \cdots + (0)x_{2p} \\ \dots \\ (.5)x_{n1} + (.5)x_{n2} + \cdots + (0)x_{np} \end{bmatrix}$$

And later we'll make use of a triple product, or a "double linear combination" such as:

$$\underset{2\times 3}{\mathbf{A}} \qquad \underset{3\times 4}{\mathbf{B}} \qquad \underset{4\times 4}{\mathbf{W}}$$

$$\begin{bmatrix}1 & -1 & 0\\ 0 & 1 & -1\end{bmatrix}\begin{bmatrix}\bar{x}_{11} & \bar{x}_{12} & \bar{x}_{13} & \bar{x}_{14}\\ \bar{x}_{21} & \bar{x}_{22} & \bar{x}_{23} & \bar{x}_{24}\\ \bar{x}_{31} & \bar{x}_{32} & \bar{x}_{33} & \bar{x}_{34}\end{bmatrix}\begin{bmatrix}.25 & .5 & 1 & 0\\ .25 & .5 & 0 & 0\\ .25 & 0 & 0 & 1\\ .25 & 0 & -1 & 0\end{bmatrix}$$

- The rows of A indicate a comparison between groups 1 and 2, and comparison between groups 2 and 3.
- B contains the means for the 3 groups or samples on the 4 variables or measures.
- Columns of W indicate the linear combination of variables. Column 1 is the mean all 4 variables; column 2 is the mean of the 1st and 2nd variables, column 3 is the difference between variable 1 and variable 4, and column 4 is just variable 3 considered alone.

Properties of Linear Transformations

We can think of a linear combination of raw data x_i's: $Y = w_1x_1 + w_2x_2$

$$\mathbf{w} = \begin{bmatrix}w_1\\ w_2\end{bmatrix} \quad \boldsymbol{X} = \begin{bmatrix}x_1\\ x_2\end{bmatrix} \quad Y = \mathbf{w}'\boldsymbol{X} = \begin{bmatrix}w_1 & w_2\end{bmatrix}\begin{bmatrix}x_1\\ x_2\end{bmatrix}$$

or of population means, μ's:

$$\mathbf{w} = \begin{bmatrix}w_1\\ w_2\end{bmatrix} \quad \boldsymbol{\mu} = \begin{bmatrix}\mu_1\\ \mu_2\end{bmatrix} \quad Y = \mathbf{w}'\boldsymbol{\mu} = \begin{bmatrix}w_1 & w_2\end{bmatrix}\begin{bmatrix}\mu_1\\ \mu_2\end{bmatrix}$$

The population means ($\boldsymbol{\mu}_x$ and $\boldsymbol{\mu}_y$) and covariance matrices ($\boldsymbol{\Sigma}_x$ and $\boldsymbol{\Sigma}_y$) for the new variables (y's) as a function of those on the old variables (x's) are defined as:

$\boldsymbol{y} = \boldsymbol{Xw}$ (w is a vector 1 linear combination or a matrix of several)

$\boldsymbol{\mu}_y = \boldsymbol{w}'\boldsymbol{\mu}_x$

$\boldsymbol{\Sigma}_y = \boldsymbol{w}'\boldsymbol{\Sigma}_x\boldsymbol{w}$

Determinant of a matrix A

The determinant of a matrix A is denoted $|\boldsymbol{A}|$, and is defined as follows:

${}_1\mathbf{A}_1 = [a_{11}]$ → $|\boldsymbol{A}| = a_{11}$

${}_2\mathbf{A}_2 = \begin{bmatrix}a_{11} & a_{12}\\ a_{21} & a_{22}\end{bmatrix}$ → $|\boldsymbol{A}| = a_{11}a_{22} - a_{12}a_{21}$

${}_3\mathbf{A}_3 = \begin{bmatrix}a_{11} & a_{12} & a_{13}\\ a_{21} & a_{22} & a_{23}\\ a_{31} & a_{32} & a_{33}\end{bmatrix}$ →

$$|\boldsymbol{A}| = a_{11}a_{22}a_{33} - a_{11}a_{23}a_{32} + a_{12}a_{23}a_{31} - a_{12}a_{21}a_{33} + a_{13}a_{21}a_{32} - a_{13}a_{22}a_{31}$$

${}_4\mathbf{A}_4 = \begin{bmatrix}a_{11} & a_{12} & a_{13} & a_{14}\\ a_{21} & a_{22} & a_{23} & a_{24}\\ a_{31} & a_{32} & a_{33} & a_{34}\\ a_{41} & a_{42} & a_{43} & a_{44}\end{bmatrix}$ $|\mathbf{A}|$ = messy

Some special relationships hold:

1) if **A** is diag $\{a_{ii}\}$, $|\boldsymbol{A}| = \prod(a_{ii})$
2) if **A** is triangular, $|\boldsymbol{A}| = \prod(a_{ii})$
3) if all the elements in a row (or in column) of a square matrix are zero, det=0
4) if **A** and **B** are both n×n (i.e., square and of the same order), then $|\boldsymbol{AB}| = |\boldsymbol{A}||\boldsymbol{B}|$
5) $|\boldsymbol{A}| = |\boldsymbol{A}'|$

Inverses

An inverse is the matrix equivalent of division. In numbers,

$\frac{1}{5} = (5)^{-1} \qquad (5)\frac{1}{5} = 5(5)^{-1} = 1 \qquad \frac{3}{5} = 3\left(\frac{1}{5}\right) = 3(5)^{-1}$

For matrices, if **C** is square and of full rank, **C** is a "regular" inverse $\boldsymbol{C}^{-1}$. Then $\boldsymbol{CC}^{-1} = \boldsymbol{C}^{-1}\boldsymbol{C} = I$. If $\boldsymbol{C}^{-1}$ exists, **C** is said to be "nonsingular" (that's a good thing). $\boldsymbol{C}^{-1}$ exists if $|\boldsymbol{C}| \neq 0$ (to get the inverse, we'll need to divide by the determinant, so it cannot be zero, otherwise that would be analogous to dividing by 0 in real numbers). (Later after learning about eigenvalues, we'll see that since $|\boldsymbol{C}| = \prod(\lambda_i)$, we need all λ_i's to be nonzero so $|\boldsymbol{C}| > 0$ to have $\boldsymbol{C}^{-1}$.)

Some properties of inverses:

1) $(\boldsymbol{C}^{-1})^{-1} = \boldsymbol{C}$ (as in scalars (numbers): $1/\left(\frac{1}{5}\right) = (5^{-1})^{-1} = 5(\boldsymbol{AB})^{-1} = \boldsymbol{B}^{-1}\boldsymbol{A}^{-1}$
2) $(\boldsymbol{C}')^{-1} = (\boldsymbol{C}^{-1})'$
3) For a diagonal matrix, $\boldsymbol{D} = diag\{d_{ii}\}, \boldsymbol{D}^{-1} = diag\{\frac{1}{d_{ii}}\}$

Orthogonal Matrices

A matrix **G** is ⊥ (orthogonal) if **G'G** is a diagonal matrix:

e.g., $\boldsymbol{G} = \begin{bmatrix} 1 & -1 \\ -1 & -1 \end{bmatrix} \quad \boldsymbol{G}'\boldsymbol{G} = \begin{bmatrix} 1 & -1 \\ -1 & -1 \end{bmatrix}\begin{bmatrix} 1 & -1 \\ -1 & -1 \end{bmatrix} = \begin{bmatrix} 2 & 0 \\ 0 & 0 \end{bmatrix}$

Orthonormal Matrices

A matrix **H** is o.n. (orthonormal) if **H'H** = **I**; **H** is orthogonal and those diagonal elements are normalized to 1. O.N. matrices have an important quality that $\boldsymbol{H}^{-1} = \boldsymbol{H}'$ (makes the computation of an inverse easier!). Both orthogonal and orthonormal matrices, **G** and **H** are of "full rank," defined in a moment.

A very cute little article (Rodgers, Nicewander, and Toothaker, 1984) describes the differences between vectors or variables that are "linearly independent," "orthogonal," and "uncorrelated." First, some definitions:

- Linear independence: we can't get one column from another (add, multiply, etc.). Geometrically, the vectors don't fall on same line; they are not (multi)collinear.
- Orthogonal: right angles, perpendicular, i.e., cos angle between =0, the sum of their cross-products=0.
- Uncorrelated implies that once the raw data are centered, the vectors are perpendicular. That is, orthogonal says the raw variables are perpendicular, and uncorrelated says the centered variables are perpendicular.
- Also (from p.134): a) 2 variables that are perpendicular become oblique once they're centered; i.e., they're orthogonal but not uncorrelated (and they're linearly independent). b) 2 variables that are not perpendicular (they're oblique) can become perpendicular once they're centered; i.e., these are uncorrelated but not orthogonal (and they're linearly independent). c) 2 variables that are both orthogonal and uncorrelated if the centering doesn't change the angle between the vectors (and they're linearly independent).
- Here are 4 pairs of vectors that we'll characterize with the terms of orthogonal, uncorrelated, and linearly independent:

$$\begin{matrix} X & Y \\ \end{matrix} \qquad \begin{matrix} X & Y \\ \end{matrix} \qquad \begin{matrix} X & Y \\ \end{matrix} \qquad \begin{matrix} X & Y \\ \end{matrix}$$

$$\begin{bmatrix} 1 & 2 \\ 1 & 3 \\ 2 & 4 \\ 3 & 5 \end{bmatrix} \quad \begin{bmatrix} 0 & 1 \\ 0 & 0 \\ 1 & 1 \\ 1 & 0 \end{bmatrix} \quad \begin{bmatrix} -1 & 1 \\ -1 & -1 \\ 1 & 1 \\ 1 & -1 \end{bmatrix} \quad \begin{bmatrix} 1 & 5 \\ -5 & 1 \\ 3 & 1 \\ -1 & 3 \end{bmatrix}$$

Figure 1: Vector Relationships

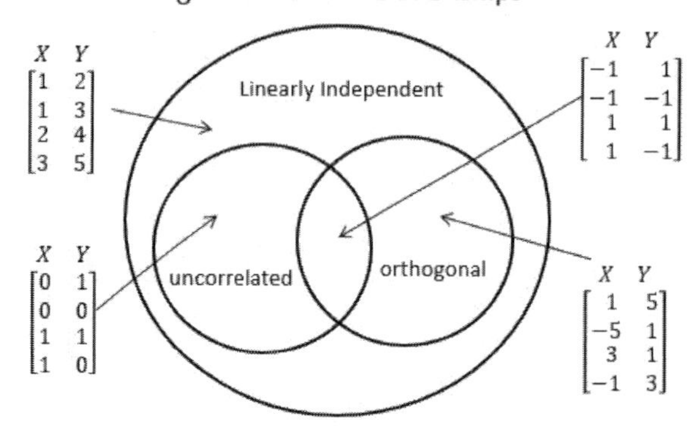

Rank of a Matrix

The rank of a matrix is roughly the amount of information contained in a matrix. The rank of ${}_n\mathbf{X}_p$ is denoted rank($\mathbf{X}$) $\leq$ min(n,p).

If n<p, and a) if **X** is of "full rank," then rank (**X**) = n, [ ]
or b) if **X** is less than full rank, rank (**X**) < n

if n>p, and a) if **X** is of "full rank," then rank (**X**) = p,

or b) if **X** is less than full rank, rank (**X**) < p

But how do we know whether **X** is of full rank or not, and if not, how do we know what rank (**X**) is?

A matrix that is "less than full rank" means there are redundancies in the matrix. That is, at least one column (or row) can be expressed as a linear combination of some or all of the other columns (or rows).

e.g., #1
$$X = \begin{bmatrix} 1 & 2 \\ 2 & 4 \\ 3 & 6 \\ 4 & 8 \end{bmatrix}$$

${}_4\mathbf{X}_2$ rank($\mathbf{X}$) $\leq$ min (4,2) ≤ 2. In this case, rank($\mathbf{X}$) = 1 because column 2 = 2.0×column 1. That is, column 2 is redundant with, or can be derived from, column 1.

e.g., #2
$$Y = \begin{bmatrix} 1 & 1 & 1 \\ 0 & 2 & 1 \\ 1 & 3 & 2 \end{bmatrix}$$

${}_3\mathbf{Y}_3$ rank($\mathbf{Y}$) $\leq$ min (3,3) ≤ 3. Here, the rank($\mathbf{Y}$) = 2 because column 3 can be derived from columns 1 and 2; that is, column 3 = .5(column 1) + .5(column 2).

e.g., #3 $\quad \boldsymbol{Z} = \begin{bmatrix} 1 & 1 \\ 0 & 2 \\ 1 & 3 \end{bmatrix}$

${}_3\mathbf{Z}_2$ rank(**Z**) ≤ min (3,2) ≤ 2. The rank(**Z**) = 2 because the columns are not redundant: one is not a linear combination of the other.

- The rank of a matrix is not an easy characteristic to determine by eye.
- The rank of a vector **Q** = 1 unless **Q**=0; the rank of a vector =1 unless the vector contains all zeros:

  e.g., rank=0: $\begin{bmatrix} 0 \\ 0 \\ 0 \end{bmatrix}$

  e.g., rank 1 ≤ min (3,1) since vector ≠ 0 vector: $\begin{bmatrix} 1 \\ 0 \\ 6 \end{bmatrix}$

Think of that vector $[1 \quad 0 \quad 6]'$ as 1 unique piece of information (e.g., one measure taken) for each subject.

The rank of this 3×2 matrix ≤ min (3,2) ≤ 2 = 1 since one column vector = 0: $\begin{bmatrix} 0 & 1 \\ 0 & 4 \\ 0 & 7 \end{bmatrix}$.

- The rank of a diagonal matrix that is k×k = k. For example, the rank of this matrix ≤ min (2,2) ≤ 2 = 2 because we can't derive 1 column from another $\begin{bmatrix} 15 & 0 \\ 0 & 7 \end{bmatrix}$.
- $r(\boldsymbol{X}) = r(\boldsymbol{X}')$. A matrix still contains the same amount of non-redundant information whether or not it has been transposed.
- $r(\boldsymbol{X}'\boldsymbol{X}) = r(\boldsymbol{X}\boldsymbol{X}') = r(\boldsymbol{X})$. No information is lost or gained (i.e., rank doesn't decrease or increase) when taking linear combinations of **X**, when those linear combinations are defined by **X** itself. That's a quality important in statistics. It means that the rank of a SSCP or covariance matrix based on **X** will = rank **X**, the data matrix.

If **X** is n×p (n<p) and **X** is of full rank, rows of **X** are linearly independent $[\quad\quad]$.

If **X** is n×p (n>p) and **X** is of full rank, columns of **X** are linearly independent $\begin{bmatrix} \\ \\ \end{bmatrix}$.

If **X** is n×n and **X** is of full rank, both columns and rows are linearly independent.

The notion of "linear independence" is that there are no redundancies, that a row (or column) cannot be written as a linear combination or the other rows (or columns). Conversely a linear combination would be a linear dependence. In one of our typical applications, **X** is an n×p data matrix where n>p (more subjects than variables), so the rank (**X**) ≤ min(n,p) or ≤p. $r(\boldsymbol{X}'\boldsymbol{X}) = r(\boldsymbol{X}\boldsymbol{X}') = r(\boldsymbol{X}) \leq p$. This rank "k" ≤ p is the number of positive "eigenvalues" we would find…so what's an eigenwhozee?

Eigenvalues and Eigenvectors

Eigenvalues and eigenvectors are derived from an eigen-structural decomposition of square, and for us, usually symmetric matrices, typically correlation or covariance matrices. Let's ease into the concept. If we want (the computer) to derive the weights in a p×1 vector $\boldsymbol{v}$, so that the newly created linear combination:

$$\boldsymbol{y} = \boldsymbol{X}\boldsymbol{v} \quad (i.e., y_i = v_1 X_{i1} + v_2 X_{i2} + \cdots + v_p X_{ip})$$

achieved the objective of explaining as much of the variance among the old variables (the X's, as in $\boldsymbol{\Sigma}_x$) as possible, we'd want to maximize $\boldsymbol{\Sigma}_y$, and we know (from properties of linear combinations): $\boldsymbol{\Sigma}_y = \boldsymbol{v}'\boldsymbol{\Sigma}_x\boldsymbol{v}$.

One way to make the entries in $\boldsymbol{\Sigma}_y$ large is to make the weights (the $\boldsymbol{v}$'s) very large, but this solution is trivial. Thus, we add the restriction that the sums of squares of the weights must be one ($\boldsymbol{v}'\boldsymbol{v} = 1$). To maximize a function (F) simultaneously with such a constraint, we need derivative calculus and a "LaGrangian multiplier" (λ):

$$F = \boldsymbol{v}'\boldsymbol{X}'\boldsymbol{X}\boldsymbol{v} - \lambda(\boldsymbol{v}'\boldsymbol{v} - 1) \;\propto\; \boldsymbol{v}'\boldsymbol{\Sigma}_x{}'\boldsymbol{v} - \lambda(\boldsymbol{v}'\boldsymbol{v} - 1) \;\propto\; \boldsymbol{v}'\boldsymbol{R}_x\boldsymbol{v} - \lambda(\boldsymbol{v}'\boldsymbol{v} - 1)$$

where **R** is the correlation matrix among the p variables ($\propto$ stands for "is proportional to"). The partial derivatives of that function F with respect to each weight v_1, v_2, through v_p must be set to zero:

$$\frac{\partial F}{\partial v} = 2\boldsymbol{R}_x\boldsymbol{v} - 2\lambda\boldsymbol{v} = \boldsymbol{R}_x\boldsymbol{v} - \lambda\boldsymbol{v} = (\boldsymbol{R}_x - \lambda\boldsymbol{I})\boldsymbol{v} = 0$$

One way to make this equation equal zero would be to multiply both sides by $(\boldsymbol{R}_x - \lambda\boldsymbol{I})^{-1}$ but this would imply $\boldsymbol{v} = 0$, which cannot be, given that $\boldsymbol{v}'\boldsymbol{v} = 1$. Thus, $(\boldsymbol{R}_x - \lambda\boldsymbol{I})$ must not have an inverse, so the determinant, $|\boldsymbol{R}_x - \lambda\boldsymbol{I}|$ must equal zero. Solving $|\boldsymbol{R}_x - \lambda\boldsymbol{I}| = 0$ results in a polynomial of power p:

$$v_p(-\lambda)^p + v_{p-1}(-\lambda)^{p-1} + v_{p-2}(-\lambda)^{p-2} + \cdots + v_1(-\lambda) + v_0 = 0$$

All λ's that satisfy this equation are called "characteristic roots" (i.e., power roots of this characteristic equation), or "eigenvalues" of **R**. There will be p λ's (though some may equal zero). Each λ_k is paired with an eigenvector, $\boldsymbol{v}_k$. (...like Noah's Ark...)

If an eigenvector pre-multiplies this equation, we have:

$$\boldsymbol{v}_k'(\boldsymbol{R} - \lambda\boldsymbol{I})\boldsymbol{v}_k = 0$$
$$\boldsymbol{v}_k'\boldsymbol{R}\boldsymbol{v}_k - \lambda\boldsymbol{v}_k'\boldsymbol{v}_k = 0$$
$$\boldsymbol{v}_k'\boldsymbol{R}\boldsymbol{v}_k = \lambda\boldsymbol{v}_k'\boldsymbol{v}_k$$

and because any $\boldsymbol{v}_k'\boldsymbol{v}_k = 1$, $\boldsymbol{v}_k'\boldsymbol{R}\boldsymbol{v}_k = \lambda$. Thus λ is the variance of the new composite variable formed from the old variables (in **X**) using the newly derived weights (in $\boldsymbol{v}_k$).

The eigenvalues are ordered $\lambda_1 \geq \lambda_2 \geq \cdots \geq \lambda_p \geq 0$, and put into a p×p diagonal matrix:

$$\boldsymbol{\Lambda} = \begin{bmatrix} \lambda_1 & 0 & 0 & \dots & 0 \\ 0 & \lambda_2 & & & \\ 0 & & \lambda_3 & & 0 \\ \dots & & & \dots & 0 \\ 0 & & 0 & \dots & \lambda_p \end{bmatrix}$$

and the eigenvectors $\boldsymbol{v}_1, \boldsymbol{v}_2, \dots, \boldsymbol{v}_k$ form the columns of a p×p matrix **V** that is orthonormal (i.e., $\boldsymbol{V}'\boldsymbol{V} = \boldsymbol{V}\boldsymbol{V}' = \boldsymbol{I}$; $\boldsymbol{V}' = \boldsymbol{V}^{-1}$):

$$V = \begin{bmatrix} \begin{bmatrix} v_1 \\ \\ \\ \end{bmatrix} & \begin{bmatrix} v_2 \\ \\ \\ \end{bmatrix} & \cdots & \begin{bmatrix} v_k \\ \\ \\ \end{bmatrix} \end{bmatrix}$$

We've seen that $\boldsymbol{V'RV} = \boldsymbol{\Lambda}$, and we know that $\boldsymbol{V'} = \boldsymbol{V}^{-1}$ ($\boldsymbol{V'}$ is inverse to $\boldsymbol{V}$, or $\boldsymbol{V}$ to $\boldsymbol{V'}$),

so, $\quad \boldsymbol{VV'RV} = \boldsymbol{V\Lambda}$

$$\boldsymbol{RV} = \boldsymbol{V\Lambda}$$
$$\boldsymbol{RVV'} = \boldsymbol{V\Lambda V'}$$
$$\boldsymbol{R} = \boldsymbol{V\Lambda V'}.$$

That is, we can decompose a correlation matrix into two unique matrices: one of eigenvalues $\boldsymbol{\Lambda}$ and one of eigenvectors $\boldsymbol{V}$ (and its transpose $\boldsymbol{V'}$).

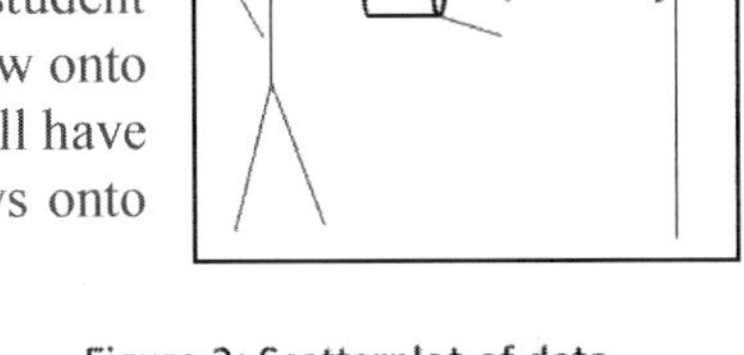

Eigenvalues and eigenvectors are abstract concepts...a more intuitive explanation follows. But first, we need the concept of a "projection." This figure shows a happy graduate student pointing a flashlight at a dot in space which projects a shadow onto the wall (at 90° or perpendicularly). In a similar manner, we'll have data in p-dimensional space and we'll project their shadows onto different walls or axes.

For eigenvalues and eigenvectors... consider a simple 2-dimensional scatterplot (i.e., p=2). The eigenvector $\boldsymbol{v}_1$ defines the direction in space in which the cloud of data points is the longest. That is, if all data points were projected onto that line, that distribution would have the greatest variance, compared with projecting the data points onto any other line in any other orientation in that p=2-dimensional space. The eigenvalue is the variance of the data points were they to be projected onto that new dimension defined by the eigenvector. The second eigenvector will be oriented orthogonally (perpendicularly), to explain the maximum of the remaining variance, and so on.

Figure 2: Scatterplot of data
Eigenvector through max dispersion
Eigenvalue = variance along that dimension

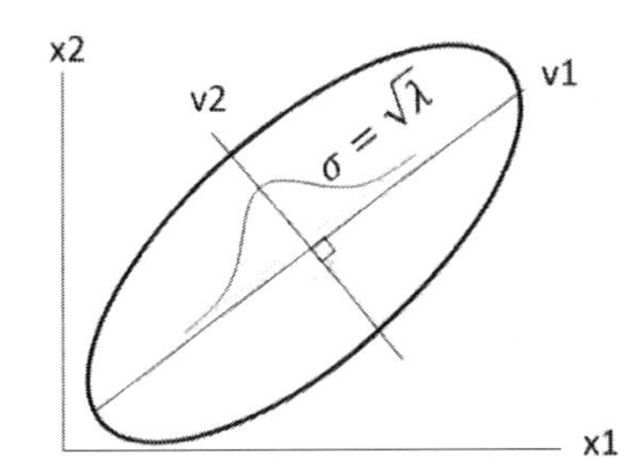

Let's do a little example by hand. We have a 2×2 matrix, and we wish to find λ to set $\det(\boldsymbol{S} - \lambda \boldsymbol{I})$ to 0:

$$|(\boldsymbol{S} - \lambda \boldsymbol{I})| = 0$$

$$\begin{bmatrix} s_{11} & s_{12} \\ s_{21} & s_{22} \end{bmatrix} - \lambda \begin{bmatrix} 1 & 0 \\ 0 & 1 \end{bmatrix} = \begin{bmatrix} s_{11} - \lambda & s_{12} \\ s_{21} & s_{22} - \lambda \end{bmatrix}$$

This matrix is symmetric ($s_{12} = s_{21}$) like our covariance and correlation matrices will be.

$$det = (s_{11} - \lambda)(s_{22} - \lambda) - s_{12}s_{12} = 0$$
$$\lambda^2 - s_{22}\lambda - s_{11}\lambda + s_{11}s_{22} - s_{12}s_{12} = 0$$

We'll solve for λ using a quadratic equation.

For example, say $\boldsymbol{S} = \begin{bmatrix} 14.19 & 10.69 \\ 10.69 & 8.91 \end{bmatrix}$

Then 2 sets of eigenvalues and vectors may be found:

$\lambda_1 = 22.56 \quad \boldsymbol{v}_1 = \begin{bmatrix} .787 \\ .617 \end{bmatrix}$

and $\lambda_2 = .54 \quad \boldsymbol{v}_2 = \begin{bmatrix} -.617 \\ .787 \end{bmatrix}$

We can verify λ_1 and λ_2 by solving the quadratic equation; showing $\boldsymbol{v}_1$ and $\boldsymbol{v}_2$ "work":

$$\boldsymbol{S}\boldsymbol{v}_1 = \begin{bmatrix} 14.19 & 10.69 \\ 10.69 & 8.91 \end{bmatrix}\begin{bmatrix} .787 \\ .617 \end{bmatrix} = \begin{bmatrix} 17.763 \\ 13.910 \end{bmatrix} = \lambda_1 \boldsymbol{v}_1 = 22.56 \begin{bmatrix} .787 \\ .617 \end{bmatrix}$$

$$\boldsymbol{S}\boldsymbol{v}_2 = \begin{bmatrix} 14.19 & 10.69 \\ 10.69 & 8.91 \end{bmatrix}\begin{bmatrix} -.617 \\ .787 \end{bmatrix} = \begin{bmatrix} -.342 \\ .416 \end{bmatrix} = \lambda_2 \boldsymbol{v}_2 = .54 \begin{bmatrix} -.617 \\ .787 \end{bmatrix}$$

We put λ's into diagonal matrix $\boldsymbol{\Lambda} = \begin{bmatrix} 22.56 & 0 \\ 0 & .54 \end{bmatrix}$

and $\boldsymbol{v}$'s as cols of $\boldsymbol{V} = \begin{bmatrix} .787 & -.617 \\ .617 & .787 \end{bmatrix}$

Then we can represent $\boldsymbol{S} = \boldsymbol{V\Lambda V'}$ (it works; try it).

With eigenvectors, $\boldsymbol{V}$ is o.n., so $\boldsymbol{V'V} = \boldsymbol{VV'} = \boldsymbol{I}$. In addition:

1) $\prod \lambda_i = |\boldsymbol{S}|$ and
2) $\sum \lambda_i = tr(\boldsymbol{S})$

Kronecker Product

The Kronecker product of 2 matrices **Y** and **Z** yield **X** such that each element in **Y** is a scaler multiplied against the entire matrix **Z**. $\boldsymbol{X} = \boldsymbol{Y} \otimes \boldsymbol{Z} \rightarrow Y_{ij}[\boldsymbol{Z}]$.

e.g., $\boldsymbol{Y} = \begin{bmatrix} 1 & 2 \\ 3 & 4 \end{bmatrix} \qquad \boldsymbol{Z} = \begin{bmatrix} 1 & 0 & 1 \\ 2 & 2 & 0 \end{bmatrix}$

$$\boldsymbol{Y} \otimes \boldsymbol{Z} = \begin{pmatrix} 1 * \begin{bmatrix} 1 & 0 & 1 \\ 2 & 2 & 0 \end{bmatrix} & 2 * \begin{bmatrix} 1 & 0 & 1 \\ 2 & 2 & 0 \end{bmatrix} \\ 3 * \begin{bmatrix} 1 & 0 & 1 \\ 2 & 2 & 0 \end{bmatrix} & 4 * \begin{bmatrix} 1 & 0 & 1 \\ 2 & 2 & 0 \end{bmatrix} \end{pmatrix} = \begin{bmatrix} 1 & 0 & 1 & 2 & 0 & 2 \\ 2 & 2 & 0 & 4 & 4 & 0 \\ 3 & 0 & 3 & 4 & 0 & 4 \\ 6 & 6 & 0 & 8 & 8 & 0 \end{bmatrix}$$

We don't use this operator often, but it's cool.

Singular Value Decomposition

We can obtain a singular value decomposition (svd) on any real matrix, and the results are related to eigensolutions (that we usually apply to real symmetric matrices). We begin with a matrix **X** which is n×p (n≥p) of real values (not all=0), like a data matrix. The svd matrix decomposition equation is $\boldsymbol{X} = \boldsymbol{PDQ'}$.

- $\boldsymbol{P}$ and $\boldsymbol{Q}$ are o.n. ($\boldsymbol{P'P} = \boldsymbol{Q'Q} = \boldsymbol{I}$), $\boldsymbol{D}$ is diagonal with "singular values."
- $\boldsymbol{P}$ is n×r, $\boldsymbol{D}$ is r×r, $\boldsymbol{Q}$ is p×r.
- $\boldsymbol{P}$ contains the eigenvectors of $\boldsymbol{XX'}$, and $\boldsymbol{Q}$ contains the eigenvectors of $\boldsymbol{X'X}$.

A svd can be useful in finding inverses. Recall that inverses are the matrix equivalent to division, and $\boldsymbol{C}^{-1}$ exists if $|\boldsymbol{C}| \neq 0$. Given that $|\boldsymbol{C}| = \prod \lambda_i$, we need all λ_i's to be nonzero, so $|\boldsymbol{C}| \neq 0$ to have $\boldsymbol{C}^{-1}$. If not, we can obtain a "generalized inverse."

Generalized Inverse

A "regular inverse," $\boldsymbol{C}^{-1}$, might not exist, but a generalized inverse, denoted $\boldsymbol{C}^{+}$, always does. ($\boldsymbol{C}^{-1}$ is a type of $\boldsymbol{C}^{+}$). The definition of a generalized inverse, C^{+}:

$$\boldsymbol{CC^{+}C} = \boldsymbol{C(C^{+}C)} = \boldsymbol{(CC^{+})C} = \boldsymbol{C}$$
$$\boldsymbol{C^{+}CC^{+}} = \boldsymbol{(C^{+}C)C^{+}} = \boldsymbol{C^{+}(CC^{+})} = \boldsymbol{C^{+}}$$

If $\boldsymbol{X} = \boldsymbol{PDQ'}$ by a svd, then $\boldsymbol{X}^{+} = \boldsymbol{QD^{-1}P'}$. Next we verify this relationship.

$\boldsymbol{XX^{+}X}$ must = $\boldsymbol{X}$ for $\boldsymbol{X}^{+}$ to be a generalized inverse:

$$(\boldsymbol{X})(\boldsymbol{X}^{+})(\boldsymbol{X}) = (\boldsymbol{PDQ'})(\boldsymbol{QD^{-1}P'})(\boldsymbol{PDQ'})$$

\ / \ /

$\boldsymbol{Q'Q}$=I $\boldsymbol{P'P}$=I

$$= \boldsymbol{PDD^{-1}DQ'}$$

\/

I

$$= \boldsymbol{PDQ'} = \boldsymbol{X}, \text{ neat, huh!!}$$

Also check the reverse: $\boldsymbol{X^{+}XX^{+}} = \boldsymbol{X}^{+}$?

$$\boldsymbol{X^{+}XX^{+}} = (\boldsymbol{QD^{-1}P'})(\boldsymbol{PDQ'})(\boldsymbol{QD^{-1}P'})$$

\ / \ /

$\boldsymbol{I}$ $\boldsymbol{I}$

$$= \boldsymbol{QD^{-1}DD^{-1}P'}$$

\/

$\boldsymbol{I}$

Thus $\boldsymbol{QD^{-1}P'} = \boldsymbol{X}^{+}$

Good news! That's it on matrix algebra until we need more.

SUMMARY

This chapter began with simple matrix algebra concepts and proceeded through more complicated manipulations. Multivariate equations are typically written in matrix language, so a minimal familiarity is helpful.

REFERENCES

Both of these books have excellent sections on matrix algebra:

1. Kirk, Roger E. (1982), *Experimental Design: Procedures for the Behavioral Sciences* (2nd ed.), Belmont, CA: Brooks/Cole (pp.778-805).
2. Morrison, D. F. (1976), *Multivariate Statistical Methods* (2nd ed.), NY: McGraw-Hill (pp.37-78).

Also see:

3. Namboodiri, Krishnan (1984), *Matrix Algebra: An Introduction*, Sage.
4. Rodgers, Joseph Lee, W. Alan Nicewander, and Larry Toothaker (1984), "Linearly Independent, Orthogonal, and Uncorrelated Variables," *The American Statistician*, 38 (2), 133-134.

HOMEWORK

This HW covers some basic matrix algebra.

Problem 1. What is a matrix? Why are we using matrices?

Problem 2. For matrices A, B, and C, conduct the following matrix algebraic calculations:

$$A = \begin{bmatrix} 1 & 2 & 1 \\ 0 & 3 & -1 \\ 2 & 1 & 4 \end{bmatrix} \qquad B = \begin{bmatrix} 3 & 2 & 1 \\ 2 & 3 & 1 \\ -1 & 2 & 3 \end{bmatrix} \qquad C = \begin{bmatrix} 1 & 2 \\ 3 & 4 \\ 5 & 6 \end{bmatrix}$$

a) Find A+B

b) Find A-2B

c) Find A’+B

d) Find A+C

e) Find (A+B)’

f) Find (2A’-B)’

Problem 3. Use the following matrices P, Q, R, and vectors x, y to calculate the products below:

$$P = \begin{bmatrix} 4 & 2 & -1 \\ 2 & 3 & -1 \\ -1 & -1 & 1 \end{bmatrix} \quad Q = \begin{bmatrix} 1 & 2 & 3 & 4 \\ 1 & 1 & 1 & 0 \\ 1 & 2 & 2 & 1 \end{bmatrix} \quad R = \begin{bmatrix} 1 & -2 \\ 0 & -2 \\ 3 & -1 \\ 2 & -1 \end{bmatrix} \quad x = \begin{bmatrix} 1 \\ 2 \\ 3 \end{bmatrix} \quad y = \begin{bmatrix} 0 \\ 1 \\ -1 \end{bmatrix}$$

a) Find PQ

b) Find PQR

c) Find QR’

d) Find yx’

e) Find x’y

f) Find x’Px

g) Find x’Py

h) Find P(x+y)

Problem 4. For a data matrix X (n×n), and the one vector j (n×1), and the one matrix W (n×n), calculate the following:

$$X = \begin{bmatrix} x_{11} & x_{12} & \cdots & x_{1n} \\ x_{21} & x_{22} & \cdots & x_{2n} \\ \cdots & & & \\ x_{n1} & x_{n2} & \cdots & x_{nn} \end{bmatrix} \quad j = \begin{bmatrix} 1 \\ 1 \\ \cdots \\ 1 \end{bmatrix} \quad W = \begin{bmatrix} 1 & 1 & \cdots & 1 \\ 1 & 1 & \cdots & 1 \\ \cdots & & & \\ 1 & 1 & \cdots & 1 \end{bmatrix}$$

a) Find j'X (the column sums of X)

b) Find Xj (the row sums of X)

c) Find WX, put values into a diagonal matrix, diag(WX)

d) Find XW and diag(XW)

e) Find j'j

f) Find Wj

g) Find W^2

Problem 5. Simple definitional question: For a matrix X, how do you know if it is orthogonal? Orthonormal?

Problem 6. Consider the following 5 matrices: ${}_5A_2$, ${}_4B_1$, ${}_3C_3$, ${}_5D_3$, ${}_3E_1$.

a) what are the dimensions of AA', CC', EE'

b) what are the dimensions of B'B, D'D

c) what is the size of a matrix that may:
   i. premultiply A (i.e., precede A), and what are the dimensions of the resulting product?
   ii. postmultiply A (i.e., follow A), & what are the dims
   iii. premultiply B, & what are the dims
   iv. postmultiply B, & what are the dims

d) A matrix S is a 3×3 variance-covariance matrix (it follows). Find $D_s^{-1/2}$ and with it construct the inter-correlation matrix.

$$\begin{bmatrix} 9 & 10 & -12 \\ 10 & 25 & 12 \\ -12 & 12 & 16 \end{bmatrix}$$

e) For the matrix A below,
   i. find A'A and tr(A'A)

ii. find AA' and tr(AA')
iii. What is the relation between the two traces? Describe why this relation should hold.

$$A = \begin{bmatrix} 1 & 2 & 3 \\ 0 & 3 & 2 \\ 6 & 3 & 44 \\ 5 & 4 & 22 \end{bmatrix}$$

f) Let c be a data vector of order n (i.e., n×1). What is (1'c)/n? Express the same quantity using the notation **1'1**. Hint: **1'1**=n.

<u>Problem 7</u>. Find the covariance matrix for the following data matrix X:

$$X = \begin{bmatrix} 1 & 7 & 5 \\ 2 & 4 & 9 \\ 3 & 6 & 2 \\ 6 & 5 & 0 \\ 8 & 3 & 9 \end{bmatrix}$$

CHAPTER 13

MULTIVARIATE NORMAL, T^2, ASSUMPTIONS AND ROBUSTNESS

Questions to guide your learning:
Q_1: What is a multivariate model?
Q_2: What are the multivariate counterparts to 1-sample and 2-sample t-tests?
Q_3: What new assumptions arise in multivariate models?

This chapter begins the transition from the univariate ANOVA, in which we model 1 dependent variable at a time, to the multivariate ANOVA, or MANOVA, in which we model several dependent variables simultaneously. We do so because this set or vector of "p" dependent variables are themselves likely to be somewhat correlated, and the resulting simultaneous models take these relationships into account, help control Type I errors, and help with the power of the statistical tests.

Univariate: t
Multivariate: T^2

This chapter begins with an overview of several classes of multivariate models to put MANOVA into perspective, then we'll focus toward MANOVA, looking at the multivariate normal distribution, the multivariate versions of a 1-sample t-test and a 2-sample t-test, both called T^2, and finally, we'll look at the now more complicated issues of model assumptions and the robustness of the model when the assumptions are violated.

Overview of Multivariate Statistical Analyses

The statistical models and analyses we have covered thus far take on the following general form:

$$X = \text{fn (factor A), or } X = \text{fn (factors A, B, ...)}$$

where X is a single dependent variable (or response measure), and it is modeled as a function of one or more predictor variables (or factors).

Simple and multiple regression models similarly consider a single dependent variable X, as a function of 1 or more predictors:

$$X = \text{fn (var V)} = b_0 + b_1V_1, \text{ or } X = \text{fn (var } V_1, V_2, \ldots) = b_0 + b_1V_1 + b_2V_2 \ldots$$

Both the ANOVA and regression models involve several variables, however, they are considered "univariate" because they model only 1 dependent variable at a time.

Techniques that are considered to be truly "multivariate" are those which attempt to model: (1) multiple dependent measures, "p" of them, simultaneously, or (2) a set of "p" variables and their interdependencies. We'll look at exemplars of each.

Multivariate Predictive Models

In multivariate models of the "prediction" sort, we model the scores on several dependent variables (X's) jointly, as a function of several explanatory predictor variables (V's):

$$(X_1, X_2, X_3, \ldots) = \text{fn } (V_1, V_2, V_3, \ldots).$$

Multivariate statistical techniques of this type include MANOVA, discriminant analysis, and canonical correlations.

In **multivariate analysis of variance (MANOVA)**, we model the set of dependent variables as a function of 1 or more factors like in the univariate ANOVA model:

$$(X_1, X_2, X_3, ...) = \text{fn (factors A, B,...)}$$

The goal is to try to understand group differences (where the groups, as usual are defined by a cross-classification of factors A, B, ...) as the study participants have been measured on a battery of dependent variables, not just the single score on one measured variable. For example, we might study how the factors of cognitive resource allocation (A) and motivational instructions (B) combine to produce effects on recall (X_1) and evaluation (X_2) together. We can, and should, run all kinds of univariate statistics, from descriptives (means and standard deviations on both X_1 and X_2, frequency distributions, etc.), correlations between them, within cells and across cells, etc. even univariate ANOVAs on X_1 and X_2 separately, to get a feel for the data, but the MANOVA takes A and B and X_1 and X_2 together into a single model. In this chapter and Chapter 14, we're working toward the MANOVA model.

In **discriminant analysis**, the goal is to find a linear combination of predictor variables to maximally distinguish groups, and to allow for the unambiguous classification of new individuals to the existing groups. If there are only 2 groups, this model can be fit via a multiple regression with the linear combination: $group\ 1\ vs. 2 = b_0 + b_1V_1 + b_2V_2 \ldots + b_pV_p$.

For example, we might ask a sample of consumers to rate our company's brand of laundry detergent on a handful of attributes (does it clean well, does it make the clothes soft, is it a good value for the money, etc.). We'd use a discriminant analysis to determine whether our customers (current users of our brand), rate the product differently from how consumers who use other brands rate it. We'll see more on discriminant analysis in Chapter 15.

A **canonical correlational analysis** is like a multivariate version of a multiple regression. The goal is to find a linear combination of predictor variables ($V = m_1V_1 + m_2V_2 \ldots m_qV_q$) that is maximally correlated with a linear combination of dep variables ($X = k_1X_1 + k_2X_2 \ldots k_pX_p$). In a regression, we have the set of V's but just 1 dependent variable. In a canonical analysis, we'll have a set of dependent variables and a set of independent variables, and the model operates like a multiple regression but on both sides of the equation.

For example, we might look into the linear combinations of subject variables (V's being demographic information and other individual differences variables), that are maximally correlated with a set of brand variables (X's being ratings of brand evaluations, semantic differentials of product characteristics, etc.).

As a brief comparison among these models:

- In multiple regression, we ask the computer to find b's to maximize the R^2 using the model: $x = \text{fn}(v_1, v_2, \ldots v_p)$.
- We know that a multiple regression can be used to fit an ANOVA if the predictors are dummy variables. (In practice, we'd simply run the ANOVA.)
- A discriminant analysis is like the regression but we're trying to predict group membership.
- In a canonical correlation analysis, we ask the computer to find weights on both sides of the equation, so the linear combination of X's is maximally correlated with a linear combination of the V's: $\text{fn}(x_1, x_2, \ldots x_q) = \text{fn}(v_1, v_2, \ldots v_p)$
- In theory, a canonical correlation can be used to fit a MANOVA if the predictors are dummy variables. (In practice, it is not; a standard MANOVA is run.)

Multivariate Interdependence Models

For multivariate models of the interdependence sort, we haven't classified our variables into "dependent variables" and "predictor variables." Instead we want to model the associative structure among our variables X_1 X_2 ... X_Z. Very often, we will begin with a correlation matrix—a matrix with rows and columns representing a set of "p" variables, and each element in the matrix is the correlation coefficient between the pair of variables in the ith row and the jth column. We'll stare at the correlation matrix and do some things to it to understand the pattern of bivariate interdependencies, but we're not trying to predict anything. Multivariate statistical techniques of this type include factor analysis, principal components analysis, multidimensional scaling, and cluster analysis.

In **factor analysis** (FA), the goal is to represent the correlational structure of our data observed on lots of variables, "p" of them, in terms of many fewer underlying constructs or factors, "r" of them, where r<<<p. As an example, we might have study participants try a new soft drink, and have them rate the soft drink on a number of attributes such as "sweetness," "carbonation," and "chemical-tasting." We'd likely find that these variables are highly inter-correlated, and so perhaps they all tap an underlying factor (sometimes called a "latent" factor) of "taste." Similarly, ratings of "expensive," "attractive can," etc. might tap another factor, perhaps having to do with image or lifestyle.

A **principal components analysis** (PCA) is a lot like factor analysis, and they're often confused. In a PCA, the goal is to maximize variance accounted for (VAF) in data on large number of variables ("p") using many fewer principal components ("r," r<<<p). The difference in focus between FA and PCA is that the components are linear combinations of the original measures to explain variance, whereas in FA, the underlying factors explain covariance.[23]

In **multidimensional scaling** (MDS), the goal is to represent objects such as brands as points in space, such that the distances in space between the brands represent the perceived similarities and differences between the brands. In marketing, these model representations have come to be known as perceptual maps. For example, brands that are close in the map would be those seen as nearly substitutable (close competitors) versus brands that seem to be very different that are mapped farther apart. The dimensions of the map are interpreted as the attributes that consumers use when distinguishing among the brands.

In **cluster analysis**, the goal is to form groups of objects (group of brands, or even groups of people as in market segmentation) so that the groups, or clusters, formed contain homogeneous objects. Usually the objective function that is maximized is to enhance between-group differences compared to within-group similarities. For example, in clustering brands in any product category, it is not unusual for consumers' perceptions to be somewhat similar among the high-end and more expensive brands, but different from the simpler, less expensive products (which themselves might get clustered together as being seen as relatively similar). Or, when clustering customers, we might see groups defined by socio-economic status, frequency of usage in the product category, levels of expertise, preferences by brands, etc.

That brief overview described several prominent multivariate models, of both the predictive and inter-dependent varieties. Next we focus on issues we'll need for the multivariate ANOVA model.

---

[23] For more on the distinction, see Iacobucci (1994, 2015) and cites therein.

Multivariate Normal Distribution and Hotelling's T^2

We begin with the multivariate version of the normal distribution. We will then see the multivariate versions of the 1-sample and 2-sample t-tests.

Univariate Inference about the Mean

At this point, let's review the univariate case of how to run a t-statistic to test the null hypothesis about a mean, mostly to have notation to generalize to the multivariate case. To test the null:

$$H_0: \mu = \mu_0 \text{ vs. } H_a: \mu \neq \mu_0$$

we estimate σ^2 by s^2, and μ by $\bar{x}$, and use both estimates to make the inference:

$$t = \frac{\bar{x} - \mu_0}{\left(\frac{s}{\sqrt{n}}\right)} \tag{1}$$

We compare our calculated t-value to the critical Student's t-statistic on n-1 df. To facilitate the generalization to more than 1 variable, let's take that t-value and look at t^2:

$$t^2 = \frac{(\bar{x} - \mu_0)^2}{\left(\frac{s^2}{n}\right)} \tag{2}$$

We wouldn't actually ever use t^2, in part because we'd have to square the critical values as well. But from the equation, we can see that if we were to use t^2, we would reject H_0 when it is big and positive (rather than when $|t|$ is big and positive, when it is not squared). Let's take it a step further, and break the numerator into 2 terms (it's squared, and to do so seems silly, but again, it will facilitate an understanding of the multivariate form):

$$t^2 = \frac{(\bar{x} - \mu_0)(\bar{x} - \mu_0)}{\left(\frac{s^2}{n}\right)} = \frac{n(\bar{x} - \mu_0)(\bar{x} - \mu_0)}{s^2}$$

Finally, writing s^2 as the reciprocal instead of in the denominator, we have:

$$t^2 = n(\bar{x} - \mu_0)(s^2)^{-1}(\bar{x} - \mu_0) \tag{3}$$

Multivariate Inference about the Mean: 1 Sample

In the multivariate world, we'll replace the mean we've computed on a single variable with a vector of means, a collection of means on each of "p" variables. The null is still:

$$H_0: \boldsymbol{\mu} = \boldsymbol{\mu}_0 \text{ vs. } H_a: \boldsymbol{\mu} \neq \boldsymbol{\mu}_0 \tag{4}$$

The vectors of μ's and hypothesized μ_0's look like this:

$$H_0: \boldsymbol{\mu} = \begin{bmatrix} \mu_1 \\ \mu_2 \\ \cdots \\ \mu_p \end{bmatrix} = \begin{bmatrix} \mu_{01} \\ \mu_{02} \\ \cdots \\ \mu_{0p} \end{bmatrix} = \boldsymbol{\mu}_0$$

p = # variables, so **μ** and $\boldsymbol{\mu}_0$ are p×1 vectors (5)

Naturally we'll estimate the μ for any variable with its $\bar{x}$. At this point, a vector of sample means is just descriptive. To pursue the inferential question of whether our guesses in the $\boldsymbol{\mu}$ vector in fact equal those hypothesized in the corresponding places in the $\boldsymbol{\mu}_0$ vector, we need something like a t-test or a z-test, and then something like a t- or normal distribution from which to obtain critical values to evaluate our calculated statistic.

In the univariate ANOVA world, we have scores observed on a single variable for each of our n subjects: x_1 x_2 ... x_n. We assume that the data points on the n subjects (the Xs) are "independent and identically distributed (i.i.d.) random variables." The "i.i.d." means that subjects are independent (the score we get from each subject does not influence the scores we get from any of the other subjects) and identically distributed (i.d.), and in this application we assume that each subject's score (x_i) is drawn from the normal distribution described by $(\mu,\ \sigma^2)$. That is, "i.i.d." simply means that we assume that x_1 x_2 ... x_n comprise a random sample of observations from a common population normal distribution. With the z-test, we say that our statistic is distributed as ("~") a normal variable (N) with population mean μ and population variance σ^2, altogether $\sim N(\mu, \sigma^2)$; i.e., the x_1 x_2 ... x_n come from the normal distribution, which has associated with it 2 parameters, μ and σ^2. The equation for the normal distribution is called the normal probability density function (p.d.f.), and notice the 2 parameters, μ and σ^2:

$$\frac{1}{\sqrt{2\pi}\sigma} exp\left\{-\frac{(X_i-\mu)^2}{2\sigma^2}\right\} \tag{6}$$

In the multivariate model, we'll say analogous things. We have p measures on each of our n subjects, so a vector of data $\mathbf{X}_i$ from each subject, but each of the n subjects are still said to be independent and identically distributed random vectors $\mathbf{X}_1$ $\mathbf{X}_2$ ... $\mathbf{X}_n$; i.e., we'll draw a random sample of n subjects and collect p pieces of information from each: $\mathbf{X}_1$ $\mathbf{X}_2$ ... $\mathbf{X}_n$ ~ MVN_p ($\boldsymbol{\mu}$, $\boldsymbol{\Sigma}$), where MVN stands for "multivariate normal," or "multinormal," with dimensionality p (i.e., p variables). It has a population mean vector:

$$\boldsymbol{\mu}' = [\mu_1 \quad \mu_2 \quad \cdots \quad \mu_p] \tag{7}$$

and a variance-covariance matrix (or just covariance matrix for short):

$$\boldsymbol{\Sigma} = \begin{bmatrix} \sigma_1^2 & & & \\ \sigma_{21} & \sigma_2^2 & sym & \\ \cdots & & & \\ \sigma_{p1} & \sigma_{p2} & \cdots & \sigma_p^2 \end{bmatrix} \tag{8}$$

The covariance matrix has the variables' variances on the diagonal. The matrix is symmetric (like a correlation matrix) and recall the relationship between a covariance and correlation coefficient is just the multiplication of the 2 variables' standard deviations: $\sigma_{ij} = \rho_{ij}\sigma_i\sigma_j =$ the covariance between variables i and j.

Each subject, $\mathbf{X}_i$ is a random observation on the p variables, distributed $MVN_p(\boldsymbol{\mu}, \boldsymbol{\Sigma})$. The multivariate normal p.d.f. for $\boldsymbol{X}_i{}' = [X_{i1} \quad X_{i2} \quad \cdots \quad X_{ip}]$ is as follows:

$$p(\boldsymbol{X}; \boldsymbol{\mu}, \boldsymbol{\Sigma}) = \frac{1}{(2\pi)^{p/2}|\boldsymbol{\Sigma}|^{1/2}} exp\left\{-\frac{1}{2}(\boldsymbol{X}_i-\boldsymbol{\mu})'\boldsymbol{\Sigma}^{-1}(\boldsymbol{X}_i-\boldsymbol{\mu})\right\} \tag{9}$$

Well, good to know, but how do we test H_0? The subjects are random vectors, each p×1, and we'll put them into a data matrix $\boldsymbol{X}$ (n×p) as rows:

$$\boldsymbol{X} = \begin{bmatrix} X_1' \\ X_2' \\ \\ \cdots \\ \\ X_n' \end{bmatrix} = \begin{bmatrix} X_{11} & X_{12} & \cdots & X_{1p} \\ X_{21} & X_{22} & \cdots & X_{2p} \\ \\ \\ \cdots & & \cdots & \cdots \\ X_{n1} & X_{n2} & & X_{np} \end{bmatrix} \qquad (10)$$

We use **X** to obtain estimates for **μ** ($\overline{\boldsymbol{x}}$) and **Σ** (**S**), first obtaining the means over subjects on each of p variables:

$$\overline{\boldsymbol{x}} = \frac{1}{n}\sum_{i=1}^{n} X_i = \begin{bmatrix} \bar{x}_1 \\ \bar{x}_2 \\ \cdots \\ \bar{x}_p \end{bmatrix} \qquad (11)$$

Then we obtain the sample covariance matrix:

$$\boldsymbol{S} = \frac{1}{n-1}\sum_{i=1}^{n} (X_i - \bar{x})\,(X_i - \bar{x})' = \begin{bmatrix} s_1^2 & & & sym \\ s_{21} & s_2^2 & & \\ \cdots & & & \\ s_{p1} & s_{p2} & \cdots & s_p^2 \end{bmatrix} \qquad (12)$$

Next, we'll use both estimates $\overline{\boldsymbol{x}}$ and **S** to make inferences about $H_0\colon \boldsymbol{\mu} = \boldsymbol{\mu}_0$. The multivariate counterpart to the 1-sample t-test is called Hotelling's T^2. It is defined as:

$$T^2 = n(\overline{\boldsymbol{x}} - \boldsymbol{\mu}_0)'\boldsymbol{S}^{-1}(\overline{\boldsymbol{x}} - \boldsymbol{\mu}_0) \qquad (13)$$

That form looks analogous to t^2 as written earlier. The numbers $\bar{x}$, μ_0, and s^2 have been replaced with vectors $\overline{\boldsymbol{x}}$ and $\boldsymbol{\mu}_0$ and the matrix **S**.

We had compared the univariate t-value to the Student's t-distribution on n-1 df. What shall we compare the multivariate T^2 to? A summer's day. There is actually no need for critical values of T^2 per se. We will transform T^2 to a statistic that follows the F-distribution under H_0:

$$F = \frac{(n-p)}{p(n-1)}T^2 \quad \text{on p and (n-p) df.} \qquad (14)$$

Notice that if p=1 (i.e., we're back in univariate-land), then this $F = t^2$. So this T^2 is truly just a multivariate extension of the simple univariate test. Let's see all this in an example.

Numerical Example of One-Sample Hotelling's T^2

In this example, we have a ridiculously small sample of 4 subjects and on each person, we have measured 3 variables. Thus, p, the number of dependent variables is 3, and n, the number of subjects is 4. Usually n >> p. Here are the n=4 subjects' vectors of data; each person provided p=3 data points:

$$x_1 = \begin{bmatrix} 2 \\ 3 \\ 11 \end{bmatrix}, \quad x_2 = \begin{bmatrix} 8 \\ 7 \\ 10 \end{bmatrix}, \quad x_3 = \begin{bmatrix} 2 \\ 6 \\ 6 \end{bmatrix}, \quad x_4 = \begin{bmatrix} 4 \\ 8 \\ 9 \end{bmatrix}$$

We put the subjects' data into the matrix **X** as rows, so the rows of **X** are subjects, 1 to n, and the columns are variables, 1 to p. The data matrix **X** is n×p:

$$X_{n\times p} = \begin{bmatrix} x_1' \\ x_2' \\ x_3' \\ x_4' \end{bmatrix} = \begin{bmatrix} 2 & 3 & 11 \\ 8 & 7 & 10 \\ 2 & 6 & 6 \\ 4 & 8 & 9 \end{bmatrix}$$

The means, 1 per variable, are easily computed, and we'll have a vector of means that is p×1:

$$\bar{\boldsymbol{x}} = \frac{1}{n}\sum_{i=1}^{n} x_i = \frac{1}{4}\left[\begin{bmatrix} 2 \\ 3 \\ 11 \end{bmatrix} + \begin{bmatrix} 8 \\ 7 \\ 10 \end{bmatrix} + \begin{bmatrix} 2 \\ 6 \\ 6 \end{bmatrix} + \begin{bmatrix} 4 \\ 8 \\ 9 \end{bmatrix}\right] = \frac{1}{4}\begin{bmatrix} 16 \\ 24 \\ 36 \end{bmatrix} = \begin{bmatrix} 4 \\ 6 \\ 9 \end{bmatrix}$$

The covariance matrix is computed next. Note the typical form in the equation, a sum (over subjects) of squared deviations from the mean (the "squaring" here is the deviations times the deviations transposed):

$$\boldsymbol{S} = \frac{1}{n-1}\sum_{i=1}^{n}(X_i - \bar{x})(X_i - \bar{x})' =$$

$$\frac{1}{3}\left[\begin{bmatrix} 2-4 \\ 3-6 \\ 11-9 \end{bmatrix}\begin{bmatrix} -2 & -3 & 2 \end{bmatrix} + \begin{bmatrix} 8-4 \\ 7-6 \\ 10-9 \end{bmatrix}\begin{bmatrix} 4 & 1 & 1 \end{bmatrix} + \begin{bmatrix} 2-4 \\ 6-6 \\ 6-9 \end{bmatrix}\begin{bmatrix} -2 & 0 & -3 \end{bmatrix} + \begin{bmatrix} 4-4 \\ 8-6 \\ 9-9 \end{bmatrix}\begin{bmatrix} 0 & 2 & 0 \end{bmatrix}\right]$$

$$= \frac{1}{3}\left[\begin{bmatrix} 4 & 6 & -4 \\ 6 & 9 & -6 \\ -4 & -6 & 4 \end{bmatrix} + \begin{bmatrix} 16 & 4 & 4 \\ 4 & 1 & 1 \\ 4 & 1 & 1 \end{bmatrix} + \begin{bmatrix} 4 & 0 & 6 \\ 0 & 0 & 0 \\ 6 & 0 & 9 \end{bmatrix} + \begin{bmatrix} 0 & 0 & 0 \\ 0 & 4 & 0 \\ 0 & 0 & 0 \end{bmatrix}\right]$$

$$= \frac{1}{3}\begin{bmatrix} 24 & 10 & 6 \\ 10 & 14 & -5 \\ 6 & -5 & 14 \end{bmatrix} = \begin{bmatrix} 8.000 & 3.333 & 2.000 \\ 3.333 & 4.667 & -1.667 \\ 2.000 & -1.667 & 4.667 \end{bmatrix}$$

In the t-test, we had a standard error term in the denominator. Recall from Chapter 12 that for a number, the reciprocal is the number raised to the (-1) power, its inverse: e.g., $5^{-1} = \frac{1}{5}$. The multivariate analog of putting a matrix in the denominator is to use the inverse $\boldsymbol{S}^{-1}$. (I didn't show you the steps from $\boldsymbol{S}$ to computing $\boldsymbol{S}^{-1}$ but you can convince yourself that $\boldsymbol{S}\boldsymbol{S}^{-1} = \boldsymbol{S}^{-1}\boldsymbol{S}=\boldsymbol{I}$.)

$$\boldsymbol{S}^{-1} = \begin{bmatrix} .321 & -.319 & -.251 \\ -.319 & .562 & .337 \\ -.251 & .337 & .442 \end{bmatrix}$$

No points for creativity here, but say we are testing the null hypothesis with values: $H_0\colon \boldsymbol{\mu} = [0\ 0\ 0]'$. In the numerator of the univariate t-test, we had $(\bar{x} - \mu_0)$, and in this multivariate world, we have that here as well, it's just that these pieces are vectors rather than single numbers:

$$(\bar{\boldsymbol{x}} - \boldsymbol{\mu}_0) = \left[\begin{bmatrix} 4 \\ 6 \\ 9 \end{bmatrix} - \begin{bmatrix} 0 \\ 0 \\ 0 \end{bmatrix} = \begin{bmatrix} 4 \\ 6 \\ 9 \end{bmatrix}\right]$$

The equation for the 1-sample Hotelling's T^2, repeated here for convenience, and plugging in our numbers:

$$T^2 = n(\bar{\boldsymbol{x}} - \boldsymbol{\mu}_0)'\boldsymbol{S}^{-1}(\bar{\boldsymbol{x}} - \boldsymbol{\mu}_0)$$

$$= 4[4 \quad 6 \quad 9]\begin{bmatrix} .321 & -.319 & -.251 \\ -.319 & .562 & .337 \\ -.251 & .337 & .442 \end{bmatrix}\begin{bmatrix} 4 \\ 6 \\ 9 \end{bmatrix}$$

$$= 4[-2.891 \quad 5.137 \quad 5.002]\begin{bmatrix} 4 \\ 6 \\ 9 \end{bmatrix} = 4(64.282) = 257.128$$

We plug the T^2 into the equation to obtain an F-statistic:

$$F = \frac{n-p}{p(n-1)}T^2 = \frac{4-3}{3(4-1)}257.128 = \frac{1}{9}257.128 = 28.57.$$

This F-test has p = 3 and n-p = 4-3 = 1 df. The critical value is $F_{.05,3,1} = 215.70$, so we do not reject H_0, which is not surprising, given the error df are so few (there's no power in the test in this tiny example). Usually n>>p so there are far more error df (n-p>>1).

Contrasts and Follow-up Tests

We will extend the 1-sample T^2 to a 2-sample T^2 shortly. Before doing so, recall that if we had a significant F-test in an ANOVA, the next thing we'd do is look at contrasts to determine the precise nature of how the null hypothesis was wrong. In this case, we only have 1 group, so group contrasts are not the issue, but there is still ambiguity in the conclusions we might draw due to the fact that there are p>1 dependent variables being modeled and tested simultaneously. Thus, if our F-value exceeds the critical $F_{\alpha;p,n-p}$, we reject H_0:

$$H_0\colon \boldsymbol{\mu} = \begin{bmatrix} \mu_1 \\ \mu_2 \\ \cdots \\ \mu_p \end{bmatrix} = \begin{bmatrix} \mu_{01} \\ \mu_{02} \\ \cdots \\ \mu_{0p} \end{bmatrix} = \boldsymbol{\mu}_0$$

Why did we reject H_0? We need to do follow-up tests to learn more about the nature of the differences between these two vectors:

- Is only 1 mean in $\boldsymbol{\mu}$ different from its corresponding theorized value in $\boldsymbol{\mu}_0$? If so, which one?
- Were 2 or more means in $\boldsymbol{\mu}$ different from their values in $\boldsymbol{\mu}_0$? Which ones?
- Was there some linear combination in $\boldsymbol{\mu}$ that was different from that in $\boldsymbol{\mu}_0$?

The tests we'll examine now are analogous to follow-up comparisons or contrasts in ANOVA designed to learn more about group differences. Here we have only 1 group, but follow-ups are needed to learn about variable differences.

Also analogous to group contrasts in ANOVA, in these variable follow-up tests, there are both *a priori* and *post hoc* approaches. In the *a priori* tests, we specify the combinations of variables we wish to test before we see the data.

If we have only 1 comparison to test, we can create a confidence interval for $\mathbf{a}'\boldsymbol{\mu}$, where **a** is a vector of coefficients that is specified *a priori*. These "a" coefficients do not need to sum to zero the way contrast coefficients comparing groups must.

The possibilities are nearly endless. For example, we might look at the mean over all p variables:

$$a'\mu = \begin{bmatrix} \frac{1}{p} & \frac{1}{p} & \dots & \frac{1}{p} \end{bmatrix} \begin{bmatrix} \mu_1 \\ \mu_2 \\ \dots \\ \mu_p \end{bmatrix} \quad (15)$$

Here we'd compare the average of the 1st two variables $(\mu_1+\mu_2)/2$ to the last mean μ_p:

$$a'\mu = \begin{bmatrix} \frac{1}{2} & \frac{1}{2} & 0 & 0 & \dots & 0 & -1 \end{bmatrix} \begin{bmatrix} \mu_1 \\ \mu_2 \\ \dots \\ \mu_p \end{bmatrix} \quad (16)$$

These coefficients compare the means on the 1st and 2nd variables, μ_1 and μ_2:

$$a'\mu = \begin{bmatrix} 1 & -1 & \dots & 0 & 0 \end{bmatrix} \begin{bmatrix} \mu_1 \\ \mu_2 \\ \dots \\ \mu_p \end{bmatrix} \quad (17)$$

We could also examine one mean μ_i by itself, as if conducting a univariate analysis:

$$a'\mu = \begin{bmatrix} 0 & 0 & \dots & 1 & 0 \end{bmatrix} \begin{bmatrix} \mu_1 \\ \mu_2 \\ \dots \\ \mu_p \end{bmatrix} \quad (18)$$

Note that MANOVA is ideal for all cases in between perfectly correlated dependent variables and totally uncorrelated dependent variables. If the dependent variables are totally uncorrelated, **Σ** would be a diagonal matrix (there would be variances as usual, on the diagonal, but all the off-diagonals, the covariances would = 0). In such a situation, a series of p ANOVAs (or t-tests), as in the last **a** vector above, would summarize the data with no loss of information in not studying the multivariate picture. At the other extreme, if the p dependent variables are perfectly correlated, the rank of **Σ** would be 1, meaning that 1 measure would suffice and no new information would be gained by modeling the other dependent variables. In this situation, 1 dependent variable might be chosen, or a summary score, such as the average of the p dependent measures (as in the first **a** vector above), would suffice in modeling the data.

A 100(1 - α)% confidence interval (e.g., 100(1-.05) = 95%) for **a'μ** is:

$$a'\bar{x} \pm \left(\sqrt{\frac{1}{n} a'Sa} \right) \left(t_{\alpha/2;n-1} \right) \quad (19)$$

If we had more than 1, say "r" follow-up tests to run on the p dependent variables, we should probably use the Bonferroni inequality and reduce α to compensate for the probability of making more Type I errors than α. Thus, for r variable comparisons defined by the vectors of coefficients $\mathbf{a}_1$, $\mathbf{a}_2$, ..., $\mathbf{a}_r$, the (100(1-α)%) confidence interval for any one of them, $\mathbf{a}_i'\boldsymbol{\mu}$ (i=1,2,...,r) would be:

$$a_i'\bar{x} \pm \left(\sqrt{\frac{1}{n} a_i'Sa_i} \right) \left(t_{\alpha/(2r);n-1} \right) \quad (20)$$

Confidence intervals for post hoc comparisons among the p variables look similar, the main difference is that there's a multiplier to the critical value, to make the interval wider (analogous to enlarging a critical value). Instead of a vector of *a priori* **a**'s, we'll call the *post hoc* coefficients **h**'s. The confidence interval for any **h**: **h'μ**:

$$\boldsymbol{h'\bar{x}} \pm \left(\sqrt{\frac{1}{n}\boldsymbol{h'Sh}}\right)\left(\sqrt{\frac{p(n-1)}{n-p}F_{\alpha;p,n-p}}\right) \tag{21}$$

For example, say n=31. As a benchmark, note that $t_{.05;30}$=2.042. In the table, it should be clear that as the number of variables, p, increases, the adjustment to the critical value will increase the size of the interval, again, to help compensate for the fact that with more variables, there are more possibilities for testing various kinds of comparisons among them.

p	n-p	F(.05,p,n-p)	right√term above
1	30	4.17	2.042
5	26	2.59	3.866
10	21	2.32	5.757

Contrasts and Follow-up Tests in the 1-Sample Example

Let's look at how a follow-up test is run on variables in our sample data. Recall that we hadn't actually rejected H_0 (because the error df were tiny), but if we had, we might suspect the difference between $\boldsymbol{\mu}$ and $\boldsymbol{\mu}_0$ was at least in part due to $\boldsymbol{\mu}_3$ ($\bar{x}_3 = 9$) because that sample mean was the most different from the hypothesized value (of 0). If we use "h" coefficients that focus on the 3rd variable, essentially turning the follow-up of the multivariate test into a univariate test, the comparison would look like this:

$$\boldsymbol{h'\mu} = \begin{bmatrix} 0 & 0 & 1 \end{bmatrix}\begin{bmatrix} \mu_1 \\ \mu_2 \\ \mu_3 \end{bmatrix} \qquad \boldsymbol{h'\bar{x}} = \begin{bmatrix} 0 & 0 & 1 \end{bmatrix}\begin{bmatrix} 4 \\ 6 \\ 9 \end{bmatrix}$$

$$\boldsymbol{h'Sh} = \begin{bmatrix} 0 & 0 & 1 \end{bmatrix}\begin{bmatrix} 8.000 & 3.333 & 2.000 \\ 3.333 & 4.667 & -1.667 \\ 2.000 & -1.667 & 4.667 \end{bmatrix}\begin{bmatrix} 0 \\ 0 \\ 1 \end{bmatrix}$$

$$= \begin{bmatrix} 2.000 & -1.667 & 4.667 \end{bmatrix}\begin{bmatrix} 0 \\ 0 \\ 1 \end{bmatrix}$$

$$\boldsymbol{h'\bar{x}} \pm \sqrt{(1/n)\boldsymbol{h'Sh}}\sqrt{\left(\frac{p(n-1)}{(n-p)}\right)F_{\alpha;p,n-p}}$$

$$= 9 \pm \sqrt{(1/4)4.667}\sqrt{\left(\frac{3(4-1)}{(4-3)}\right)215.70}$$

$$= 9 \pm (1.0802)(44.0602) = 9 \pm 47.592$$

→[-38.592, 56.592]

As we might have anticipated, this is a huge (useless) confidence interval because the error df were so few due to the tiny sample size, thus we forfeit precision. The confidence interval also spans the value of 0, which corresponds to our null hypothesis, which is information consistent with the fact that we had a small F-value and had not rejected the null.

Multivariate Inference about the Mean: 2 Samples

Just as there are univariate t-tests for hypotheses of 1 population mean ($H_0: \mu = \mu_0$) or 2 population means ($H_0: \mu_1 = \mu_2$), there are also multivariate T^2 tests for hypotheses of 1 population vector ($H_0: \boldsymbol{\mu} = \boldsymbol{\mu}_0$) and 2 population mean vectors ($H_0: \boldsymbol{\mu}_1 = \boldsymbol{\mu}_2$).

Recall the univariate 2-sample t-test to test $H_0: \mu_1 = \mu_2$ or $H_0: \mu_1 - \mu_2 = 0$. (We could specify $\mu_1 - \mu_2 = k$, or $\mu_1 = \mu_2 * k$, etc., but usually we test whether $\mu_1 - \mu_2 = 0$.)

$$t = \frac{(\bar{x}_1 - \bar{x}_2) - (\mu_1 - \mu_2)}{\sqrt{s_p^2\left(\frac{1}{n_1} + \frac{1}{n_2}\right)}}$$

This t-value is compared to a critical value on $((n_1 - 1)+(n_2 - 1)) = (n_1 + n_2 - 2)$ df. We had assumed homogeneity of variances, $\sigma_1^2 = \sigma_2^2$ in order to pool:

$$s_p^2 = \frac{\sum(X_{i1} - \bar{x}_1)^2 + \sum(X_{i2} - \bar{x}_2)^2}{(n_1 - 1) + (n_2 - 1)}$$

In multivariate-land, the hypothesis is: $H_0: \boldsymbol{\mu}_1 = \boldsymbol{\mu}_2$

$$\boldsymbol{\mu}_1 = \begin{bmatrix} \mu_{11} \\ \mu_{12} \\ \cdots \\ \mu_{1p} \end{bmatrix} = \begin{bmatrix} \mu_{21} \\ \mu_{22} \\ \cdots \\ \mu_{2p} \end{bmatrix} = \boldsymbol{\mu}_2$$

μ_{ij} i = 1, 2 group
j = 1, 2, ..., p variables

In both samples, we obtain data matrices and means:

	sample 1	sample 2
sample size:	$n_1 \times p$	$n_2 \times p$
sample data matrix:	$\begin{bmatrix} \boldsymbol{X}_1 & \\ & \end{bmatrix}$	$\begin{bmatrix} \boldsymbol{X}_2 & \\ & \end{bmatrix}$

sample mean vectors: $\bar{x}_1 = [\bar{x}_{11}\, \bar{x}_{12} \ldots \bar{x}_{1p}]'$ $\bar{x}_2 = [\bar{x}_{21}\, \bar{x}_{22} \ldots \bar{x}_{2p}]'$
/ \
group variable

In each sample, we compute the sample covariance matrices:

$$\boldsymbol{S}_1 = \frac{1}{n_1 - 1}\boldsymbol{A}_1 \qquad \boldsymbol{S}_2 = \frac{1}{n_2 - 1}\boldsymbol{A}_2 \tag{22}$$

Where the $\boldsymbol{A}$'s are the SSCPs matrices:

$$\boldsymbol{A}_1 = \sum_{i=1}^{n_1}(X_{i1} - \bar{x}_1)(X_{i1} - \bar{x}_1)' \text{ and } \boldsymbol{A}_2 = \sum_{i=1}^{n_2}(X_{i2} - \bar{x}_2)(X_{i2} - \bar{x}_2)' \tag{23}$$

The pooled covariance matrix is: $\boldsymbol{S}_p = \frac{1}{(n_1-1)+(n_2-1)}(\boldsymbol{A}_1 + \boldsymbol{A}_2)$ (24)

And of course, by pooling, we've assumed $\boldsymbol{\Sigma}_1 = \boldsymbol{\Sigma}_2$.
Then, to test the null: $H_0: \boldsymbol{\mu}_1 = \boldsymbol{\mu}_2$, we calculate:

$$T^2 = \frac{n_1 n_2}{n_1 + n_2}(\bar{\boldsymbol{x}}_1 - \bar{\boldsymbol{x}}_2)'\boldsymbol{S}_p^{-1}(\bar{\boldsymbol{x}}_1 - \bar{\boldsymbol{x}}_2) \tag{25}$$

and multiply T^2 by the proper values to get a variable that follows the F-distribution:

$$F = \frac{(n_1+n_2-p-1)}{p(n_1+n_2-2)} T^2 \tag{26}$$

We compare that F-value to a critical F-statistic on p and $(n_1 + n_2 - p - 1)$ degrees of freedom (for some α, e.g., .05).

If we reject H_0, we'd want to do follow-up testing to learn on which variable(s) the two populations differ:

A 100 * (1 - α)% confidence interval for $\boldsymbol{h}'(\boldsymbol{\mu}_1 - \boldsymbol{\mu}_2)$:

$$h'(\bar{\boldsymbol{x}}_1 - \bar{\boldsymbol{x}}_2) \pm \sqrt{\boldsymbol{h}'\boldsymbol{S}_p\boldsymbol{h}\frac{n_1+n_2}{n_1 n_2}}\sqrt{\frac{p(n_1+n_2-2)}{(n_1+n_2-p-1)}F_{\alpha;p,(n_1+n_2-p-1)}} \tag{27}$$

For example, comparing two populations variable by variable:

$\boldsymbol{h}_1' = [1\ 0\ 0\ ...\ 0]$
$\boldsymbol{h}_2' = [0\ 1\ 0\ ...\ 0]$
$\boldsymbol{h}_3' = [0\ 0\ 1\ ...\ 0]$
etc.

e.g., $$\boldsymbol{h}_2'(\bar{\boldsymbol{x}}_1 - \bar{\boldsymbol{x}}_2) - [0\ 1\ 0\ ...\ 0]\begin{bmatrix} \bar{x}_{11} - \bar{x}_{21} \\ \bar{x}_{12} - \bar{x}_{22} \\ \cdot \\ \cdot \\ \cdot \\ \bar{x}_{1p} - \bar{x}_{2p} \end{bmatrix} = \bar{x}_{12} - \bar{x}_{22} \tag{28}$$

Numerical Example of Two-Sample Hotelling's T^2

Let's see the 2-sample $\boldsymbol{T}^2$ in an example. We'll use our previous data set as sample 1 ($n_1 = 4$) and we'll draw 4 new observations from sample 2 ($n_2 = 4$), measuring the same 3 variables. The data sets and mean vectors follow:

$$\boldsymbol{X}_1 = \begin{bmatrix} x_{11}' \\ x_{12}' \\ x_{13}' \\ x_{14}' \end{bmatrix} = \begin{bmatrix} 2 & 3 & 11 \\ 8 & 7 & 10 \\ 2 & 6 & 6 \\ 4 & 8 & 9 \end{bmatrix} \qquad \boldsymbol{X}_2 = \begin{bmatrix} x_{21}' \\ x_{22}' \\ x_{23}' \\ x_{24}' \end{bmatrix} = \begin{bmatrix} 10 & 9 & 11 \\ 11 & 9 & 10 \\ 6 & 9 & 12 \\ 9 & 9 & 11 \end{bmatrix}$$

$$\bar{\boldsymbol{x}}_1 = \begin{bmatrix} \bar{x}_{11} \\ \bar{x}_{12} \\ \bar{x}_{13} \end{bmatrix} = \begin{bmatrix} 4 \\ 6 \\ 9 \end{bmatrix} \qquad \bar{\boldsymbol{x}}_2 = \begin{bmatrix} \bar{x}_{21} \\ \bar{x}_{22} \\ \bar{x}_{23} \end{bmatrix} = \begin{bmatrix} 9 \\ 9 \\ 11 \end{bmatrix}$$

/\
group variable

The SSCP matrices:

$$\boldsymbol{A}_1 = \sum_{i=1}^{n_1}(X_{i1} - \bar{x}_1)(X_{i1} - \bar{x}_1)'$$

$$= \left[\begin{bmatrix} 2-4 \\ 3-6 \\ 11-9 \end{bmatrix} [-2 \quad -3 \quad 2] + \begin{bmatrix} 8-4 \\ 7-6 \\ 10-9 \end{bmatrix} [4 \quad 1 \quad 1] + \begin{bmatrix} 2-4 \\ 6-6 \\ 6-9 \end{bmatrix} [-2 \quad 0 \quad -3] \right.$$

$$\left. + \begin{bmatrix} 4-4 \\ 8-6 \\ 9-9 \end{bmatrix} [0 \quad 2 \quad 0] \right]$$

$$= \begin{bmatrix} 24 & 10 & 6 \\ 10 & 14 & -5 \\ 6 & -5 & 14 \end{bmatrix} \quad \leftarrow \text{group 1 SSCP}$$

$$\boldsymbol{A}_2 = \sum_{i=1}^{n_2} (X_{i2} - \bar{x}_2)(X_{i2} - \bar{x}_2)'$$

$$= \left[\begin{bmatrix} 10-9 \\ 9-9 \\ 11-11 \end{bmatrix} [1 \quad 0 \quad 0] + \begin{bmatrix} 11-9 \\ 9-9 \\ 10-11 \end{bmatrix} [2 \quad 0 \quad -1] + \begin{bmatrix} 6-9 \\ 9-9 \\ 12-11 \end{bmatrix} [-3 \quad 0 \quad 1] \right.$$

$$\left. + \begin{bmatrix} 9-9 \\ 9-9 \\ 11-11 \end{bmatrix} [0 \quad 0 \quad 0] \right]$$

$$= \begin{bmatrix} 14 & 0 & -5 \\ 0 & 0 & 0 \\ -5 & 0 & 2 \end{bmatrix} \quad \leftarrow \text{group 2 SSCP}$$

We'll aggregated these SSCPs to obtain the pooled covariance matrix:

$$\boldsymbol{S}_p = \frac{1}{(n_1-1)+(n_2-1)} (\boldsymbol{A}_1 + \boldsymbol{A}_2)$$

$$= \frac{1}{3+3} \begin{bmatrix} 38 & 10 & 1 \\ 10 & 14 & -5 \\ 1 & -5 & 16 \end{bmatrix} = \begin{bmatrix} 6.333 & 1.667 & .167 \\ 1.667 & 2.333 & -.833 \\ .167 & -.833 & 2.667 \end{bmatrix} \leftarrow \text{pooled covariance matrix}$$

The inverse of the pooled covariance matrix, and the vector of mean differences follow:

$$\boldsymbol{S}_p^{-1} = \begin{bmatrix} .204 & -.169 & -.066 \\ -.169 & .623 & .205 \\ -.066 & .205 & .443 \end{bmatrix}$$

$$(\bar{\boldsymbol{x}}_1 - \bar{\boldsymbol{x}}_2) = \begin{bmatrix} 4 \\ 6 \\ 9 \end{bmatrix} - \begin{bmatrix} 9 \\ 9 \\ 11 \end{bmatrix} = \begin{bmatrix} -5 \\ -3 \\ -2 \end{bmatrix}$$

The test statistic then is: $T^2 = \frac{n_1 n_2}{n_1+n_2} (\bar{\boldsymbol{x}}_1 - \bar{\boldsymbol{x}}_2)' \boldsymbol{S}_p^{-1} (\bar{\boldsymbol{x}}_1 - \bar{\boldsymbol{x}}_2)$

$$= \frac{4 \times 4}{4+4} [-5 \quad -3 \quad -2] \begin{bmatrix} .204 & -.169 & -.066 \\ -.169 & .623 & .205 \\ -.066 & .205 & .443 \end{bmatrix} \begin{bmatrix} -5 \\ -3 \\ -2 \end{bmatrix}$$

$$= \frac{16}{8} [-.381 \quad -1.434 \quad -1.171] \begin{bmatrix} -5 \\ -3 \\ -2 \end{bmatrix} = 2(8.549) = 17.098$$

And finally, the F-test is: $F = \frac{n_1+n_2-p-1}{p(n_1+n_2-2)} T^2 = \frac{4+4-3-1}{3(4+4-2)} 17.098 = \frac{4}{18} 17.098 = 3.80$, which we'd compare to a critical F-statistic on p=3 and $(n_1 + n_2 - p - 1)$ =4 df, here that $F_{.05;3;4} =$ 6.59. Our conclusion is that we do not reject H_0, again probably due to the tiny sample sizes, which reduce the power of the test.

Assumptions Underlying the 2-Sample T^2 Test

We'll close this chapter by looking at the assumptions underlying the multivariate T^2 test, and seeing what happens when the assumptions are violated. The univariate models (t-tests, ANOVA) are so robust that when we assume $\sigma_1^2 = \sigma_2^2$ to pool error variability, even if that equality doesn't hold, the results of the t-test or F-test are close to what they should be and they'll tell us to reject the null or not, typically appropriately. In multivariate models, however, it's worth looking at the assumptions because the models are a tad touchier—with p variables, there are just many more combinations of ways that assumptions can be off. Let's begin with assumptions.

First, for group 1, we draw a random sample $\mathbf{X}_{11}$ $\mathbf{X}_{12}$ ... $\mathbf{X}_{1n}$ from MVN$_p(\boldsymbol{\mu}_1, \Sigma)$, and for group 2, we draw a random sample $\mathbf{X}_{21}$ $\mathbf{X}_{22}$ ... $\mathbf{X}_{2n}$ from MVN$_p(\boldsymbol{\mu}_2, \Sigma)$. We're assuming the $\mathbf{X}_{ij}$'s are sampled from multivariate normal populations, with a common population covariance matrix, $\boldsymbol{\Sigma}$ (that's the same as saying we assume $\Sigma_1 = \Sigma_2 = \Sigma$). We compute a pooled estimate of the shared covariance matrix, since both $\mathbf{S}_1$ and $\mathbf{S}_2$ are thought to be sample realizations (with just random fluctuations) of $\boldsymbol{\Sigma}$.

So when we say that we're testing the null hypothesis, H_0: $\boldsymbol{\mu}_1 = \boldsymbol{\mu}_2$, we're really testing the null hypothesis:

H_0: $\boldsymbol{\mu}_1 = \boldsymbol{\mu}_2$ <u>and</u> $\Sigma_1 = \Sigma_2$ <u>and</u> populations 1 and 2 are both MVN. (29)

Usually, we are interested in testing $\boldsymbol{\mu}_1 = \boldsymbol{\mu}_2$, so we assume the parts of the statement about $\Sigma_1 = \Sigma_2$ and MVN are true. When we reject H_0, it could be that $\boldsymbol{\mu}_1 \neq \boldsymbol{\mu}_2$and $\Sigma_1 = \Sigma_2$ and MVN, as we hope, but it could be alternatively that $\boldsymbol{\mu}_1 = \boldsymbol{\mu}_2$ but $\Sigma_1 \neq \Sigma_2$, etc.

The good news is that T^2, like t, is more sensitive to the differences between mean vectors than it is to the differences between covariance matrices. There is a basic, general, umbrella robustness to the statistic.

More specifically, when researchers have examined the behavior of T^2 with respect to departures from multivariate normality (the MVN), they find that T^2, like t, is fairly robust (i.e., the statistic results in the correct decision about rejecting or not rejecting H_0) most often, even in moderately-sized samples, unless the distribution is very skewed. The T^2, like t, is more sensitive to skewness vs. a symmetric normal, than to kurtosis vs. normal peak). That is, if our data are really skewed (the two figures at the left below), we should worry more about the validity of the T^2 test than if the data are platy- (flat) or leptokurtic (peaked) (the two figures at the right).[24]

Furthermore, researchers have also examined how well T^2 holds up with respect to violations of equal covariance matrices. After all, we wouldn't obtain pooled estimate $\mathbf{S}_p$ unless we thought $\boldsymbol{\Sigma}_1 = \boldsymbol{\Sigma}_2$. The multivariate situation is more complicated than the univariate situation due to both the fact that there are more variances (p>1) as well as the fact that there

[24] For more information, see references Chase and Bulgren (1971) and Mardia (1975).

are now also covariances that we're assuming to be equal. That is, in the 2-sample t-test, we only had to worry whether $\sigma_1^2 = \sigma_2^2$ (or $s_1^2 = s_2^2$ probabilistically). Now, we have more than 1 element in a matrix to be equal, e.g., even with p only equal to 2:

$$\mathbf{\Sigma}_1 = \begin{bmatrix} \sigma_{1(1)}^2 & \sigma_{12(1)} \\ \sigma_{12(1)} & \sigma_{2(1)}^2 \end{bmatrix} = \ ? \ = \mathbf{\Sigma}_2 = \begin{bmatrix} \sigma_{1(2)}^2 & \sigma_{12(2)} \\ \sigma_{12(2)} & \sigma_{2(2)}^2 \end{bmatrix} \quad (30)$$

It's easy to imagine experiments that include treatments that would be expected to have an influence on the means (rating changes) as well as the variances (more individual differences, less agreement in ratings), and there are more variables on which these effects might occur.

Thus, before conducting the test (T^2 for H_0: $\boldsymbol{\mu}_1 = \boldsymbol{\mu}_2$), we might test the assumption about whether $\mathbf{\Sigma}_1 = \mathbf{\Sigma}_2$. There are several such tests, and here is one. We calculate the covariance matrix in each sample:

$$\boldsymbol{S}_1 = \frac{1}{n_1 - 1}\boldsymbol{A}_1 \qquad \boldsymbol{S}_2 = \frac{1}{n_2 - 1}\boldsymbol{A}_2 \quad (31)$$

and we create the pooled covariance matrix:

$$\boldsymbol{S} = \frac{1}{(n_1 - 1) + (n_2 - 1)}(\boldsymbol{A}_1 + \boldsymbol{A}_2)$$

The test statistic compares them:

$$M = \sum_{i=1}^{2}(n_i - 1)\ell n|\boldsymbol{S}| - \sum_{i=1}^{2}(n_i - 1)\ell n|\boldsymbol{S}_i| \quad (32)$$

This test is somewhat like comparing an "observed" data point to an "expected" data point (o-e) because if H_0 is true, $\mathbf{\Sigma}_1 = \mathbf{\Sigma}_2 = \mathbf{\Sigma}$. We'd compare M/C to χ^2 on $(.5)(df_{effect})p(p+1)$ df , where $\frac{1}{c} = 1 - \frac{(2p^2 + 3p - 1)(n+1)}{6(p+1)(2n)}$ when $n_1 = n_2 = "n"$ if we have large sample sizes and if p ≤ 5. (For p>5 or smaller n's, we'd have to find another test statistic; e.g., we can also test H_0: $\mathbf{\Sigma}_1 = \mathbf{\Sigma}_2$ in structural equations programs such as Lisrel or Eqs.)

However, here's the tricky thing. The testing of H_0: $\mathbf{\Sigma}_1 = \mathbf{\Sigma}_2$ comes with its own concerns. The test is "overly powerful"; we'd be very likely to reject H_0: $\mathbf{\Sigma}_1 = \mathbf{\Sigma}_2$ even in cases where the departure from homogeneity of the covariance matrices would not have much of an effect on the validity of the T^2 in the subsequent testing of H_0: $\boldsymbol{\mu}_1 = \boldsymbol{\mu}_2$. For example, the test is quite sensitive to departures from nonnormality, whereas the T^2 test is more robust against departures from nonnormality. In addition, if were were to reject H_0: $\mathbf{\Sigma}_1 = \mathbf{\Sigma}_2$, alternative procedures for testing H_0: $\boldsymbol{\mu}_1 = \boldsymbol{\mu}_2$ are not as well understood, and certainly not as familiar to other researchers (e.g., journal reviewers).

Finally, there is good news, however. If we have equal sample sizes ($n_1 = n_2$) and those sample sizes are large, we can assume T^2 is robust against violations of equal covariance matrices. That is, modest differences of $\mathbf{\Sigma}_1 \neq \mathbf{\Sigma}_2$ will have little effect. If the sample sizes are unequal ($n_1 \neq n_2$), then the probability of making Type I errors may be > or < α depending on the eigenvalues of $\mathbf{\Sigma}_1\mathbf{\Sigma}_2^{-1}$ (which reflect the degree and pattern of heterogeneity between $\mathbf{\Sigma}_1$ and $\mathbf{\Sigma}_2$).[25]

[25] For more information, see references Hakstian, Roed, and Lind (1979), and Ito and Shull (1964).

A Simulation Testing the Robustness of T^2

When researchers offer conclusions about a model being robust to violations of its assumptions, their confidence usually comes from having run a simulation or "Monte Carlo" study. In such studies, a true set of population parameters is created, and samples are drawn repeatedly on the computer that violate the assumptions, and the test is whether the statistic leads to the proper conclusions about rejecting the null hypothesis or not, and seeing how close (or unbiased) the sample statistics are compared to the known population parameters.

A great example is an article by Hakstian, Roed, and Lind (1979). The paper is relatively easy to read as these things go, and for our purposes, it's a fairly comprehensive simulation study. Data are drawn from MVN distributions with determined population means and covariance matrices. They set the null to be true, that is, H_0: $\boldsymbol{\mu}_1 = \boldsymbol{\mu}_2$. So we know ahead of time that we should never be rejecting the null, except around the 5% mark, which would be the tolerable Type I error rate.

The study then varies the extent to which the covariance matrices $\boldsymbol{\Sigma}_1$ and $\boldsymbol{\Sigma}_2$ are unequal. The simulation also manipulates other variables, including sample size (is the model more robust with more observations?), and the number of variables (is the model more robust if it is simpler, i.e., p is small?). Given that experimental design, a sample of data is drawn, the T^2 is computed, and this is done many, many times, to build up an observed, empirical sampling distribution. Doing so will allow us to see how many times we'd reject H_0: $\boldsymbol{\mu}_1 = \boldsymbol{\mu}_2$, even though we know it's true.

The factors varied in their Monte Carlo study were (p.1256):

1. p = number of variables = 2, 6, 10 (3 levels)
2. relative sample sizes: $n_1 = n_2$, or $n_1 = 2 \times n_2$, or $n_1 = 5 \times n_2$ (3 levels)
3. average sample size = $3 \times p$ or $10 \times p$ (2 levels)
4. when $n_1 \neq n_2$, was the larger sample drawn from $\boldsymbol{\Sigma}_1$ or $\boldsymbol{\Sigma}_2$ (2 levels)
5. degree of heterogeneity of $\boldsymbol{\Sigma}_1$ and $\boldsymbol{\Sigma}_2$ (4 levels) illustrated here:
   a. case 1: "mild" departure from homogeneity:

$$\boldsymbol{\Sigma}_1 = \begin{bmatrix} 1.00 & 0 & 0 & 0 \\ 0 & 1.00 & 0 & 0 \\ \cdots & & & \\ 0 & 0 & \cdots & 1.00 \end{bmatrix} \qquad \boldsymbol{\Sigma}_2 = \begin{bmatrix} 1.44 & 0 & 0 & 0 \\ 0 & 1.44 & 0 & 0 \\ \cdots & & & \\ 0 & 0 & \cdots & 1.44 \end{bmatrix}$$

   b. case 2: "moderate to substantial" departure:

$$\boldsymbol{\Sigma}_1 = \begin{bmatrix} 1.00 & 0 & 0 & 0 \\ 0 & 1.00 & 0 & 0 \\ \cdots & & & \\ 0 & 0 & \cdots & 1.00 \end{bmatrix} \qquad \boldsymbol{\Sigma}_2 = \begin{bmatrix} 2.25 & 0 & 0 & 0 \\ 0 & 2.25 & 0 & 0 \\ \cdots & & & \\ 0 & 0 & \cdots & 2.25 \end{bmatrix}$$

   c. case 3: heterogeneity on some (half) variables:

$$\boldsymbol{\Sigma}_1 = \begin{bmatrix} 1.00 & 0 & 0 & 0 \\ 0 & 1.00 & 0 & 0 \\ \cdots & & & \\ 0 & 0 & \cdots & 1.00 \end{bmatrix} \qquad \boldsymbol{\Sigma}_2 = \begin{bmatrix} 1.00 & 0 & 0 & 0 \\ 0 & 1.00 & 0 & 0 \\ \cdots & \cdots & & \\ 0 & 0 & 2.25 & 0 \\ 0 & 0 & 0 & 2.25 \end{bmatrix}$$

   d. the benchmark case of complete homogeneity of covariance matrices: $\boldsymbol{\Sigma}_1 = \boldsymbol{\Sigma}_2 = \boldsymbol{I}$.

In that 3×3×2×2×4 factorial, 2000 samples were drawn from each of the combinations of factors and their resulting MVN distribution. The T^2 test was computed, set aside, done 1999 more time. The researchers observed the proportion of times that the T^2 values exceeded the critical T^2 value, even though we know that $\boldsymbol{\mu}_1 = \boldsymbol{\mu}_2$. That incidence is the set of Type I errors, and if the model is robust, we'd expect to see them only approximately α percent of the time (e.g., .05 if we had used α=.05 in determining the critical values). The findings in Hakstian et al. (1979) are complex, but a selection of them are presented in the following tables. The first 2 tables show the results for the conditions when the larger samples are drawn from the sample with the larger covariance values.

The results indicate that when there is in fact homogeneity (in the 2nd column, $\Sigma_1 = \Sigma_2$), as we assume, the Type I error rates are indeed approximately .05. Equal sample sizes (1st rows) also keep the Type I error rates about where they should be. Slight heterogeneity (case 1) is better than moderate heterogeneity (case 2), and the greater heterogeneity (case 2) operates to make the tests more conservative rather than more liberal (i.e., $\alpha < .05$), especially for large discrepancies between the 2 sample sizes.

For $p = 2$, and $\alpha = 0.05$:

		heterogeneity		
n_1:n_2	homogeneity	case 1	case 2	case 3
Average n per sample = 6:				
6:6	.049	.052	.050	.054
8:4	.047	.030	.031	.035
10:2	.047	.027	.014	.029
Average n per sample = 20:				
20:20	.055	.050	.065	.046
27:13	.049	.045	.026	.034
33:7	.060	.025	.007	.031

For $p = 10$, and $\alpha = 0.05$:

		heterogeneity		
n_1:n_2	homogeneity	case 1	case 2	case 3
Average n per sample = 30:				
30:30	.059	.049	.056	.058
40:20	.057	.029	.013	.023
50:10	.050	.015	.005	.027
Average n per sample = 100:				
100:100	.048	.049	.050	.050
133:67	.049	.022	.010	.036
167:33	.057	.008	.001	.025

The next 2 tables show the results for the conditions when the larger samples are drawn from the group with the smaller covariance matrix values. These results are more problematic, given the usual tendencies among researchers in academic publishing to seek low levels of Type I errors. In many cells in the following 2 tables, the effective Type I error rate greatly exceeds 5%, especially when working with 10 variables (vs. 2).

For $p = 2$, and $\alpha = 0.05$:

		heterogeneity		
n_1:n_2	homogeneity	case 1	case 2	case 3
Average n per sample = 6:				
6:6	.049	.052	.050	.054
8:4	.047	.073	.104	.077
10:2	.047	.084	.148	.111
Average n per sample = 20:				
20:20	.055	.050	.065	.046
27:13	.049	.073	.087	.073
33:7	.060	.097	.163	.107

For $p = 10$, and $\alpha = 0.05$:

		heterogeneity		
n_1:n_2	homogeneity	case 1	case 2	case 3
Average n per sample = 30:				
30:30	.059	.049	.056	.058
40:20	.057	.088	.158	.106
50:10	.050	.124	.337	.203
Average n per sample = 100:				
100:100	.048	.049	.050	.050
133:67	.049	.085	.163	.100
167:33	.057	.156	.369	.211

Given these results, we come to several conclusions. First, the equality of covariance matrices, $\Sigma_1 = \Sigma_2$ is better than the heterogeneity cases. Second, sample sizes that are roughly equal are better than sample sizes with greater imbalance. Third, if we could anticipate which group would likely yield greater variability, we should draw larger samples from it, in the hopes that the larger n might help compensate for the greater heterogeneity. Fourth, we might want to run simpler studies with fewer variables rather than many.

The 1-sample and 2-sample t-tests and ANOVA are so robust that we hadn't spent much time worrying about assumptions or robustness. In multivariate models, there are just more things happening, so it's not particularly surprising that there is potential for more issues going astray.

The next step is to consider what we do if we have more than 2 groups. Just as we proceeded from 2 sample t-tests to ANOVA, in multivariate data, we'll move from the 2-sample T^2 to a MANOVA. In ANOVA, if we have k groups measured on 1 variable and we wish to test the null H_0: $\mu_1 = \mu_2 = \mu_k$, we cannot test the means in pairs (via a t-test or F-test) because doing so would inflate the Type I error rates. Similarly, in the multivariate case, with k groups measured on p variables, we cannot test the null H_0: $\boldsymbol{\mu}_1 = \boldsymbol{\mu}_2 = \boldsymbol{\mu}_k$ in pairs of mean vectors (via T^2), because this also would inflate α. In addition, it wouldn't capture the inter-relationships among the p variables. Thus we turn to MANOVA in the next chapter.

SUMMARY

This chapter presents the multivariate 1-sample and 2-sample Hotelling's T^2 statistics. The multivariate normal was presented, and assumptions underlying the tests were considered. With this preparation, we are ready for the fuller MANOVA model, next in Chapter 14.

REFERENCES

1. Algina, James and Takako C. Oshima (1990), "Robustness of the Independent Samples Hotelling's T^2 to Variance-Covariance Heteroscedasticity When Sample Sizes are Unqueal and in Small Ratios," *Psychological Bulletin*, 108 (2), 308-313.
2. Chase, G. R. and William G. Bulgren (1971), "A Monte Carlo Investigation of the Robustness of T^2," *Journal of the American Statistical Association*, 66 (335), 499-502.
3. Cole, David A., Scott E. Maxwell, Richard Arvey and Eduardo Salas (1994), "How the Power of Manova Can Both Increase and Decrease as a Function of the Intercorrelations Among the Dependent Variables," *Psychological Bulletin*, 115 (3), 465-474.
4. Hakstian, A. Ralph, J. Christian Roed, and John C. Lind (1979), "Two-Sample T^2 Procedure and the Assumption of Homogeneous Covariance Matrices," *Psychological Bulletin*, 86 (6), 1255-1263.
5. Harris (1985), "Chapter 3: Hotelling's T^2: Tests on One or Two Mean Vectors," in his book, *A Primer of Multivariate Statistics*.
6. Iacobucci, Dawn (2015), *Marketing Models: Multivariate Statistics and Marketing Analytics*, 2nd ed., Nashville, TN: Earlie Lite Books, Inc.
7. Iacobucci, Dawn (1994), "Classic Factor Analysis," in Richard Bagozzi (ed.), *Principles of Marketing Research*, Cambridge, MA: Blackwell, 279-316.
8. Ito, Koichi and William J. Schull (1964), "On the Robustness of the T^2_0 Test in Multivariate Analysis of Variance when Variance-Covariance Matrices are Not Equal," *Biometrika*, 51 (1 and 2), 71-82.
9. Manly, Bryan F. J. (1986), *Multivariate Statistical Methods: A Primer*, London & NY: Chapman and Hall.
10. Mardia, K. V. (1975), "Assessment of Multinormality and the Robustness of Hotelling's T^2 Test," *Journal of the Royal Statistical Society, Series C (Applied Statistics)*, 24 (2), 163-171.

CHAPTER 14

MULTIVARIATE ANALYSIS OF VARIANCE

Questions to guide your learning:
Q_1: What is the multivariate ANOVA, or MANOVA model?
Q_2: What are the H and E matrices?
Q_3: What are the 4 main test statistics that replace the ANOVA F-test?

In univariate statistics, we've seen 1-sample and 2-sample t-tests, and we've seen how ANOVA is a generalization of those models for multiple groups, and if the ANOVA F-tests are significant, we've seen how to conduct contrasts to understand the precise nature of the group differences. Last chapter, we saw the multivariate 1-sample and 2-sample T^2 tests, in this chapter we'll see their generalization to the multivariate analysis of variance, or MANOVA, and next chapter, we'll see how to conduct multivariate contrasts.

The ANOVA model may have 1 or more predictor factors, but we model 1 dependent variable at a time. In the MANOVA, we model multiple indicators, $p>1$ dependent variables simultaneously.

One application of MANOVA would be if we measured several indicators of some construct. If each of those p variables really taps the same construct, it's likely those dependent variables would be correlated. We could run a factor analysis, find that all the variables load on the same factor, create a score averaging the items, and then run a single ANOVA. That would be an acceptable and simple way to proceed (and indeed it is probably the most common thing to do). Yet to understand why we might use MANOVA, consider that even if the p variables are correlated, they're not likely to be perfectly correlated, and if we wished to retain some of their individual personality and nuance, we might wish to keep the p variables separate, not aggregated into a single score. So, with p variables, one approach for analyzing possible group differences would be to model each of the p variables one at a time in a series of p ANOVAs. That approach would also give us some good, basic information. However, analyzing the p variables separately, by definition wouldn't take into account the correlations among the p variables. The MANOVA model does because it analyzes the p variables simultaneously.

Advantages of MANOVA over a series of p ANOVAS

There are at least 4 advantages of running 1 MANOVA vs. running p ANOVAs. First, we might well be interested in some behaviors or perceptions as they occur on several dependent variables jointly, as a unit. If all p variables are uncorrelated (i.e., **S** is diagonal), then we could conduct p separate ANOVAS and not lose any information with respect to the correlations between the variables. However, we'd still need to adjust α for making the series of p tests. We could use α/p or MANOVA. (And rarely are p variables uncorrelated.)

Second, when the null hypothesis is true, $H_0\colon \boldsymbol{\mu}_1 = \boldsymbol{\mu}_2 = \cdots = \boldsymbol{\mu}_k$, running a single MANOVA can help protect us from making too many Type I errors. That's quite simply because we'd be running 1 MANOVA vs. running p ANOVAs. It's also true because of the series of contrasts we're likely to run. Let's think of ANOVA or MANOVA as a "2-staged" procedure, where in Stage I we evaluate an overall test, and if it's significant, we go to Stage II and conduct follow-up tests and contrasts. If H_0 is true, the probability of not conducting any follow-up tests is $1 - \alpha$ (e.g., .95), because we wouldn't proceed to the next stage, so of course there is less of an opportunity (.05) to make type I errors in the 2nd stage.

On the other hand, if H_0 is false, we logically cannot commit a Type I error at the 1st stage in the overall testing. A Type I error is one that occurs when rejecting H_0 while H_0 is true, and here, we've said H_0 is false, so we cannot incorrectly reject it. So while MANOVA protects us from making too many Type I errors in the 1st stage of the analysis when H_0 is true, to be fair, it doesn't protect us from Type I errors in the 2nd stage. That is, if we reject the null, and we proceed to testing a handful of contrasts, we can certainly commit Type I errors in the follow-up stage, and this rate can exceed α (as for any technique). Consider the following example. Say the population mean vectors are as follows: $\begin{bmatrix}\mu_1\\ \mu_2\\ \mu_3\end{bmatrix} \neq \begin{bmatrix}\mu_4\\ \mu_2\\ \mu_3\end{bmatrix} \neq \begin{bmatrix}\mu_5\\ \mu_2\\ \mu_3\end{bmatrix}$. That is, the 3 groups differ only on the 1st variable. Any tests involving variables 2 or 3 that showed significance would be committing Type I errors. So MANOVA does not protect us from Type I errors in the follow-up testing if H_0 is false (and presumably, we wouldn't have collected the data if we thought H_0 was true).[26]

An alternative to using MANOVA to protect us from too many Type I errors would be the simpler approach of conducting p separate ANOVAS with a Bonferroni adjustment (i.e., use α/p). Between the two, the technique that is more powerful depends on sample sizes, effect sizes, correlations, etc. (cf., Ramsey, 1982). The series of ANOVAs still does not take into account the intercorrelations among the p variables.

A third advantage in favor of MANOVA is that it can be more powerful than a series of p ANOVAs. In this figure, 2 scatterplots represent the data for 2 samples measured on 2 variables, X_1 and X_2. In the scatter plot ovals, that is, when using the information on both variables, it is clear that the groups are distinct—our statistics should say that they are significantly different. That picture of how the groups are separable is the clearest view in the figure. By comparison, looking at either of the marginal, univariate distributions, the one for X_1 or the one for X_2 do not show the group differences as clearly. A univariate F-test on X_2 might very well be significant, but a univariate F-test on X_1 would certainly not be. Thus, the multivariate MANOVA view of these data would show that these two groups differ when both variables are considered simultaneously, but the groups might not be seen as significantly different on either variable considered alone.

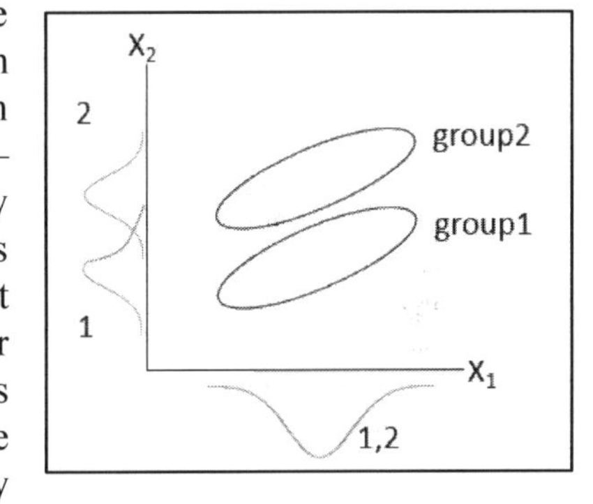

Along these lines, Bray and Maxwell (1985, p. 31) offer a nice discussion of the effects of the correlations between variables on results in MANOVA. They present figures like "a" and "b" below and point out that the 2 groups are more clearly distinguished in b) than in a), even though there is the same amount of overlap in the marginal, univariate distributions. However, don't be misled into thinking that the trick is to have negative correlations; rather, it has to do with the amount of overlap.

[26] For more information, see Bird and Hadzi-Pavlovic (1983).

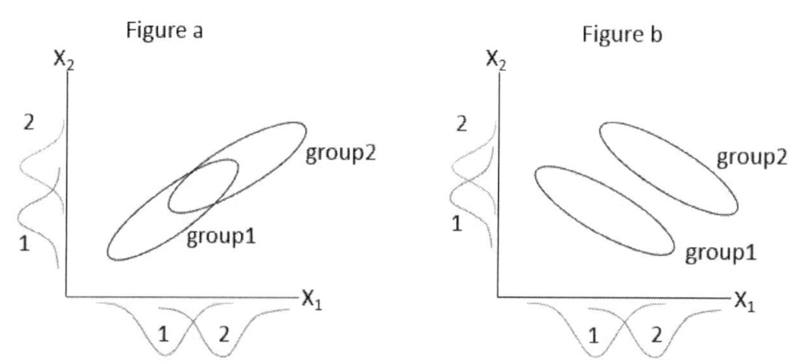

That is, even if the data were oriented as in Figures c and d, we could see that the groups were more clearly distinguishable in d) than in c).

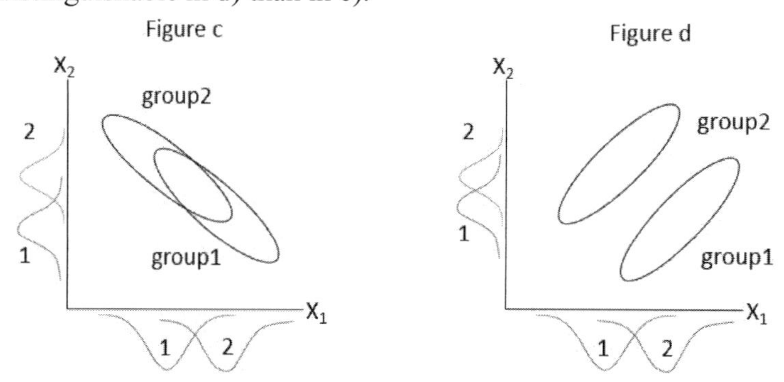

These figures are some of the ways to understand how the multivariate modeling takes into account the correlations among the p dependent variables, and can help distinguish the groups, finding them to be significantly different through decreasing the amount of overlap between them. In any of the figures a) through d), the univariate F-tests on X_1 and X_2 might be significant, but the MANOVA tests, in at least b) and d), would surely be significant.

The fourth advantage of MANOVA is that sometimes it can help us understand our data better. Perhaps the largest disadvantage of using the MANOVA model over a series of p ANOVAs is that it is more complicated. However, this chapter is intended to ameliorate that challenge.

The Four Major MANOVA Test Statistics

As just previewed, MANOVA like ANOVA can best be thought of as a two-staged procedure. In stage I, we test the overall null hypothesis; that there is no difference in the mean centroids (vectors) between the k groups: $H_0\colon \boldsymbol{\mu}_1 = \boldsymbol{\mu}_2 = \cdots = \boldsymbol{\mu}_k$.

$$\begin{bmatrix} \mu_{11} \\ \mu_{12} \\ \cdots \\ \mu_{1p} \end{bmatrix} = \begin{bmatrix} \mu_{21} \\ \mu_{22} \\ \cdots \\ \mu_{2p} \end{bmatrix} = \cdots = \begin{bmatrix} \mu_{k1} \\ \mu_{k2} \\ \cdots \\ \mu_{kp} \end{bmatrix}$$

If we reject the overall null hypothesis, we proceed to stage II and conduct follow-up tests and contrasts to understand the nature of the group differences; e.g., which groups differ, and on which variables?

There are two issues or debates in MANOVA: how to do the first stage, and how to do the second stage. In the first stage, there is a choice among overall test statistics, and in the second stage, there is a choice among methods of follow-up testing. In this chapter, we'll focus on stage I and the choice among the overall test statistics. In the next chapter, we'll look at the issues of the follow-up tests in stage II.

The nature of the problem of the choices in the overall test in stage I is simple. In the univariate ANOVA, there is one test statistic, the F-test for testing any overall H_0, e.g., the main effect for A, H_0: $\alpha_1 = \alpha_2 = \cdots = 0$, the interaction between A and B, H_0: $(\alpha\beta)_{ij} = 0$; etc. However, in MANOVA, there are several tests to choose from. Why? It's a bit analogous to deriving estimates of μ and σ^2; we can obtain estimates of either parameter using different statistical criteria (e.g., maximum likelihood estimation, least squares, unbiased estimates, etc.). For estimates of μ, these all converge to $\bar{x} = \Sigma x_i/n$, but for σ^2, the maximum likelihood estimate is $\Sigma(x_i - \bar{x})^2/n$ and an unbiased estimate is $\Sigma(x_i - \bar{x})^2/(n-1)$. Similarly when deriving test statistics in ANOVA, even using different criteria, the derivations converge to same test statistic, that F-test. However, different criteria result in different test statistics in MANOVA. Again, why? Simply due to the very multivariate nature of problem, things are more complicated (we have p variables and variances, and $\frac{p(p-1)}{2}$ covariances).

In ANOVA, we rely on the F-test: $F = \frac{\frac{SS_A}{a-1}}{\frac{SS_{S(A)}}{a(n-1)}}$. In MANOVA, we don't have single numbers for terms like SS_A, $SS_{S(A)}$. We have p variables, so we have the terms SS_A and $SS_{S(A)}$ for each variable, as well as both terms SS_A and $SS_{S(A)}$ for the cross products of the pairs of p variables. These interrelationships are how the MANOVA model will "take into account" the correlations between the p variables.

The equivalent to $SS_{between}$ in MANOVA is the matrix "**H**," the between groups SSCP matrix. The "**H**" stands for hypothesis matrix—the effect we want to test. For example, for a 1-way design with "a" groups:

Univariate ANOVA:

$$SS_{between} = \sum_{i=1}^{a}\sum_{j=1}^{n}(\bar{x}_i - \bar{x})^2$$

Multivariate ANOVA:

$$\boldsymbol{H} = \sum_{i=1}^{a}\sum_{j=1}^{n}(\bar{\boldsymbol{x}}_i - \bar{\boldsymbol{x}})(\bar{\boldsymbol{x}}_i - \bar{\boldsymbol{x}})'$$

with mean vectors in group i for the p variables: $\bar{\boldsymbol{x}}_i' = [\bar{x}_{i1} \quad \bar{x}_{i2} \quad \cdots \quad \bar{x}_{ip}]$, and the grand mean vector for the p variables: $\bar{\boldsymbol{x}}' = [\bar{x}_1 \quad \bar{x}_2 \quad \cdots \quad \bar{x}_p]$. This source (uni-or multi-variate) has associated with it (a-1) df.

Analogously, the equivalent to SS_{error} or SS_{within} in MANOVA is the matrix "**E**", the error, or within group SSCP matrix:

Univariate: $$SS_{within} = \sum_{i=1}^{a}\sum_{j=1}^{n}\left(x_{ij} - \bar{x}_i\right)^2$$

Multivariate: $$\boldsymbol{E} = \sum_{i=1}^{a}\sum_{j=1}^{n}\left(x_{ij} - \bar{x}_i\right)\left(x_{ij} - \bar{x}_i\right)'$$

This source (uni- or multi-variate) has a(n-1) df.

And to be complete, the total variability terms are:

Univariate: $SS_{total} = \sum_{i=1}^{a}\sum_{j=1}^{n}\left(x_{ij} - \bar{x}\right)^2$

Multivariate: $\boldsymbol{T} = (\boldsymbol{E} + \boldsymbol{H}) = \sum_{i=1}^{a}\sum_{j=1}^{n}\left(x_{ij} - \bar{x}\right)\left(x_{ij} - \bar{x}\right)'$

The total df are (an-1).

Notice that the df for ***H***, ***E***, and ***T***, are determined by the number of subjects and groups. The numbers of variables, p, doesn't play a direct role in df. The number of variables p will enter into the statistics later.

This fairly straightforward generalization holds in higher-order designs too. For example, in a between-subjects design, with both factors A and B fixed:

Univariate

source	df	SS	MS	F
A	a-1	$\sum_i^a\sum_j^b\sum_k^n(\bar{x}_i - \bar{x})^2$	SS/df	$MS_A/MS_{S(AB)}$
B	b-1	$\sum\sum\sum\left(\bar{x}_j - \bar{x}\right)^2$		$MS_B/MS_{S(AB)}$
A×B	(a-1)(b-1)	$\sum\sum\sum\left(\bar{x}_{ij} - \bar{x}_i - \bar{x}_j + \bar{x}\right)^2$		$MS_{AB}/MS_{S(AB)}$
S(AB)	ab(n-1)	$\sum\sum\sum\left(x_{ijk} - \bar{x}_{ij}\right)^2$		
total	abn-1	$\sum\sum\sum\left(x_{ijk} - \bar{x}\right)^2$		

Multivariate:

source	df	SSCP matrices	to test H_0, compare to	
$\mathbf{H}_A$	a-1	$\sum_i^a\sum_j^b\sum_k^n(\bar{x}_i - \bar{x})(\bar{x}_i - \bar{x})'$	$\mathbf{H}_A$	**E**
$\mathbf{H}_B$	b-1	$\sum\sum\sum(\bar{x}_j - \bar{x})(\bar{x}_j - \bar{x})'$	$\mathbf{H}_B$	**E**
$\mathbf{H}_{A\times B}$	(a-1)(b-1)	$\sum\sum\sum(\bar{x}_{ij} - \bar{x}_i - \bar{x}_j + \bar{x})(\bar{x}_{ij} - \bar{x}_i - \bar{x}_j + \bar{x})'$	$\mathbf{H}_{A\times B}$	**E**
$\mathbf{E}_{S(AB)}$	ab(n-1)	$\sum\sum\sum(x_{ijk} - \bar{x}_{ij})(x_{ijk} - \bar{x}_{ij})'$		
T(total)	abn-1	$\sum\sum\sum(x_{ijk} - \bar{x})(x_{ijk} - \bar{x})'$		

As a second example, take a 3-factor factorial, between-subjects design, with factors A and B as fixed, and C is random:

Univariate

source	df	SS	F
A	a-1	$\sum_i^a\sum_j^b\sum_k^c\sum_\ell^n(\bar{x}_i - \bar{x})^2$	MS_A/MS_{AC}
B	b-1	$\sum\sum\sum\sum\left(\bar{x}_j - \bar{x}\right)^2$	MS_B/MS_{BC}
C	c-1	$\sum\sum\sum\sum(\bar{x}_k - \bar{x})^2$	$MS_C/MS_{S/ABC}$
AB	(a-1)(b-1)	$\sum\sum\sum\sum\left(\bar{x}_{ij} - \bar{x}_i - \bar{x}_j + \bar{x}\right)^2$	MS_{AB}/MS_{ABC}
AC	(a-1)(c-1)	$\sum\sum\sum\sum(\bar{x}_{ik} - \bar{x}_i - \bar{x}_k + \bar{x})^2$	$MS_{AC}/MS_{S/ABC}$
BC	(b-1)(c-1)	$\sum\sum\sum\sum\left(\bar{x}_{jk} - \bar{x}_j - \bar{x}_k + \bar{x}\right)^2$	$MS_{BC}/MS_{S/ABC}$
ABC	(a-1)(b-1)(c-1)	$\sum\sum\sum\sum\left(\bar{x}_{ijk} - \bar{x}_{ij} - \bar{x}_{ik} - \bar{x}_{jk} + \bar{x}_i + \bar{x}_j + \bar{x}_k - \bar{x}\right)^2$	$MS_{ABC}/MS_{S(ABC)}$
S(ABC)	abc(n-1)	$\sum\sum\sum\sum\left(x_{ijk\ell} - \bar{x}_{ijk}\right)^2$	
T	abcn-1	$\sum\sum\sum\sum\left(x_{ijk\ell} - \bar{x}\right)^2$	

Multivariate: source	df	SSCP matrices	to test H_0, compare	to
$\mathbf{H}_A$	same as	$\sum_i^a \sum_j^b \sum_k^c \sum_l^n (\bar{x}_i - \bar{x})(\bar{x}_i - \bar{x})'$	$\mathbf{H}_A$	$\mathbf{H}_{AC}$
$\mathbf{H}_B$	in univar.	$\Sigma\Sigma\Sigma\Sigma(\bar{x}_j - \bar{x})(\bar{x}_j - \bar{x})'$	$\mathbf{H}_B$	$\mathbf{H}_{BC}$
$\mathbf{H}_C$		$\Sigma\Sigma\Sigma\Sigma(\bar{x}_k - \bar{x})(\bar{x}_k - \bar{x})'$	$\mathbf{H}_C$	$\mathbf{E}$
$\mathbf{H}_{AB}$		$\Sigma\Sigma\Sigma\Sigma(\bar{x}_{ij} - \bar{x}_i - \bar{x}_j + \bar{x})(\bar{x}_{ij} - \bar{x}_i - \bar{x}_j + \bar{x})'$	$\mathbf{H}_{AB}$	$\mathbf{H}_{ABC}$
$\mathbf{H}_{AC}$		$\Sigma\Sigma\Sigma\Sigma(\bar{x}_{ik} - \bar{x}_i - \bar{x}_k + \bar{x})(\bar{x}_{ik} - \bar{x}_i - \bar{x}_k + \bar{x})'$	$\mathbf{H}_{AC}$	$\mathbf{E}$
$\mathbf{H}_{BC}$		$\Sigma\Sigma\Sigma\Sigma(\bar{x}_{jk} - \bar{x}_j - \bar{x}_k + \bar{x})(\bar{x}_{jk} - \bar{x}_j - \bar{x}_k + \bar{x})'$	$\mathbf{H}_{BC}$	$\mathbf{E}$
$\mathbf{H}_{ABC}$		$\Sigma\Sigma\Sigma\Sigma(\bar{x}_{ijk} - \bar{x}_{ij} - \bar{x}_{ik} - \bar{x}_{jk} + \bar{x}_i + \bar{x}_j + \bar{x}_k - \bar{x})$ $(\bar{x}_{ijk} - \bar{x}_{ij} - \bar{x}_{ik} - \bar{x}_{jk} + \bar{x}_i + \bar{x}_j + \bar{x}_k - \bar{x})'$	$\mathbf{H}_{ABC}$	$\mathbf{E}$
$\mathbf{E}$		$\Sigma\Sigma\Sigma\Sigma(x_{ijk\ell} - \bar{x}_{ijk})(x_{ijk\ell} - \bar{x}_{ijk})'$		
$\mathbf{T}$		$\Sigma\Sigma\Sigma\Sigma(x_{ijk\ell} - \bar{x})(x_{ijk\ell} - \bar{x})'$		

Everything generalizes from the univariate to the multivariate. For balanced designs and equal cell sizes, all the effects (main effects and interactions) are orthogonal (independent, partialling out all other effects). If we do not reject H_0: no main effect for A, then no follow-ups are necessary on A. If we reject H_0: no main effect for B, we'll do the follow-ups, contrasting groups and trying different combinations of variables. If we reject H_0: no main effect for C, we'll do follow-ups, and the contrasts for groups and the variable combinations may well be different from those for factor B. If an interaction is significant, we'll do follow-ups, studying multivariate simple effects (ooo! or eew!?).

The matrix analog of $SS_{between}/SS_{within}$ is $\boldsymbol{HE}^{-1}$. And while the ratio $\frac{SS_{between}/df_{between}}{SS_{within}/df_{within}}$ is a single number, the F-test, now we have an entire matrix of such numbers. For example, with just p=2, if $\boldsymbol{H} = \begin{bmatrix} 210 & -90 \\ -90 & 90 \end{bmatrix}$ and $E = \begin{bmatrix} 88 & 80 \\ 80 & 126 \end{bmatrix}$, then $HE^{-1} = \begin{bmatrix} 7.18 & -5.27 \\ -3.95 & 3.23 \end{bmatrix}$. The value 7.18 is the $SS_{between}/SS_{within}$ for the 1st variable, and 3.23 is that ratio for the 2nd variable. Note that while $\boldsymbol{H}$ and $\boldsymbol{E}$ are both symmetric, in general, $\boldsymbol{HE}^{-1}$ is not. So $\boldsymbol{HE}^{-1}$ has p^2 elements, not p(p-1)/2) because it's not symmetric:

$$\boldsymbol{HE}^{-1} = \begin{bmatrix} (he^{-1})_{11} & (he^{-1})_{12} & \cdots & (he^{-1})_{1p} \\ (he^{-1})_{21} & (he^{-1})_{22} & \cdots & (he^{-1})_{2p} \\ \cdots & & & \\ (he^{-1})_{p1} & (he^{-1})_{p2} & \cdots & (he^{-1})_{pp} \end{bmatrix}$$

We need to derive test statistics, the equivalents of F-tests using all the information in this matrix. To do so, we combine the p^2 numbers to test H_0 by solving for the eigenvalues of $\boldsymbol{HE}^{-1}$: $\lambda_i = \lambda_1 \geq \lambda_2 \geq \cdots \geq \lambda_s$. Think of any one λ_i of $\boldsymbol{HE}^{-1}$ as roughly $SS_{between}/SS_{within}$ (we'll see in Chapter 15 that it is $SS_{between}/SS_{within}$ for the "discriminate function" variate i), so each λ_i is roughly "F-like." All the test statistics are functions of the λ_i's.

The matrix $\boldsymbol{HE}^{-1}$ has rank s, where s=min (p=#vars, k-1=df_{effect}). For example, in a 1-factor design with a=3 groups and say p=4 measures, then s=min(4,2)=2. So we can extract s=2 λ_i's. There are 4 main test statistics that are like the ANOVA F-test, and we'll need the λ_i's to calculate them.

Wilks' Likelihood Ratio Test Statistic Λ

The Wilks' likelihood ratio test statistic Λ (sometimes also called U or W) is defined as the product over all "s" of the λ_i's of the reciprocal of 1 plus λ_i:

$$\Lambda = \prod_{i=1}^{s} \frac{1}{1+\lambda_i} = \frac{|\boldsymbol{E}|}{|\boldsymbol{E}+\boldsymbol{H}|} = \frac{1}{|\boldsymbol{I}+\boldsymbol{H}\boldsymbol{E}^{-1}|} = \prod_{i=1}^{s} (1+\lambda_i)^{-1} \quad (1)$$

If it's the case that we can think of λ_i as $SS_{between}/SS_{within}$, then $1/(1+\lambda_i)$ is like a SS_{within}/SS_{total}, or a variance not explained:

$$\frac{1}{1+\lambda_i} = \frac{\frac{SS_w}{SS_w}}{\frac{SS_w}{SS_w}+\frac{SS_b}{SS_w}} = \frac{\frac{SS_w}{SS_w}}{\frac{SS_t}{SS_w}} = \frac{SS_w}{SS_t} \quad (2)$$

Using this statistic, we'll reject H_0 for small values of Λ (because it's capturing σ^2 not accounted for).

We need critical values so we know when to reject H_0. The distribution of Λ is tough, so the most frequent advice is to use approximations. In general, a good approximation is:

$$F = \frac{\left(1-\Lambda^{\left(\frac{1}{q}\right)}\right)\left(mq - .5p\left(df_{effect}+1\right)\right)}{\Lambda^{\left(\frac{1}{q}\right)}(p)\left(df_{effect}\right)} \quad (3)$$

where $m = \left[df_{error} - \left(p+1-df_{effect}\right)/2\right]$ and $q = \sqrt{\frac{(p^2)\left(df_{effect}\right)^2-4}{p^2+\left(df_{effect}\right)^2-5}}$

The df_{effect} = a-1 for main effect A; (a-1)(b-1) for AB interaction, etc., that hasn't changed. Similarly, the df_{error} = a(n-1) for a between-subjects, 1-factor design, or (a-1)(c-1) if we're testing factor A in a 3-factor design where A and B are fixed, and C is random, etc. Again, the df have not changed.

The F-value in equation (3) is approximately distributed as an F-statistic on p(df_{effect}) and (mq-.5 p(df_{effect})+1) df. If k, the number of groups = 2 or 3, that is, df_{effect}=1 or 2, or p=1 or 2, the F-test is exactly distributed as an F random variable. Another approximation has a simpler expression, but we'd need very large samples for it to hold:

$$X^2 = -m\, \ell n(\Lambda), \quad \text{m as defined above} \quad (4)$$

This X^2 is approximately distributed χ^2 on (p×df_{effect}) df.

Pillai-Bartlett Trace V

The Pillai-Bartlett trace V statistic is defined as the sum over all the λ_i's of $\lambda_i/(1+\lambda_i)$:

$$V = \sum_{i=1}^{s} \frac{\lambda_i}{1+\lambda_i} = tr[\boldsymbol{H}(\boldsymbol{E}+\boldsymbol{H})^{-1}] \quad (5)$$

In terms of interpreting V, again, think simply of λ_i being like $\frac{SS_{between}}{SS_{within}}$, then

$$\frac{\lambda_i}{1+\lambda_i} = \frac{\frac{SS_b}{SS_w}}{\frac{SS_w}{SS_w}+\frac{SS_b}{SS_w}} = \frac{\frac{SS_b}{SS_w}}{\frac{SS_t}{SS_w}} = \frac{SS_{between}}{SS_{total}} \tag{6}$$

One approximate test statistic is computed as:

$$F = \frac{(df_{error} - p + s)V}{b(s - V)} \tag{7}$$

where b=max(df_{effect},p). This F-test is approximately distributed F on (sb) and $(s(df_{error} - p + s))$ df. Alternatively, in large samples, we can calculate:

$X^2 = (df_{error})V$

and that is approximately distributed χ^2 on p(df_{effect}) df. (8)

In a moment, we'll compare the 4 test statistics. For now, we note that V is the most robust of the 4 (Olson, 1974, 1976), and it is the focus (with R) in Bird and Hadzi-Pavlovic (1983).

Hotelling-Lawley Trace T

The 3rd of the major test statistics is the Hotelling-Lawley trace, T. It is quite simply the sum of the λ_i's:

$$T = \sum_{i=1}^{s} \lambda_i = tr(\boldsymbol{HE}^{-1}) \tag{9}$$

Again, if we think of λ_i as $SS_{between}/SS_{within}$ then obviously T is also a between-to-within ratio, much like F-tests in ANOVA (which are MS rather than SS). The F-test is $\frac{1}{C}T$, and it is approximately distributed as an F-statistic on a and b df, with: $a = p(df_{effect})$, $b = 4 + [(a+2)/(B-1)]$, $C = [a(b-2)]/[b(df_{error} - p - 1)]$, and $B = \frac{(df_{error}+df_{effect}-p-1)(df_{error}-1)}{(df_{error}-p-3)(df_{error}-p)}$.

Roy's Greatest Characteristic Root (GCR), R

The 4th statistic is called Roy's greatest characteristic root (GCR), R. (This "R" is not the regression R. This R stands for Roy or root.) The term "greatest" here means largest, and given that the s λ_i's are ranked from largest to smallest, the R statistic involves only the 1st λ_i, that is, λ_1:

$$R = \lambda_1 \tag{10}$$

Given that the λ_i's are ranked from largest to smallest, the R statistic relies on the information only in the 1st λ_i, so, λ_1, which captures more information than those that follow, however, given that the remaining roots or eigenvalues are not included in the computation, the R statistic ignores all the information that may be contained in the later eigenvalues $\lambda_2 \ldots \lambda_s$. The focus on λ_1 wastes information unless s = 1 literally, or s is "almost" 1, as when $\lambda_1 \gg \lambda_2 > \lambda_3 > \cdots > \lambda_s$. Note that the statistic $T = \sum_{i=1}^{s} \lambda_i$ will equal R when s=1 (i.e., when p=1 or k=2). Critical values have been tabulated (Heck, 1960; Pillai, 1956) but of course SAS and other statistical computing packages will calculate F-tests and p-values for these R values.

Occasionally (in other references), these 4 test statistics will be defined not as functions of the eigenvalues of $\boldsymbol{HE}^{-1}$, but as functions of $\boldsymbol{H}(\boldsymbol{H}+\boldsymbol{E})^{-1}$ or $\boldsymbol{E}(\boldsymbol{H}+\boldsymbol{E})^{-1}$. Olson

(1976, p.580) provides a table showing how they are related (changing the gcr row to be consistent with the literature):[27]

Statistic:	Defining matrix (i.e., eigenvalues come from this matrix): $\boldsymbol{HE}^{-1}$	$\boldsymbol{H}(\boldsymbol{H}+\boldsymbol{E})^{-1}$	$\boldsymbol{E}(\boldsymbol{H}+\boldsymbol{E})^{-1}$
Roy's gcr R	λ_1	$\frac{\theta_1}{(1-\theta_1)}$	$\frac{1-m_1}{m_1}$
Hotelling-Lawley trace T	$\sum_{i=1}^{s}\lambda_i$	$\sum\frac{\theta_i}{(1-\theta_i)}$	$\sum\frac{(1-m_i)}{m_i}$
Wilks' Lrt Λ	$\prod_{i=1}^{s}\frac{1}{1+\lambda_i}$	$\prod(1-\theta_i)$	$\prod m_i$
Pillai-Bartlett trace V	$\sum_{i=1}^{s}\frac{\lambda_i}{1+\lambda_i}$	$\sum\theta_i$	$\sum(1-m_i)$
	λ_i's are the eigenvalues of $\boldsymbol{HE}^{-1}$	θ_i's are the eigenvalues of $\boldsymbol{H}(\boldsymbol{H}+\boldsymbol{E})^{-1}$	m_i's are the eigenvalues of $\boldsymbol{E}(\boldsymbol{H}+\boldsymbol{E})^{-1}$

The eigenvalues are related across the 3 matrices as follows: $\lambda_i = \frac{\theta_i}{(1-\theta_i)} = \frac{(1-m_i)}{m_i}$. We've used $\boldsymbol{HE}^{-1}$ because it is most closely analogous to the $SS_{between}/SS_{within}$ logic of the univariate ANOVA.

Comparing the 4 Test Statistics R, T, Λ, and V with respect to Power

These 4 test statistics are printed by SAS and other statistical computing packages every time a MANOVA is run. For the special case of $s = 1$ (s being the min(df_{effect},p)), all the F-tests will be the same. That's nice; there would be no ambiguity as to which results to pay attention to, interpret and report. However, when $s \neq 1$, the 4 test statistics will yield slightly different information, sometimes disagreeing on whether the effect being tested is significant or not. We have to know more about how these 4 statistics compare to know when to pay attention to which ones in preference over the other ones. In this section, we'll compare the statistics with regard to their power, and in the next section, we'll compared them with regard to their robustness.

Power, recall, is the likelihood of rejecting the null when we should, when it is false. When the null hypothesis H_0 is false, and the group mean vectors differ, the "non-centrality structure" is the information that indicates how the group means differ. We'll consider this issue simply, as in whether the group means differ along 1 "dimension" ("a concentrated structure") or along several dimensions ("a diffuse structure"). Specifically, when we have a

---

[27] Some authors (e.g., Bray and Maxwell, 1985; Olson, 1974) build R as $\frac{\lambda_1}{1+\lambda_1}$, the largest eigenvalue of $\boldsymbol{H}(\boldsymbol{H}+\boldsymbol{E})^{-1}$, which, if s=1, makes R equal to our V rather than T, however more references, including SAS's computation, take $R = \lambda_1$, the largest eigenvalue of $\boldsymbol{HE}^{-1}$, thus when s=1, R=T. R's closeness to T also makes better sense given the ranking of the test statistics with respect to power and robustness, discussed next.

concentrated structure, the groups differ along a single underlying dimension, it will be the case that the 1st eigenvalue will be much larger than the others: $\lambda_1 >>> \lambda_2 > \lambda_3 > \cdots > \lambda_s$). For this situation, the 4 tests rank from most powerful—R, T, Λ, V—to least powerful. In this structure, most of the information in the matrix $\boldsymbol{HE}^{-1}$ is contained in λ_1 given its dominant size (and $\lambda_2, \dots, \lambda_s$ add little). That's all the information that is in $R = \lambda_1$. In the other statistics, bringing in the other λ_i's, doesn't help much. They are very small and perhaps just noise (and could diminish the statistic, especially in multiplication: e.g., .9×.2×.1 = .018 <<< .9).

In the opposite scenario, when we have a diffused structure, where the groups differ along several dimensions, the eigenvalues are still ordered by size, but they're more comparable in magnitude, i.e., $\lambda_1 \cong> \lambda_2 \cong> \lambda_3 \cong> \cdots \cong> \lambda_s$). Naturally, for this condition, the rank of the 4 tests in terms of power is the reversed from that above. The most powerful would be V, then Λ, then T, and R would be least powerful.

Some researchers have tried to argue that we should expect diffused structures in behavioral sciences, that is groups differing along more than 1 dimension (Olson, 1976), though others would argue we'd expect unidimensionality. Seems to me, surely that depends on the field and phenomena being studied. The point is, none of these 4 test statistics is "uniformly most powerful," meaning none of the tests is always more likely to reject the H_0 when the H_0 is false; it depends on how the H_0 is false.

Regarding power in general, that is, for all 4 tests, some results have been established.

1) As the number of variables, p, increases, we need to increase the sample size to retain comparable power. (This is just like in regression. If we want to fit regression models with many predictors, we need larger samples.) So again, the lesson is that we can't just add variables to the analysis without thought or without cost.
2) Power increases as the effect size increases, of course—if the groups differ by a lot, we'll more easily detect it.
3) Power increases as sample size increases (of course).
4) The intercorrelations among the p dependent variables has an effect on power, increasing it or decreasing it, relative to running p separate ANOVAs. The MANOVA could be significant and none of the ANOVAs are, or the ANOVAs could be significant but the MANOVA is not. This issue is somewhat related to the concentrated vs. diffuse issue.[28]
5) Power is decreased over all 4 statistics in the presence of kurtosis (platykurtic, flat distributions, or leptokurtic, peaked distributions; see Olson 1974, 1976).

Having said all that, the differences between the 4 tests in each of the rankings is not large, and even so, the difference is for small to moderately large samples. When N=100 or 200, there seem to be no substantial power differences of any importance; that is, it's essentially the case that R, T, Λ, and V are asymptotically equivalent. And of course they are equal when s=1.

So, in sum, if we have to choose between interpreting and reporting Λ, V, R, and T, what shall we do? Look at the pattern of results. The clearest case (after they're being equal when s=1) is if all 4 tests say "significant" or all say "not significant" even if the p-values are slightly different. If the p-values are not unanimous in all being above or below .05, then we do not simply count and say, "ok, the majority rules, 3 of the 4 are significant, so it must be significant." Instead, we'll keep in mind their relative power, e.g., if we have a concentrated structure ($\lambda_1 >>> \lambda_2 > \lambda_3 > \cdots$), we'd expect R to be more powerful than V, Λ, or T, so if

[28] For more information, see Stevens (1980), who describes how to compute power for MANOVA.

R is significant, but the others are not, the results are still sensible (we'd go with R and significance, even though the other "3 out of 4" say not).

We need to also keep in mind the relative robustness of the 4 test statistics, especially with respect to characteristics that might be possible violations in our data. For example, if the data are quite skewy, we should probably weight the results based on V most heavily because V is generally the most robust, as we'll see in the next section.

Comparing the 4 Test Statistics R, T, Λ, and V with respect to Robustness

The test statistics may also be compared with respect to their robustness. Robustness is evaluated in regard to how a statistic behaves when there are violations of the assumptions, so first, let's look at the 3 major MANOVA assumptions.

First, we assume that we have a random sample and independent observations. If we depict a data matrix **X** that is N×p, where N is the total number of study participants. For example, if we had run a 2-factor factorial, then N=a×b×n, and of course p is the number of variables:

$$\text{data matrix:} \begin{bmatrix} X\ is \\ \\ N \times p \\ \end{bmatrix}$$

Each row in **X** is a vector $\boldsymbol{x}_{ij}$, for person i=1…N and variable j=1…p. (We can divvy the N subjects into their experimental groups later.) For the model to be appropriate, we assume each vector $\boldsymbol{x}_{ij}$ is drawn from a multivariate normal distribution, $MVN_p(\boldsymbol{\mu}, \boldsymbol{\Sigma})$, with some mean vector and covariance matrix.

Neither ANOVA nor MANOVA is robust if the assumption about independence doesn't hold. The N observations must be independent of each other (one subject's data cannot affect another's), and we must select a random sample (not one that is systematic or biased in any way).

Second, after the concept of independence among the observations, we mentioned MVN. This assumption is a little more forgiving in that if the p dependent variables are not quite distributed as MVN_p in each population, we could do variable transformations to make them more normal-ish. Variable transformations are especially useful if we're expecting skewed data or heterogeneity of variances across groups (or upon viewing our data, see that either the skewing or heterogeneity is true). Some classes of transformations are widely acceptable, even expected when researchers are handling certain kinds of (univariate) data:

type of data	transformation accepted/expected
latencies, reaction times	natural log
proportions	arcsine
frequencies, counts (e.g., #errors)	square root

Variable transformations carry the cost that the interpretation is affected. It is helpful to transform the data, conduct the ANOVA, obtain the F-tests and p-values, and then reverse-transform the data to obtain means that are more interpretable in the original unit of the data. If one is nervous about using variable transformations, one can simply hope that the methods will be robust.

One way that a distribution could be non-normal is with kurtosis (platykurtic, flat, or leptokurtic, peaked). With peaked data (positive kurtosis), tests become slightly more conservative (though the effect is mild). Ranking the 4 tests statistics on their likely Type I error rates: V would be the closest to the nominal α rate, and then Λ, T, and R will produce fewer Type I errors than the chosen α rate (i.e., R is more conservative). One exception is that, for R, the likelihood of Type I errors can be much > α if only 1 group shows kurtosis; that is, if there is also heterogeneity of covariances.

Third, just as we saw in the last chapter with Hotelling's T^2, statisticians are usually mostly interested in the assumption of homogeneity of covariance matrices, and what happens to the model parameters and conclusions if the assumption does not hold. The assumption of the covariance matrices being statistically equal across all "k" groups is quite daunting because now, in addition to the variances that must be at least roughly equal, so must be all the many covariances.

To make an assumption of the equality of covariance matrices, all k groups must share $\mathbf{\Sigma}$; that is, $\mathbf{\Sigma}_1 = \mathbf{\Sigma}_2 = \mathbf{\Sigma}$. In the figure, the variances for all p variables on the diagonal must be homogeneous across all groups (from group-to-group, but not necessarily variable-to-variable), and the (p(p-1)/2) covariances on the off-diagonal cells in the matrix must also be equal across all groups.

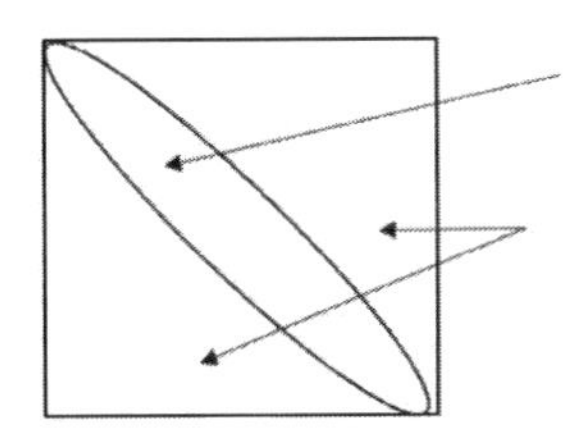

homogeneity of variances for each of the p variables

<u>and</u> homogeneity of covariances for the (p(p-1)/2) variables

Back in the '70s, when computers were getting powerful enough and fast enough to investigate some of these statistical issues, there was a series of papers that investigated the robustness of the 4 test statistics R, T, V, Λ against violations of normality and of equal covariance matrices.

Olson (1974, 1976) examined the 4 test statistics R, T, V, Λ (and others) in a simulation study. The number of dependent variables varied (p = 2, 3, 6, 10), as did the number of groups in the MANOVA (k = 2, 3, 6, 10), along with cell sample sizes (n = 5, 10, 15, and the n's were always equal across all cells). The simulation also included several factors affecting the degree of heterogeneity, the effect size (the degree to which the groups differed from H_0), and the number of underlying dimensions along which the group differed. Olson (1974, 1976) found that in the presence of heterogeneous covariance matrices, Type I errors can be very high for R (and then T and Λ). V was found to be less affected; that is, V is the most robust statistic against violations of the homogeneity of covariance matrices assumption.

In general, across all 4 statistics: 1) Robustness was shown to improve with smaller p (also, smaller p is easier to interpret, so increasing the number of dependent variables comes at a cost). 2) Robustness improves as the number of groups decreases (again, probably due to simplicity). 3) Robustness improves as sample size decreases, but this is (greatly) offset by loss of power.

Olson (1974, 1976) concluded that using R results in too many Type I errors (as well as poor power in diffuse structures). The test statistics V, Λ, and T are asymptotically equivalent (when $df_{error} \geq 10 \times p \times df_{effect}$). In smaller samples, we should avoid using Λ and T if there are heterogeneous covariance matrices. The statistic V is recommended for general use—it is the most robust and it is powerful in diffuse structures and powerful enough in concentrated structures.

Stevens (1979) criticized Olson's Monte Carlo work for creating heterogeneity conditions that Stevens believes were too severe (i.e., $\sigma_1^2 = 36 * \sigma_2^2$). This extreme heterogeneity was thought to not be realistic for real data. In testing the effects of less extreme heterogeneity, V, T, and Λ performed about the same. V was nevertheless acknowledged to be slightly more robust, and slightly more powerful for diffuse structures, and T and Λ slightly were more powerful for concentrated structures.

Olson (1979) responded. V's robustness advantage was demonstrated to hold up even for less extreme heterogeneity (i.e., $\sigma_1^2 = 4\sigma_2^2$ or $9\sigma_2^2$, rather than $36\sigma_2^2$). So there.

Regarding robustness in general, that is, for all 4 tests, some results have been established:

1) When sample sizes are all equal, and those n's are large, and p is relatively small, and the covariance matrices aren't too different, then all 4 test statistics are robust.
2) In general, the Pillai-Bartlett trace V is the most robust of all.
3) When sample sizes (the n's) are not equal, not 1 of the 4 is robust. Depending on other factors, the Type I errors can greatly exceed α (so the test is dangerous), or Type I errors can occur much less frequently than α (due to a lessened power of the test, so the test is not useful).

SUMMARY

This chapter introduced MANOVA. The logic of comparing $SS_{between}$ to SS_{error} in the univariate ANOVA is carried over and executed by comparing an **H** matrix to an **E** matrix in MANOVA. There are 4 main test statistics that can be calculated as a function of the eigenvalues of (the information contained in) these 2 matrices. SAS and other computing packages will print the values of the 4 tests, along with their corresponding F-tests (and df, and p-values).

In closing, it is worth mentioning that there are 2 special cases of MANOVA that occur when s=1, which is when the number of groups=2 or p=1. When the number of groups=2, the Hotelling's T^2 is the uniformly most powerful test (and all 4 MANOVA test statistics will equal it). When p=1, the univariate F-test is the uniformly most powerful test (and all 4 MANOVA test statistics will equal it).

REFERENCES

Books:

11. Bray, James H. and Scott E. Maxwell (1985), *Multivariate Analysis of Variance*, Sage.
12. Most general "multivariate stats" books also cover MANOVA, albeit briefly.

Articles:

1. Bird, Kevin D. and Dusan Hadzi-Pavlovic (1983), "Simultaneous Test Procedures and the Choice of a Test Statistic in MANOVA," *Psychological Bulletin*, 93 (1), 167-178.
2. Hakstian, A. Ralph, J. Christian Roed, and John C. Lind (1979), "Two-Sample T^2 Procedure and the Assumption of Homogeneous Covariance Matrices," *Psychological Bulletin*, 86 (6), 1255-1263.

3. Heck, D. L. (1960), "Charts of Some Upper Percentage Points of the Distribution of the Largest Characteristic Root," *Annals of Mathematical Statistics*, 31, 625-642.
4. Olson, Chester L. (1979), "Practical Considerations in Choosing a MANOVA Test Statistic: A Rejoinder to Stevens," *Psychological Bulletin*, 86 (6), 1350-1352.
5. Olson, Chester L. (1976), "On Choosing a Test Statistic in Multivariate Analysis of Variance," *Psychological Bulletin*, 83 (4), 579-586.
6. Olson, Chester L. (1974), "Comparative Robustness of Six Tests in Multivariate Analysis of Variance," *Journal of the American Statistical Association*, 69 (348), 894-908.
7. Pillai, K. C. S. (1956), "On the Distribution of the Largest or Smallest Root of a Matrix in Multivariate Analysis," *Biometrika*, 43, 122-127.
8. Ramsey, Philip H. (1982), "Empirical Power of Procedures for Comparing 2 Groups on p Variables," *Journal of Educational Statistics*, 7, 139-156.
9. Stevens, James P. (1980), "Power of Multivariate Analysis of Variance Tests," *Psychological Bulletin*, 88 (3), 728-737.
10. Stevens, James, P. (1979), "Comment on Olson: Choosing a Test Statistic in Multivariate Analysis of Variance," *Psychological Bulletin*, 86 (2), 355-360.

APPENDIX A

This appendix shows the SAS inputs required to run MANOVAs, and excerpts from the output files.

```
*file: anova-sas3-manova-2grp-2dv.docx;
*Hotelling's T2 (or MANOVA) on 2 groups;
data easyex; input grp x1 x2; datalines;
1 2.0 3.0
1 2.0 2.0
1 3.0 3.0
1 3.0 4.0
1 1.5 1.5
1 1.5 2.0
1 2.5 2.0
1 2.0 1.5
1 2.0 2.5
1 2.5 2.5
1 2.5 3.0
1 3.0 3.5
1 2.5 3.5
1 2.5 4.0
1 3.5 4.0
1 3.0 4.5
1 3.5 4.5
2 6.0 4.5
2 6.0 3.5
2 7.0 4.5
2 7.0 5.5
2 5.5 3.0
```

```
2 5.5 3.5
2 6.5 3.5
2 6.0 3.0
2 6.0 4.0
2 6.5 4.0
2 6.5 4.5
2 7.0 5.0
2 6.5 5.0
2 6.5 5.5
2 7.5 5.5
2 7.0 6.0
2 7.5 6.0
proc print data=easyex; proc plot; plot x2*x2 = grp;
proc glm data=easyex; class grp; model x1 x2 = grp;
manova h=grp / printe printh summary; *this manova statement is new. H=grp says test the hypothesis for the factor called grp. The options of printe and printh will print the E and H matrices, in case you're curious (they're not typically reported). Summary will print the univariate ANOVAs on each of the dependent variables;
contrast 'grp 1 vs 2' grp 1 -1;
means grp; run;
*Normally we would not do the analysis that follows (it is just to make a point);
data new; set easyex; newv = (.49347 * x1) + (-.21827*x2); *we have created a new variable, its source and purpose to be explained shortly;
proc glm data=new; class grp; model newv = grp; run;
```

Excerpts from the SAS output follow.

In this **E** matrix, the values of 12 and 32 are the SS_{error} for X_1 and X_2 respectively. The value 16 (note the symmetry in the matrix) is the the covariance between the two dependent variables' errors, a concept that is not hugely intuitive, but it's how the inter-correlation between the dependent variables is "taken into account" that makes the MANOVA different from, and usually more powerful than, the individual ANOVAs.

```
                 The GLM Procedure
          Multivariate Analysis of Variance
               E = Error SSCP Matrix
                          X1                  X2
      X1                  12                  16
      X2                  16                  32
```

This **H** matrix contains the SS_{effect} (grp) values for X_1 (136) and X_2 (19.125), as well as their cross-product SS_{effect} value (51):

```
          H = Type III SSCP Matrix for GRP
                          X1                  X2
      X1                 136                  51
      X2                  51              19.125
```

Next, the "characteristic roots" (an old-fashioned word for eigenvalues) are printed. We have p=2, so there are 2 lines, but s=1 (recall, $s=\min(df_{effect},p)$, here $s=\min(1,2)=1$), so λ_1

will be nonzero (here it is 23.0429688, and it is associated with the eigenvector 0.493, -0.218), but any additional λ's will equal 0 (the vector affiliated with λ_2 is just computer vomit—ignore the -0.0805 and 0.2147 values).[29]

```
              Characteristic Roots and Vectors of: E Inverse * H
Characteristic                          Characteristic Vector  V'EV=1
          Root       Percent                     X1                X2
    23.0429688        100.00             0.49347325       -0.21826702
     0.0000000          0.00            -0.08052422        0.21473125
```

Next, those eigenvalues are combined into the 4 test statistics we have seen for the MANOVA.

```
MANOVA Test Criteria and Exact F Statistics for the Hypothesis of No GRP Effect
Statistic                          Value    F Value    Num DF    Den DF    Pr > F
Wilks' Lambda                 0.04159220     357.17         2        31    <.0001
Pillai's Trace                0.95840780     357.17         2        31    <.0001
Hotelling-Lawley Trace       23.04296875     357.17         2        31    <.0001
Roy's Greatest Root          23.04296875     357.17         2        31    <.0001
```

The values for the 4 test statistics differ, but given that s=1, their F-tests converge. Often the "values" are not reported, but the F-value, df, and p-values would be.

Recall in the SAS commands, we had created a new variable (p=1) as a function of the old variables (p=2), defined as: newv = (.49347 * x1) + (-.21827*x2). Running an ANOVA on this single composite yields useful information. In defining the 4 test statistics, we continually referred to the interpretation of λ_i as $SS_{between}/SS_{within}$, and it is, for the "ith discriminant function." That function is defined by the eigenvector (.49347 -.21827) printed above next to its eigenvalue λ_1, and those weights were used to create the new variable. In the ANOVA of this new variable, the $SS_{between}$=23.04248 and the SS_{within}=.99998. Their ratio is $SS_{between}/SS_{within}$=23.04248/.99998 = 23.04294, which equals λ_1 (within rounding error, 23.0429688). Cool, yes?

```
*file: anova-sas4-manova-2grp-3dv.docx;
*Two sample Hotelling T2 (or MANOVA on 2 groups), now with p=3;
data two; input samp v1 v2 v3; datalines;
1    3    4    11
1    9    8    10
1    3    7    6
1    5    9    9
2    10   9    11
2    11   9    10
2    6    9    12
2    9    9    11
proc print data=two;
proc glm data=two; class samp; model v1 v2 v3 = samp;
manova h = samp / printh printe summary; means samp; run;
```

[29] See Appendix B to this chapter for a note about the matrix $\boldsymbol{HE}^{-1}$ and SAS.

```
*The following shows how to define a linear combination of the p dependent variables (the m= statement). In this case, the linear combination equals simple V3. Ordinarily you would really want a combo of more than 1 (up to p) of the variables. This example, though, shows that these results would be equal to running a univariate ANOVA on V3;
proc glm data=two; class samp; model v1 v2 v3 = samp;
manova h = samp m = 0*v1 + 0*v2 + 1*v3 / printh printe summary;
means samp; run;
*Normally do not do analysis that follows (just for learning);
proc discrim simple pool=test wcov wcorr pcov;
class samp; var v1 v2 v3; run;
```

Excerpts from the SAS output follow. The **E** and **H** matrices are printed, as are the eigenvalues (s=1, so $\lambda_1 = 1.804$, and the other λ's are 0). Wilks' Lambda = 0.357, Pillai's Trace = 0.643, the Hotelling-Lawley Trace is $T = 1.804$, and Roy's Greatest Root is 1.804 (note that $T = R$ because s=1). The F-value for all 4 test statistics is 2.41 on 3 and 4 df, p=0.2078. What's new is the definition and analysis of a linear combination of the p=3 variables. SAS very helpfully provides feedback as to the transformation that we defined, so we can check.

M Matrix Describing Transformed Variables

	V1	V2	V3
MVAR1	0	0	1

Then in the analysis, the E "matrix" for this linear combination, which is essentially only the univariate examination of the 3rd variable, is the value 16, and that is the SS_{error} for V3 in an ANOVA on V3. Similarly, the H "matrix" is the value 8, which is the SS_{effect} for V3, so this "linear combination" thing "works."

In SAS, if we wanted to create more than 1 linear combination of our old p=3 variables, we'd say:

```
manova h=samp m=0*v1 +  0*v2 +  1*v3,
              .33*v1 + .33*v2 + .33*v3,
               1*v1    -1*v2 +  0*v3;
```

That would define coefficients in the columns of this matrix.

$$C'\bar{X}A:\quad A = \begin{bmatrix} 0 & .33 & 1 \\ 0 & .33 & -1 \\ 1 & .33 & 0 \end{bmatrix} \begin{matrix} var1 \\ var2 \\ var3 = p \end{matrix}$$

Linear combination: 1 2 3=q

The last procedure we had included was a "discriminant" analysis. We'll say more about these in Chapter 15. For our purposes here, it was simply to test for the homogeneity or the equality of the within-group covariance matrices. The procedure produces a lot of output, but for our data, the bottom line is this X^2:

Chi-Square	DF	Pr > ChiSq
30.245012	6	<.0001

Given the p-value, we reject the null hypothesis that $\mathbf{\Sigma}_1 = \mathbf{\Sigma}_2$. Unfortunately, that puts us in a weird zone where we know an assumption does not hold, yet there aren't many great alternative statistics to pursue for this analysis. We can pray for robustness; we do the MANOVA and hold our breath.

```
*file: and anova-sas5-manova-2factors.docx; *if the experiment has 2 or more independent
variables, each effect needs its own manova statement;
data try; input a b v1 v2; datalines;
1 1 30 35
1 2 25 33
1 1 21 41
1 2 15 21
1 1 27 32
1 2 25 34
1 1 35 34
1 2 31 36
1 1 20 37
1 2 14 21
1 1 23 38
1 2 19 25
2 1 20 15
2 2 26 25
2 1 11  3
2 2 18 19
2 1  7  2
2 2 11  8
2 1 21 15
2 2 24 22
2 1 15 11
2 2 17 13
2 1 13 12
2 2 20 15
proc glm; class a b; model v1 v2 = a b a*b;
manova h=a     / printh printe summary;
manova h=b     / printh printe summary;
manova h=a*b   / printh printe summary;
means a b a*b; run;
```

Excerpts from the SAS output follow. First, the main effect for A is tested. SAS prints the **E** and **H** matrices required for this test, the eigenvalues, and the 4 test statistics and their F-values (and df and p-values).

E = Error SSCP Matrix

	v1	v2
v1	674.33333333	456
v2	456	643.5

H = Type III SSCP Matrix for a

	v1	v2
v1	280.16666667	775.58333333
v2	775.58333333	2147.0416667

Statistic	Value	F Value	Num DF	Den DF	Pr > F
Wilks' Lambda	0.19707231	38.71	2	19	<.0001
Pillai's Trace	0.80292769	38.71	2	19	<.0001
Hotelling-Lawley Trace	4.07427964	38.71	2	19	<.0001
Roy's Greatest Root	4.07427964	38.71	2	19	<.0001

Given these results, we'd report: "The MANOVA showed that the A main effect was significant, F(2,19) = 38.71, p<.0001."

Next, SAS tests the main effect for B (notice that the **E** matrix has not changed, and notice that SAS labels the **H** matrices so we know we're testing "a" above and "b" here below):

E = Error SSCP Matrix

	v1	v2
v1	674.33333333	456
v2	456	643.5

H = Type III SSCP Matrix for b

	v1	v2
v1	0.1666666667	-0.25
v2	-0.25	0.375

Statistic	Value	F Value	Num DF	Den DF	Pr > F
Wilks' Lambda	0.99740440	0.02	2	19	0.9756
Pillai's Trace	0.00259560	0.02	2	19	0.9756
Hotelling-Lawley Trace	0.00260235	0.02	2	19	0.9756
Roy's Greatest Root	0.00260235	0.02	2	19	0.9756

For B, we'd report: "The MANOVA showed that the B main effect was not significant, F(2,19) = 0.02, p=0.9756."

Finally, for the A×B interaction:

E = Error SSCP Matrix

	v1	v2
v1	674.33333333	456
v2	456	643.5

H = Type III SSCP Matrix for a*b

	v1	v2
v1	130.66666667	212.33333333
v2	212.33333333	345.04166667

Statistic	Value	F Value	Num DF	Den DF	Pr > F
Wilks' Lambda	0.64735958	5.17	2	19	0.0161
Pillai's Trace	0.35264042	5.17	2	19	0.0161
Hotelling-Lawley Trace	0.54473654	5.17	2	19	0.0161
Roy's Greatest Root	0.54473654	5.17	2	19	0.0161

For the A×B interaction, we'd report: "The MANOVA showed that the A×B interaction was significant, F(2,19) = 5.17, p=0.0161."

APPENDIX B

This second appendix addresses an issue about $\boldsymbol{HE^{-1}}$, the fact that it is not symmetric, and SAS. Multivariate texts uniformly refer to the $\boldsymbol{HE^{-1}}$ matrix, yet SAS refers to $\boldsymbol{E^{-1}H}$. Both forms put the between-groups variability in the "numerator" and the within-groups variability in the "denominator" (via $\boldsymbol{E^{-1}}$), yet we know from basic matrix algebra that even when 2 matrices conform to multiplication in both orders, $\boldsymbol{AB}$and $\boldsymbol{BA}$, their products are not generally equal, $\boldsymbol{AB} \neq \boldsymbol{BA}$. In this case, SAS's labeling (or even the computation) is not critical, because we do not use the full $\boldsymbol{HE^{-1}}$ matrix, just its eigenvalues. And the eigenvalues of $\boldsymbol{HE^{-1}}$ equal those of $\boldsymbol{E^{-1}H}$.

This illustration demonstrates the equality for a 2×2 matrix. We'll define 2 matrices, with 4 cell elements a, b, c, d, but the off-diagonals have reversed locations, as would be the case for $\boldsymbol{HE^{-1}}$ and $\boldsymbol{E^{-1}H}$, given that one matrix ($\boldsymbol{HE^{-1}}$) is the transpose of the other ($\boldsymbol{E^{-1}H}$): $\boldsymbol{HE^{-1}} = (\boldsymbol{E^{-1}H})' = \boldsymbol{H}'(\boldsymbol{E^{-1}})' = \boldsymbol{HE^{-1}}$ (in turn, because $\boldsymbol{H}$, $\boldsymbol{E}$, and $\boldsymbol{E^{-1}}$ are all symmetric). (The transposition will be the essence as to why the eigenvalues are the same.) For two 2×2 matrices:

$$Matrix_1 = \begin{bmatrix} a & b \\ c & d \end{bmatrix} \qquad Matrix_2 = \begin{bmatrix} a & c \\ b & d \end{bmatrix}$$

The eigen equations are the same:

$$(a-\lambda)(d-\lambda) - cb \qquad \text{and} \qquad (a-\lambda)(d-\lambda) - bc$$

as of course will be their eigenvalues derived via the quadratic equations:

$$\lambda^2 - a\lambda - d\lambda + ad - cb \qquad \text{and} \qquad \lambda^2 - a\lambda - d\lambda + ad - bc$$

So whether we're working with $\boldsymbol{HE^{-1}}$ or SAS is working with $\boldsymbol{E^{-1}H}$, in both cases, we are working with the same eigenvalues in the 4 test statistics.

Consider the numeric example presented earlier in this chapter. We had p=2, with $\boldsymbol{H} = \begin{bmatrix} 210 & -90 \\ -90 & 90 \end{bmatrix}$ and $\boldsymbol{E} = \begin{bmatrix} 88 & 80 \\ 80 & 126 \end{bmatrix}$, then $\boldsymbol{HE^{-1}} = \begin{bmatrix} 7.18 & -5.27 \\ -3.95 & 3.23 \end{bmatrix}$. The eigenvalues of $\boldsymbol{HE^{-1}}$ are 10.179 and 0.226. If we were to compute $\boldsymbol{E^{-1}H}$, we'd obtain the same eigenvalues of 10.179 and 0.226. The 4 test statistics are a function of the eigenvalues, and they are indeed equal to each other when computing them from either source $\boldsymbol{HE^{-1}}$ or $\boldsymbol{E^{-1}H}$. So SAS can call the matrix $\boldsymbol{E^{-1}H}$ and we (and most sources) will continue to call it $\boldsymbol{HE^{-1}}$. Done.

As an aside that is perhaps categorically TMI, while the eigenvalues of $\boldsymbol{HE^{-1}}$ and $\boldsymbol{E^{-1}H}$ are the same, the eigenvectors are not. Eigen decompositions are done on some (real) square matrices, often symmetric, as when we derive factors from a correlation matrix ($\boldsymbol{R} = \boldsymbol{V\Lambda V'}$). The more general decomposition operation for rectangular (real or complex) matrices is the singular value decomposition (SVD, see Chapter 11), whereby a matrix $\boldsymbol{A}$ is decomposed into left-singular-vectors ($\boldsymbol{L}$), singular values ($\boldsymbol{Q}$), and right-singular-vectors ($\boldsymbol{R}$): $\boldsymbol{A} = \boldsymbol{LQR'}$. The singular values of $\boldsymbol{A}$ are the square roots of the eigenvalues of $\boldsymbol{AA'}$ and $\boldsymbol{A'A}$. For $\boldsymbol{AA'}$ we have $\boldsymbol{AA'} = (\boldsymbol{LQR'})(\boldsymbol{LQR'})' = (\boldsymbol{LQR'})(\boldsymbol{RQL'}) = \boldsymbol{LQQL'}$ and for $\boldsymbol{A'A}$, we have $\boldsymbol{A'A} = (\boldsymbol{LQR'})'(\boldsymbol{LQR'}) = (\boldsymbol{RQL'})(\boldsymbol{LQR'}) = \boldsymbol{RQQR'}$. Furthermore, the left-singular-vectors $\boldsymbol{L}$ are the eigenvectors of $\boldsymbol{AA'}$, and the right-singular-vectors $\boldsymbol{R}$ are the eigenvectors of $\boldsymbol{A'A}$. In this context, $\boldsymbol{A}$ is the equivalent of $\boldsymbol{HE^{-1}}$ and $\boldsymbol{A'}$ that of $\boldsymbol{E^{-1}H}$. While the eigenvectors of $\boldsymbol{HE^{-1}}$ and $\boldsymbol{E^{-1}H}$ do not easily map onto each other, it is the case that the left-singular-vectors derived from a SVD of $\boldsymbol{HE^{-1}}$ match the right-singular-vectors of $\boldsymbol{E^{-1}H}$ (and the right-vectors of $\boldsymbol{HE^{-1}}$ match the left-vectors of $\boldsymbol{E^{-1}H}$), again, all due to the transposition.

HOMEWORK

This HW asks a couple of questions about the MANOVA test statistics.

Problem 1.

You've run a 3-factor between-subjects design

Factor A has a=7 levels and is a random factor.
Factor B has b=5 levels and is a fixed factor.
Factor C has c=2 levels and is a fixed factor.
n=4 subjects were run in each (abc) cell.
4 variables were measured on each subject.

You are testing for a main effect for factor B ($\boldsymbol{H}_B$), and the eigenvalues of $\boldsymbol{HE}^{-1}$ (for B being tested in the ***H*** and the proper term being used as ***E***) follow: $\lambda_1 = 5.2,\ \lambda_2 = 3.75,\ \lambda_3 = 2.00,$ $\lambda_4 = .75$.

Answer the following questions:

1. Describe ***E***
2. λ_5 =?
3. p = ?
4. df_{effect} = ?
5. df_{error} = ?
6. Derive the 4 test stats, Λ, V, R and T
7. Compute the F approximations for Λ, V, and T and list the appropriate critical F (again, for $\alpha = .05$).
8. What is H_0?
9. Would you reject H_0 (assume multi-normality and equality of covariance matrices hold)? Defend your decision.
10. Which of the 4 test statistics should you pay the most attention to, and why?

CHAPTER 15

MANOVA CONTRASTS AND FOLLOW-UP TESTING

Questions to guide your learning:
Q_1: How do we do follow-up testing, contrasts among groups and linear combination among dependent variables, if we have a significant MANOVA?

In this chapter, we'll discuss follow-up tests for MANOVA. We'll assume that we have run a MANOVA and we've found 1 or more significant result(s)—main effect(s) or interaction(s). To learn more about the nature of our data, why we rejected a null hypothesis, we'll conduct follow-up statistical tests, just like the contrasts and simple effects we saw in Chapter 5 for ANOVA. The complication in this chapter of course is that we now have p dependent variables, rather than 1, so in addition to drawing multiple group comparisons, we must consider how to examine the relationships among the p variables.

There is certainly joy in significance, but follow-up procedures can be complicated. We have many choices, both with respect to determining which groups differ, and how or on which variables. We'll discuss 4 major methods of conducting follow-up tests: 1) We could follow a MANOVA on p variables by running p univariate ANOVAs on each of the dependent variables separately. 2) We could run a 1 df contrast between groups on all p variables, as in a Hotelling's T^2, or a 1 df contrast between groups on 1 variables, as in a Scheffé test like we did in ANOVA. 3) We could run a simultaneous test procedure, a multivariate version of a contrast in which we investigate groups and dependent variables, doing both at the same time. 4) We could run a discriminant analysis. We'll see what this is and how it might help to clarify our understanding of how our groups differ.

1) Univariate ANOVAs

The first natural option for investigating details of our data upon rejecting some null hypothesis in the MANOVA is to conduct univariate ANOVAs on each of the p variables separately. Historically, this has been, and continues to be the most frequently used approach. Its prevalence is no doubt due to the fact that it is the simplest thing to do—we know our data differ on the p variables, to get more detail, we inquire as to whether our groups differ on the measures, variable by variable.

In addition, our research questions might be univariate, of the form, "Do our groups differ on X_1?" and "Do our groups differ on X_2?," etc., rather than of a multivariate, and admittedly somewhat odd and complex form, i.e., "Do our groups differ on some linear combination of X_1, X_2, ..., and X_p?" It's fine to run ANOVAs upon finding significance in MANOVA, however, doing so does not take into account the correlations among the p variables.

Hummel and Sligo (1971) compared this approach of running p ANOVAs to STPs (the 3rd approach, discussed shortly) on Type I experiment-wise error rates. They varied sample size (10, 30, 50), and p (3, 6, 9), as well as the correlations among the p variables. They found that this approach, running p ANOVAs, behaved well with respect to Type I errors. The α rates did not exceed .05 for a nominal $\alpha = .05$. By comparison, the STPs (below) were found to be very conservative (with Type I error rates around .000-.017), increasingly so for higher p. That is, using STPs, we're not likely to make a Type I error, but we're also not likely to reject the null even when H_0 is false (i.e., if we're rarely rejecting the null, the approach is also not sensitive to real differences). We prefer conservative tests to liberal ones—if a test is conservative so a statistic is not found to be significant, it only hurts the researcher and we tend to have faith that if the phenomenon is real, it will be discovered and established in subsequent research, but if a statistic is liberal and found to be significant when in fact nothing is going on in the data, it hurts the field in leading us all astray. Still, it's worth noting that one concern with a follow-up approach that is too conservative is that we could very well reject the overall H_0: $\boldsymbol{\mu}_1 = \boldsymbol{\mu}_2 = \cdots = \boldsymbol{\mu}_k$ in the MANOVA, and yet not be able to reject H_0: $\mu_{1j} = \mu_{2j} = \cdots = \mu_{kj}$ for any particular variable j=1, 2, ..., p, or variable combindation. All, in all, particularly because the p ANOVAs did not inflate (or decrease) Type I errors, Hummel and Sligo (1971) recommended this method, using p ANOVAs, as the means of investigating significant results found in a MANOVA.

The p ANOVAs are also an easy way to go. Nevertheless, they might not answer all our research questions. The basic way that a MANOVA differs from p ANOVAs is that the MANOVA takes into account the correlations among the p dependent variables. Part of what that means is that if we were to conduct p ANOVAs, the F-test for any of the p variables will be the same no matter what order we run the p ANOVAs (e.g., X_1, X_2, X_3... or X_3, X_2, X_1,...). Each ANOVA model is being computed on its single dependent variable, so by definition, the series of p ANOVAs is insensitive to the correlations between the p variables. Yet the p variables are correlated (usually), so the F-tests cannot be considered to be somehow "partial" tests of the MANOVA. For example, the sums of squares on each ANOVA won't add to the sums of squares in the MANOVA, and the MANOVA's $\boldsymbol{H}$ matrix has sums of squares for cross-products between pairs of the p dependent variables, and of course the univariate ANOVAs do not.

2) Single DF Contrasts among Groups

The second option for pursuing and investigating significant results from a MANOVA follows the first in its simplicity and familiarity. We can either run a single df contrast among the experimental groups for all p variables, or we can run a single df contrast among the groups on just a single variable.

For the first of these, running a 1 df contrast among groups for all p variables, we'd have begun, in the first stage of the data analysis, by estimating the MANOVA model, and after rejecting H_0: $\boldsymbol{\mu}_1 = \boldsymbol{\mu}_2 = \boldsymbol{\mu}_3 = \boldsymbol{\mu}_4$ for some effect (the main effect for B, the A×B interaction, etc.), we might test some group contrast, e.g., perhaps we'd test whether groups 1 and 2 differed from group 4, on all p variables. We'd use contrast coefficients: H_0: $.5\boldsymbol{\mu}_1 + .5\boldsymbol{\mu}_2 + 0\boldsymbol{\mu}_3 - \boldsymbol{\mu}_4$ and given that this 1 df follow-up contrast involves all p variables, we'd be running a Hotelling's T^2. That would be fine.

For the second version of this option, after running the MANOVA and rejecting H_0: $\boldsymbol{\mu}_1 = \boldsymbol{\mu}_2 = \boldsymbol{\mu}_3 = \boldsymbol{\mu}_4$, we might wish to test whether H_0: $\mu_{25} = \mu_{35}$, that is, run a 1 df contrast among the groups on just a single variable. In this example, we'd compare groups 2 and 3 on dependent variable number 5. Pursuing this approach is merely the tact of conducting a typical ANOVA follow-up, such as a Scheffé F test, business as usual.

3) Multivariate versions of Contrasts and Comparisons, on both Groups and Variables

In this section, we'll see a technique that seems nicely complementary to the MANOVA as a procedure for conducting follow-up tests. We'll be using the logic of contrasts to compare across groups, and we'll create and examine linear combinations of the p dependent variables to take into account their inter-correlations, much as the MANOVA itself does. Given that the groups and variables are analyzed simultaneously (unlike an ANOVA, which might compare the groups but does so on only a single variable at a time), this approach was long ago termed, a "simultaneous test procedure" (STP). The label is apt, however it is no longer used, so don't bother to refer to a STP when writing up results, for example; no one would know the reference.

One quality we would like in the two-staged analysis, where in stage I we run a MANOVA, and in stage II we conduct follow-up tests is "coherence." Coherence here means that if the overall test of $H_0\colon \boldsymbol{\mu}_1 = \boldsymbol{\mu}_2 = \cdots = \boldsymbol{\mu}_k$ is not significant, then there won't be any contrasts to reject in the follow-up stage. Coherence is handy: "In a coherent two-stage analysis, the overall test can be regarded as a theoretically unnecessary but a practically useful starting point in the exploration of the data: theoretically unnecessary because the overall test could be omitted without affecting the outcomes of tests on contrasts; useful in practice because a nonsignificant overall test indicates that there would be no point in carrying out follow-up tests" (Bird and Hadzi-Pavlovic, 1983, p.168). We might disagree with whether the 1st stage is unnecessary, after all, it is what leads us to the 2nd stage, and it is expected to be reported. However, the 2nd part of the quote is what points out why coherence is helpful—if the overall test in the MANOVA is not significant, we stop; there is no need to try to find significance in the follow-up tests.

Okay, so we want coherence: how is it achieved? In the 1st stage, running the MANOVA, we saw we had a choice among 4 test statistics (R, V, Λ, T). Each is also used as the basis for a follow-up test procedure. If the follow-up test procedure is based on the MANOVA statistic used for the overall test, then the statistical decisions are usually coherent (it's a necessary but not sufficient condition). Thus, if the F-test (and df and p-value) based on the R (or Λ, etc.) is reported for the MANOVA stage, then the contrasts based on R (or Λ, etc.) must be reported in the follow-up stage. In the last chapter, we saw how those test statistics compared with respect to power and robustness, and when they were all equal. After making a decision about rejecting or not rejecting a null hypothesis in the MANOVA stage, (Bird and Hadzi-Pavlovi, 1983, p.168) remind us that "when choosing a MANOVA test statistic for a coherent analysis, it is necessary to take into account the performance of [the follow-up simultaneous test procedures as well], rather than simply choosing a statistic that performs well in the first stage of testing."

The point is that we cannot be opportunistic. We'll run a MANOVA in stage I and report an F-test based on 1 of the 4 test statistics. Next we'll run follow-up tests in stage II and we'll report their F-tests based on the same test statistic.

Bird and Hadzi-Pavlovic (1983) made an argument for using R because its follow-up tests will be "consonant"; that is, if we reject the overall H_0 in the MANOVA, we can always define at least 1 follow-up contrast that will be significant. The statistic V can be dissonant; if the overall test is significant, we might not be able to find any contrast to declare significant in follow-up tests.

More observations from Bird and Hadzi-Pavlovic (1983): when p=2, V and R both behave ok, but for more dependent variables, p>2 (i.e., p=4 or p=6 in their study), they consider V as too conservative, and R is better (but their results show R is rather conservative also). The paper didn't investigate the T or Λ statistics, because the V test is better on many criteria such as robustness (or power in a diffuse structure).[30] Some researchers have suggested that, because the follow-up tests are so conservative, perhaps a slightly more liberal α (e.g., $\alpha = .10$) should be used in the follow-up stage.

The Multivariate Contrast—Groups and Variables

How are the groups and dependent variables tested simultaneously in the follow-up procedure? The multivariate contrast is of the form:

$$\psi = c'\boldsymbol{\mu}a \tag{1}$$

The vector c' is the usual set of contrast coefficients to test and compare differences among the groups. The means, for all groups on all p dependent variables are in $\boldsymbol{\mu}$, which of course we estimate using the sample means $\overline{\boldsymbol{x}}$. The piece that is new is the vector a, which will contain the weights to combine the p dependent variables.

We'll estimate each multivariate contrast in (1) by $\hat{\psi} = c'\overline{\boldsymbol{x}}a$. As always, the contrast coefficients must sum to 0: $c' = [c_1\ c_2\ \dots\ c_k]$, $\sum_{i=1}^{k} c_i = 0$.

The means are compiled into a k×p matrix, with 1 row for each group, and 1 column for each dependent variable:

$$\overline{\boldsymbol{x}} = \begin{bmatrix} \bar{x}_{11} & \bar{x}_{12} & \dots & \bar{x}_{1p} \\ \bar{x}_{21} & \bar{x}_{22} & \dots & \bar{x}_{2p} \\ \dots & & & \\ \bar{x}_{k1} & \bar{x}_{k2} & \dots & \bar{x}_{kp} \end{bmatrix}$$

$\bar{x}_{ij}$ is mean over subjects in group i on variable j
(groups × variables), i.e., a (k × p) matrix

We put the "a" coefficients into a vector. Recall these a's do not need to sum to 0 like the c's for the contrasts between groups. If we wanted to compare the mean on the 1st two variables to the mean on the last two variables, for example, the vector would be: $\boldsymbol{a}' = [.5\ .5\ 0\ 0 \dots 0 - .5 - .5]$.

$$\boldsymbol{a}' = [a_1\ a_2\ \dots\ a_p] \text{ variable comparison coefficients} \tag{2}$$

To test the multivariate contrast, H_0: $\psi = 0$, we reject H_0 if:

$$\frac{n(\hat{\psi})^2}{(c'c)(a'Ea)} > \text{its critical value} \tag{3}$$

(When p=1 (1 dependent variable), this is essentially the Scheffé follow-up test.)

That's how it's done; it's very simple (and we saw SAS commands for doing so in the appendix to Chapter 14). Let's expand the scenario to testing several contrasts among the groups, not just 1, but "L" of them, and several combinations of variables, "m" of them. What had been vectors of contrasts (ψ), contrast coefficients (c), and weights to form linear combinations (a) are now matrices:

$$_L\boldsymbol{\psi}_m = {}_L\boldsymbol{c}_a\overline{\boldsymbol{x}}_p\boldsymbol{a}_m \tag{4}$$

[30] The dismissal of T and Λ due to their relatively poor performance at the overall stage is a little strange given their earlier argument to evaluate a statistic's performance at both its overall and follow-up stages.

$$\begin{bmatrix} \boldsymbol{\psi}_{L\times m} \end{bmatrix} = \begin{bmatrix} \boldsymbol{c}_{L\times a} \end{bmatrix} \begin{bmatrix} \overline{\boldsymbol{x}}_{a\times p} \end{bmatrix} \begin{bmatrix} \boldsymbol{a}_{p\times m} \end{bmatrix}$$

row = group contrast column = combination of variables	***c*** = a group contrast in each row	group by vars matrix	***a*** = 1 variable combination in each column

These follow-up tests allow us to study group differences and variable combinations. The elements in the multivariate matrix $\boldsymbol{\psi}$ are $\{\psi_{ij}\}$, the contrast between groups defined by c_i' and combination of variables defined by a_j.

To get accustomed to that equation, consider the following. What if the matrix **a** (p×m) was the identity matrix **I** (p×p)? Then the tests conducted would be the same as if we were doing p separate, univariate ANOVAs.

Just as we created an F-test from a univariate $\hat{\psi}$, we'll do so here. We'll test H_0: $\psi = 0$, $\hat{\psi} = c'\overline{\boldsymbol{x}}a$ and reject H_0 if:

$$\frac{n(\hat{\psi})^2}{(c'c)(a'Ea)} > \text{critical value} \tag{5}$$

with a corresponding confidence interval (λ, the eigenvalue from the MANOVA):

$$c'\bar{x}a \pm \sqrt{\textstyle\sum_j \frac{c_j^2}{n_j}(a'Ea)\lambda} \tag{6}$$

It's times like these that we give thanks for computers and SAS and such.

4) Discriminant analysis

The fourth technique that is typically included as a means to do follow-up tests after finding significant results in a MANOVA is a discriminant analysis. The discriminant analysis model is used to achieve several kinds of goals; in this section, we'll see how it applies to MANOVA. What is it and why is it relevant?

In a discriminant analysis, we're looking to form a new variable as a linear combination (or linear "function") of our p original variables, X_1 X_2, ..., X_p that maximally discriminates among k groups. That goal should sound vaguely familiar—in the MANOVA, we have multiple groups and we're seeking to understand how they differ on p variables.

Let's see what this model is trying to do. Say we have 2 groups (k=2) and 2 dependent variable (p=2), and the data look like this figure:

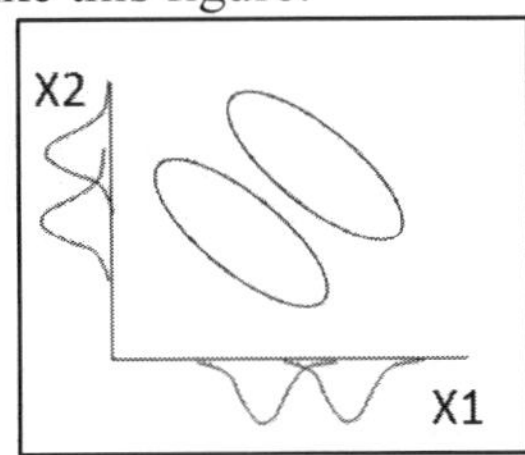

In discriminant analysis, we're trying to find the linear combination of variables that would result in the groups being maximally distinguished. One example might be for us to define a linear combination: new variable = $1 \times X_1 + 0 \times X_2$; that is, we'd be using only X_1 and not X_2. We can project the 2 groups' distributions from the scatterplot in the previous figure onto this new "function," which of course is merely the X_1 axis. (Recall the concept of a projection as the notion of light casting shadows from Chapter 12.) We'd see that the groups are fairly distinguishable on X_1. There is some overlap, but a good deal of clarity.

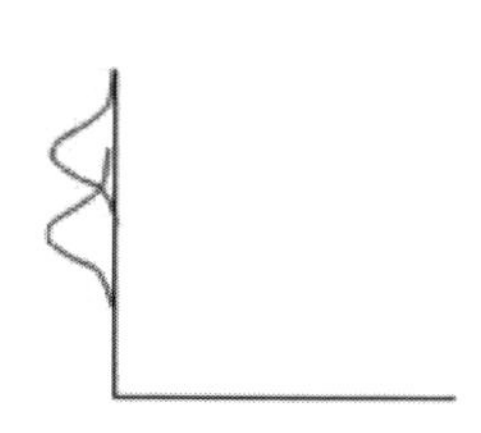

Alternatively we could try the other variable, defining a linear combination as $0 \times X_1 + 1 \times X_2$. Then, projecting the 2 groups' scatterplots onto the new "function" which is the X_2 axis, we'd see the distribution of the new variable. It also shows some differentiation, with only a little overlap.

As a 3rd example, we can try a summing or averaging function, $1 \times X_1 + 1 \times X_2$ (below, left), which results in the group projections as depicted in the figure below, to the right. This function does a really poor job of distinguishing the groups.

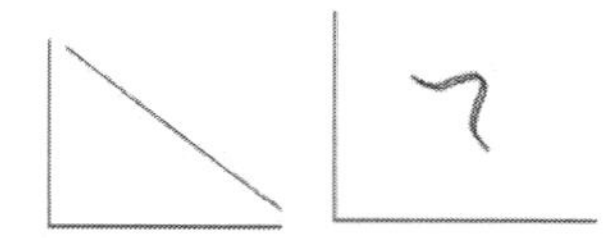

As a 4th and final example, we can try the linear combination function, $-1 \times X_1 + 1 \times X_2$, which looks like the figure to the left and results in projections like the figure to the right. Now the 2 groups have been maximally distinguished:

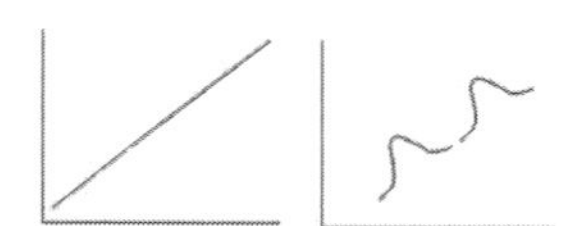

Examples 1 and 2 have some overlap and some differentiation. Example 3 shows no distinction ("distinction" or group "differences" being the goal in a technique designed to be able to "discriminate" between members in 1 sample and members in another). Example 4 is the linear combination that maximally (best) differentiates the groups. The group differences on this newly defined variable are the greatest—it is the discriminant function.

How do we find the discriminant function/axis? We'll derive it by first starting with the familiar t-statistic on 1 of our p dependent variables: $t = (\bar{x} - \mu)/\sqrt{(s^2/n)}$ on n-1 df. With all p variables, we could do p univariate t-tests, but we know that doing so would increase Type I error rate, it would ignore the fact that the p variables are correlated, etc. So, we'll define a new variable that combines all the p variables:

$$y = a_1X_1 + a_2X_2 + \cdots + a_pX_p \qquad (7)$$

$$\boldsymbol{y} = \boldsymbol{Xa}$$

$\boldsymbol{y}$ = n×1 vector scores on new calculated variable
$\boldsymbol{X}$ = n×p scores of n subjects on old variables
$\boldsymbol{a}$ = p×1 weights that define the linear combination

Keeping in mind the properties of linear transformations, that if we define $\boldsymbol{y} = \boldsymbol{Xa}$, then we know $\boldsymbol{\mu}_y = \boldsymbol{a}'\boldsymbol{\mu}_x$ and $s_y^2 = \boldsymbol{a}'\boldsymbol{Sa}$. So we can test hypotheses about this new variable, e.g., H_0: $\boldsymbol{\mu}_y = \boldsymbol{a}'\boldsymbol{\mu}_x = 0$:

$$\text{t-test on this new variable: } t = \frac{a'\bar{x} - a'\mu_0}{\sqrt{a'Sa/n}} = \frac{a'(\bar{x} - \mu_0)}{\sqrt{a'Sa/n}} \tag{8}$$

Usually, in deriving parameter estimates for models, we talk about LSE (least squares estimation), that is, minimizing errors in data-fitted values, yet alternatively, we could discuss our goals as one of maximizing t, or t^2 (the maximum for t^2 will be the maximum for t, and the squared-term is easier to handle than maximizing |t|):

$$t^2 = \frac{\left(a'(\bar{x} - \mu_0)\right)^2}{a'Sa/n} = \frac{na'(\bar{x} - \mu_0)(\bar{x} - \mu_0)'a}{a'Sa} \tag{9}$$

Now our goal is to find the values in **a** to maximize t^2. We add the side constraint that we maximize t^2 with the condition that the **a** also satisfy $\boldsymbol{a'Sa} = 1$ (making the variance of new variable, the linear combination =1). Doing so normalizes the problem. Without the constraint, we could just choose **a** to make $\boldsymbol{a'Sa}$ real small, so t becomes real big. In addition, the problem now becomes simpler because it's essentially reduced to the goal of maximizing the numerator of t^2, not the whole ratio (we're fixing the denominator in (9) to 1).

To choose **a** that maximizes the numerator, we take the derivative of (9) with respect to **a**, using the LaGrangian multiplier to capture the side condition ($\boldsymbol{a'Sa} - 1 = 0$):

$$[n(\bar{\boldsymbol{x}} - \boldsymbol{\mu}_0)(\bar{\boldsymbol{x}} - \boldsymbol{\mu}_0)' - \lambda \boldsymbol{S}]\boldsymbol{a} = 0$$

$$\text{or: } [n\boldsymbol{S}^{-1}(\bar{\boldsymbol{x}} - \boldsymbol{\mu}_0)(\bar{\boldsymbol{x}} - \boldsymbol{\mu}_0)' - \lambda \boldsymbol{I}]\boldsymbol{a} = 0 \tag{10}$$

This form is recognizable—our solution will be based on the eigensolution (eigenvalues and eigenvectors) of the matrix: $n\boldsymbol{S}^{-1}(\bar{\boldsymbol{x}} - \boldsymbol{\mu}_0)(\bar{\boldsymbol{x}} - \boldsymbol{\mu}_0)'$. The eigenvector: $\boldsymbol{a} = \boldsymbol{S}^{-1}(\bar{\boldsymbol{x}} - \boldsymbol{\mu}_0)$ is the discriminant function. The eigenvalue is the maximized solution for $T^2 = n(\bar{\boldsymbol{x}} - \boldsymbol{\mu}_0)'\boldsymbol{S}^{-1}(\bar{\boldsymbol{x}} - \boldsymbol{\mu}_0)$. That is, this discriminant function (leads to the max t^2) is the "variable" or dimension where the data differ the most from H_0.

Let's look at this a little less technically. Consider the example in the figure below (left), in which p=2, there are equal variances ($\sigma^2_{X1} = \sigma^2_{X2}$), and no covariances or correlations. That is, $\boldsymbol{\Sigma}_1 = \boldsymbol{\Sigma}_2$ = diagonal, scalar matrix, e.g., $\boldsymbol{\Sigma} = \begin{bmatrix} 3.1 & 0 \\ 0 & 3.1 \end{bmatrix}$. In this very special (rarely seen) case, the discriminant function would be the line that connects the 2 groups' centroids (the vector of means, or the means in p-d).

In the 2nd example in the figure (above, right), we've kept the correlation between X_1 and X_2 at 0 as in the 1st example ($\rho_{12} = 0$), but now the σ^2's are unequal ($\sigma^2_{X1} < \sigma^2_{X2}$). In this case, a different linear composite might be more helpful. When we project (think of the lights casting shadows) the distributions of the 2 groups' data onto the axis labeled "a" which connects those 2 group centroids (the multidimensional means), the groups are somewhat separable, but when we project the 2 groups' data onto the axis labeled "b" (which is just X_1), then the data are even more clearly distinguishable.

Let's add one final complication, to be most like real data. We have 2 groups, p=2, different variances ($\sigma_{X1}^2 < \sigma_{X2}^2$) and now also a correlation between the variables ($\rho_{12} > 0$; which would be typical in data). For such data, if we projected the data onto the axis labeled "a," the groups would have some overlap (below, left). But if we project the data onto the axis labeled "b" (below, right), then the groups are maximally distinguished or our discriminant function has been maximized.

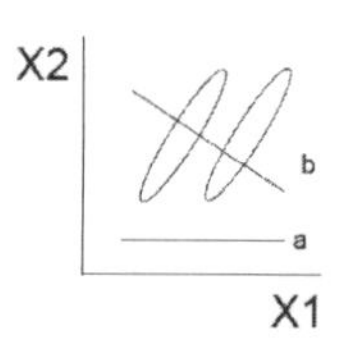

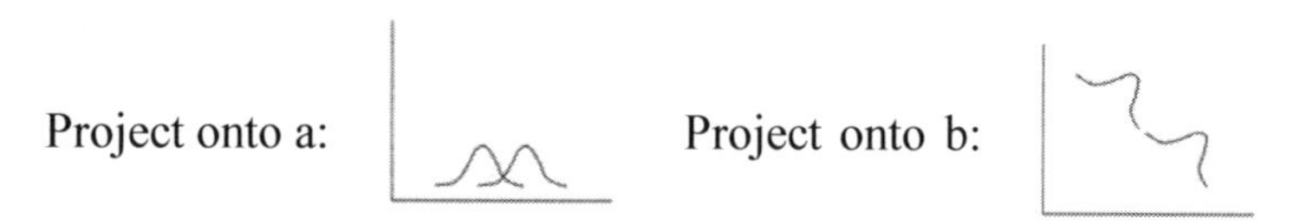

So what is that magical axis "b"—it's no longer the simple case of connecting the centroids. Instead, the discriminant function or axis is the line perpendicular to the line that connects the intersection points of the groups' contours. At the left, the data are in the inner ellipses, and image them radiating out. Connect those dots, as in the right, and the axis that is orthogonal (from the north-northwest to the south-southeast) is the discriminant function or axis:

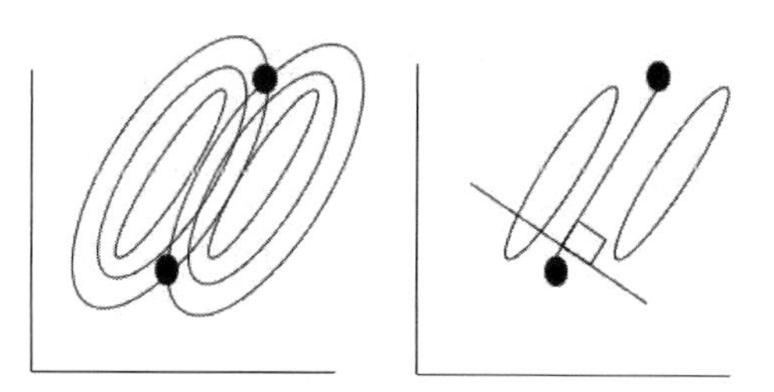

Discriminant analyses bring lots of terminology, such as the discriminant scores being the scores of original data using the linear combination weights, and the discriminant structure coefficients being the correlations of each variable with the new linear combination variables, and more. The terms are often used interchangeably, but they can be different things in different applications. In our application to MANOVA, it's more precise to simply refer to the eigenvectors. So let's return to (a 1-factor) MANOVA. We'll extend the notion of maximizing t^2 to having many (p) variables and having multiple (k) groups. In this more complicated contcxt, to find **a** to maximum value: $\frac{\boldsymbol{a'Ha}}{\boldsymbol{a'Ea}}$, we solve the eigenproblem: $(\boldsymbol{H} - \lambda\boldsymbol{E})\boldsymbol{a} = 0$, or $(\boldsymbol{E}^{-1}\boldsymbol{H} - \lambda\boldsymbol{I})\boldsymbol{a} = 0$.

The solution to obtain the maximum value of $\frac{\boldsymbol{a'Ha}}{\boldsymbol{a'Ea}}$ is the 1st eigenvalue, that is, λ_1. The coefficients that define the new linear combination, $\boldsymbol{a}_1$ are contained in the 1st eigenvector.

Remember that with a rank of s, then s eigenvalues: $\lambda_1 \geq \lambda_2 \geq \cdots \geq \lambda_s$ will be >0, and they are associated with the eigenvectors $\boldsymbol{a}_1, \boldsymbol{a}_2, \ldots, \boldsymbol{a}_s$. Each eigenvector $\boldsymbol{a}_i$ is orthogonal to the previous $\boldsymbol{a}_1 \ldots \boldsymbol{a}_{i-1}$, so each discriminant function maximizes the differences between groups along a dimension that is orthogonal to (uncorrelated with) the previous linear combinations.

So, the ratio of between-group to within-group variance is maximized on this new variable, and this is the discriminant variable. That's why, earlier we said "think of λ_i as $SS_{between}/SS_{within}$" because it is exactly that, for ith discriminant function.

In a study comparing the choices of first running a MANOVA and then either running a series of p ANOVAs or a discriminant analysis, Spector (1977) concluded: 1) If we want linear combination of variables where groups are the most different (and other byproducts of the analysis such as the ability of discriminant analysis to predict and classify new subjects), we'd use discriminant analysis. 2) If we want to look at the nature of group differences, especially in hypothesis testing and experiments, we should do the p ANOVAs. (I would add we should also use follow-up methods 2 and 3 discussed previously as well!)

Finally, let's return to the statement made previously that if the MANOVA R overall test is significant, we can always find at least 1 contrast that's significant in the follow-up testing. That contrast would be defined by the 1st discriminant function. The solution would be $\boldsymbol{c'\bar{x}a}$, where **a** is the eigenvector associated with the first λ_1, and **c** is the discriminant function mean ($\boldsymbol{c} = \boldsymbol{\bar{x}a}$). We would try to interpret $c_1{}'$ group contrasts as well as the $a_1{}'$ variables combinations. Then perhaps we might test simplified (implied), versions of **c** and **a** that might be more theoretically interpretable. For example (from Bird and Hadzi-Pavlovic), when the computer derives very precisely, the 1st contrast vector as **c**' [-.453 .815 -.362] and the 1st variate coefficient vector: **a**' = [-.074 .248 .634 .729], those solutions are difficult to explain. Yet the values in **c** are suggestive of contrasting group 2 vs. groups 1 and 3, as in $\boldsymbol{c}_2{}'$ = [-1/2 1 -1/2], and the values in the derived **a** are suggestive of computing an average on the last 2 variables: $\boldsymbol{a}_2{}'$ = [0 0 1/2 1/2] (again, the a's do not define a contrast, so Σa doesn't have to = 0). Note the original ***c*** and ***a*** vectors are the maximum, precise solution. But if the simplified ***c*** and ***a*** coefficients are also significant, they would be easier to interpret. We would test $H_0\colon \psi_2 = \boldsymbol{c}_2'\boldsymbol{\mu a}_2 = 0$ with $\psi = \boldsymbol{c}_2{}'\boldsymbol{\bar{x}a}_2$. Even if a statistical computing package doesn't allow entering **c** or **a**, it will produce $\boldsymbol{H}, \boldsymbol{E}, \boldsymbol{HE}^{-1}$, etc., and we can compute ψ's ourselves. (SAS enables c's through the contrast statement and a's through the m= option on the manova statement; see the appendix to Chapter 14.)

As final remark, don't forget that common sense is allowed. If 2 or more of the p dependent variables are obviously already correlated (e.g., they are items that measure the same or similar constructs), it is helpful to combine them, usually by averaging them, before running even the MANOVA. The average will be more reliable and stable and that helps to yield more precise findings, and finally the whole set of p is reduced and simplified, both for the model (e.g., with no phenomenon equivalent to multicollinearity) and for interpretation (cf., Bray and Maxwell, 1982, p.35).

SUMMARY

This chapter describes the follow-up statistical tests to conduct once a MANOVA has yielded significant results. Procedures are analogous to contrasts and simple effects in ANOVA, but are somewhat more complicated due to the multivariate nature of the data, p>1.

REFERENCES

1. Bird, Kevin D. (1975), "Simultaneous Contrast Testing Procedures for Multivariate Experiments," *Multivariate Behavioral Research*, 10 (3), 343-351.
2. Bird, Kevin D. and Dusan Hadzi-Pavlovic (1983), "Simultaneous Test Procedures and the Choice of a Test Statistic in Manova," *Psychological Bulletin*, 93 (1), 167-178.

3. Borgen, Fred H. and Mark J. Seling (1978), "Uses of Discriminant Analysis Following Manova: Multivariate Statistics for Multivariate Purposes," *Journal of Applied Psychology*, 63 (6), 689-697.
4. Bray, James H. and Scott E. Maxwell (1982), "Analyzing and Interpreting Significant Manovas," *Review of Educational Research*, 52 (3), 340-367.
5. Harris, Richard J. (2013), *A Primer of Multivariate Statistics*, 3rd ed., NY: Psychology Press.
6. Huberty, Carl J. (1984), "Issues in the Use and Interpretation of Discriminant Analysis," *Psychological Bulletin*, 95 (1), 156-171.
7. Huberty, Carl J. and John D. Morris (1989), "Multivariate Analysis Versus Multiple Univariate Analysis," *Psychological Bulletin*, 105 (2), 302-308.
8. Hummel, Thomas J. and Joseph R. Sligo (1971), "Empirical Comparison of Univariate and Multivariate Analysis of Variance Procedures," *Psychological Bulletin*, 76 (1), 49-57.
9. Ramsey, Philip H. (1980), "Choosing the Most Powerful Pairwise Multiple Comparison Procedure in Multivariate Analysis of Variance," *Journal of Applied Psychology*, 65 (3), 317-326.
10. Spector, Paul E. (1977), "What to Do with Significant Multivariate Effects in Multivariate Analysis of Variance," *Journal of Applied Psychology*, 62 (2), 158-163.
11. Wilkinson, Leland (1975), "Response variable Hypotheses in the Multivariate Analysis of Variance," *Psychological Bulletin*, 82 (3), 408-412.

HOMEWORK

This HW prepares you for your take-home final (and will make you a quant rock star).

Problem 1. What follows is an example of what a computational MANOVA problem might look like on your final exam. An example of the instruction of what you should do with the data follows the listing of the data. Or, if I had requested simply that you "analyze these data," the analyses listed are those I'd be looking for.

For the purposes of this homework assignment, run the data through SAS. Summarize what results are significant via the univariate and multivariate analyses of variance. In both models, follow-up significant effects with appropriate contrasts. Find out what's going on in these data.

The following data are the results of a two-factor factorial experiment. Each of 12 males and each of 12 females was randomly assigned to an experimental condition in which 1 of 3 drugs (especially selected for their expected effect on performance and learning) was administered (i.e., drugs A, B, and C). Two different types of measures of performance (X_1 and X_2) were taken on each subject.

The data follow (borrowed from Morrison, D. F. (1976), *Multivariate Statistical Methods*, 2nd ed., NY: McGraw-Hill, p.190, with permission):

Drug:		A		B		C	
Measure:		X_1	X_2	X_1	X_2	X_1	X_2
Gender:	male	5	6	7	6	21	15
		5	4	7	7	14	11
		9	9	9	12	17	12
		7	6	6	8	12	10
	female	7	10	10	13	16	12
		6	6	8	7	14	9
		9	7	7	6	14	8
		8	10	6	9	10	5

Analyze these data and summarize the results.

CHAPTER 16

MANOVA APPLICATION TO REPEATED MEASURES

Questions to guide your learning:
Q_1: What is the MANOVA approach to within-subjects (or repeated measures) designs?
Q_2: How are its assumptions less restrictive than the univariate ANOVA approach to analyzing such data?

In Chapter 8, we saw the univariate ANOVA model for within-subjects or repeated measures data. In this chapter, we revisit the issue of within-subjects data, but we'll use a MANOVA to analyze them. The univariate ANOVA model requires assumptions that are more restrictive than the MANOVA model for such data, and the ANOVA model is not very robust to violations of those assumptions. Thus, the MANOVA model is a very useful model for analyzing repeated measures or within-subjects data.

> If we have "within-subjects" = "repeated measures" data, it's probably best to analyze them via MANOVA.

Repeated measures data can arise in many situations. A classic example is that a subject's performance is measured over multiple trials:

x_1 = measure subjects' performance at time 1
x_2 = measure performance at time 2
x_3 = measure performance at time 3

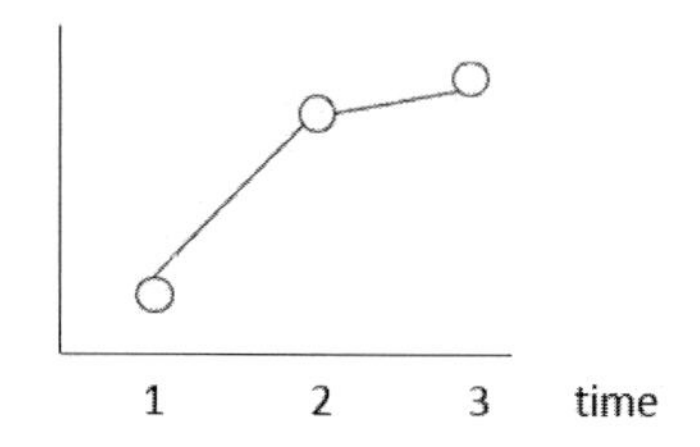

or x_1 = consumers' propensity to buy "X" at time 1
x_2 = consumers' propensity to buy "X" at time 2
x_3 = consumers' propensity to buy "X" at time 3

or x_1 = measure taken in "condition" 1
x_2 = measure taken in "condition" 2
x_3 = measure taken in "condition" 3

The key to these scenarios is that we're measuring the same variable at several points in time. We could characterize the data as multivariate, after all, there are multiple data points on each study participant. Those multiple variables might all be measuring the same construct, but there would be p of them, 1 for each of p time points. In the MANOVA model of Chapter 14 (or for the Hotelling's T^2 in Chapter 13), when we used the term "multivariate," we had in mind that we had captured different variables with different content, variables X, Y, Z. For example, the 3 variables might measure different consumer perceptions about an ad and the brand it features (all measured at the same point in time):

x_1 = attitude toward an ad
y_1 = attitude toward the brand in the ad
z_1 = likelihood to purchase the brand

We could even combine these perspectives, measuring "p" variables at more than one point in time, making the data "doubly multivariate"; e.g., variables "a" and "b" measured at times 1, 2, and 3:

Xa1 = attitude toward the ad, time 1
Xa2 = attitude toward the ad, time 2
Xa3 = attitude toward the ad, time 3
Xb1 = attitude toward the brand, time 1
Xb2 = attitude toward the brand, time 2
Xb3 = attitude toward the brand, time 3

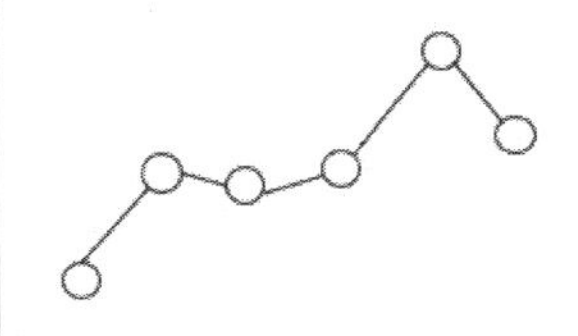

Of course, all the MANOVA model knows is that it's been asked to analyze p variables (anthromorphizing a bit here). It doesn't know whether the content of the p variables is p=3 different variables X, Y, and Z, or variable X measured at times 1, 2, and 3. So, we could certainly use MANOVA to analyze repeated measures data. Let's see why that might be a good idea.

We'll begin with a simple example of 1 variable measured at 3 points in time (the figure below, left). That would give us a repeated measures data set of this form (below, right):

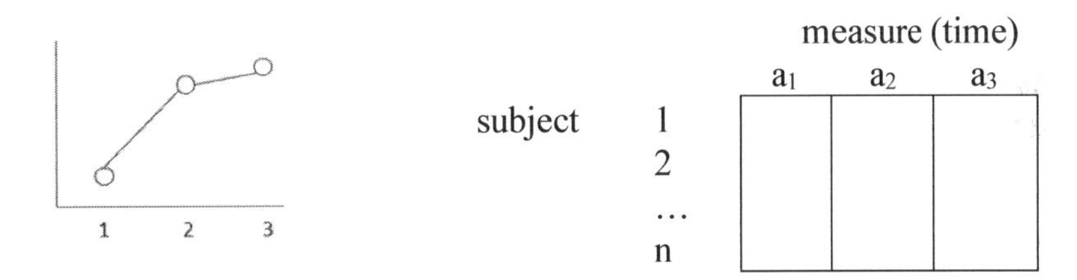

For comparison, recall that in Chapter 8, we had considered the traditional, univariate ANOVA approach to repeated measures. The design just depicted is a 1-factor within-subject design (A×S), and to test the null hypothesis, $H_0: \mu_1 = \mu_2 = \cdots = \mu_a$, we had constructed the following F-test.

Source	df	SS	MS	F
A	a-1	$\sum\sum(\bar{x}_i - \bar{x})^2$	$SS_A/(a-1)$	$MS_A/MS_{A\times S}$
S	n-1	$\sum\sum(\bar{x}_j - \bar{x})^2$	$SS_S/(n-1)$	
A×S = error	(a-1)(n-1)	$\sum\sum(\bar{x}_{ij} - \bar{x}_i - \bar{x}_j + \bar{x})^2$	$SS_{A\times S}/[(a-1)(n-1)]$	
Total	an-1	$\sum\sum(x_{ij} - \bar{x})^2$		

The univariate ANOVA carried several assumptions:

1) That observations (subjects) are independent (required of both between- and within-subjects designs).
2) Observations are normally distributed (also required of both kinds of designs, if we want the p-value associated with the F-test to mean anything).
3) Homogeneity of variances.
    a) For between-subjects designs, each group of subjects has a distribution, each distribution has a variance, σ^2. The variances are assumed to be approximately (statistically) equal.

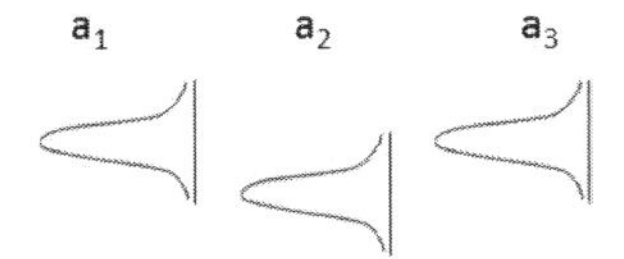

b) For within-subjects designs, that assumption translates into the homogeneity of treatment differences variances ("h.o.t.d.v.").

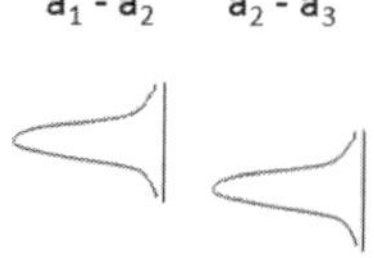

c) The F-test for between-subjects designs is somewhat robust to violations of this assumption (in 3a). But the F-test for within-subject designs is not robust to violations of the h.o.t.d.v. assumption (3b).

4) In addition, for within-subjects designs, we have to worry about the homogeneity of treatment differences' covariances. If we have, for example a=4 conditions: a_1, a_2, a_3, a_4, we would be modeling 3 treatment difference scores, say: $d_1 = a_2 - a_1$, $d_2 = a_3 - a_2$, and $d_3 = a_4 - a_3$. With p=3 difference score variables, *d*'s, we'd have 3 variances to assume are equal: $\sigma_1^2 = \sigma_2^2 = \sigma_3^2 = \sigma$. We'd also have 3 covariances we'd have to assume are equal: $\sigma_{12} = \sigma_{23} = \sigma_{13}$ (or, we could state the equality in terms of correlations, given that $\sigma_{ij} = \rho_{ij}\sigma_i\sigma_j$ and we've already assumed $\sigma_i = \sigma_j$: $\rho_{12} = \rho_{13} = \rho_{23}$). (Notice of course this covariance issue does not arise if a=2. There would be 2 conditions, data in a_1 and a_2 and only 1 difference score, $d_1 = a_1 - a_2$. The d score d_1 has variance σ^2 but no other difference scores with which to covary.)

Together, for within-subjects data, assumptions (3b) and (4) stipulate that the difference scores should have a covariance matrix with the following form, called "compound symmetry":[31]

$$\boldsymbol{\Sigma} = \underbrace{\overbrace{\begin{bmatrix} \sigma^2 & \rho\sigma^2 & \rho\sigma^2 & \cdots & \rho\sigma^2 \\ & \sigma^2 & \rho\sigma^2 & & \rho\sigma^2 \\ & & & & \\ & sym & & & \\ & met & & & \sigma^2 \\ & ric & & & \end{bmatrix}}^{a-1}}_{a-1}$$

For "a" variables, we would model "a-1" difference scores, hence $\boldsymbol{\Sigma}$ is (a-1)×(a-1). The diagonal variances are all equal, and the off-diagonal covariances are all equal.

Specifically, to continue with an example in which the repeated measures design involved the measure of the same variable at times 1 through 4, there would be 3 difference scores, so $\boldsymbol{\Sigma}$ is 3×3:

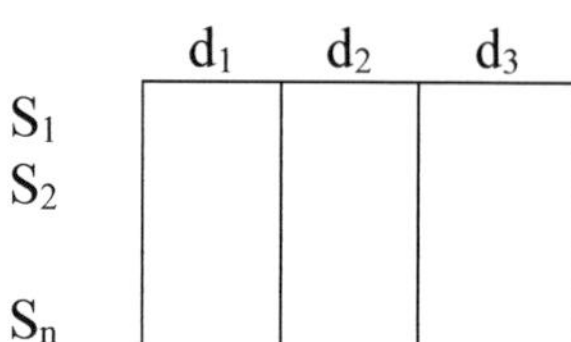

[31] There seems to be a difference (but I cannot find a source reference) in that the h.o.t.d.v assumption, also known as sphericity, requires the variances of the difference scores be equal, and that compound symmetry is a structure of assumptions on the original variables (not the difference scores) that the variances are equal, and the covariances are equal as well.

$$\boldsymbol{\Sigma} = \begin{bmatrix} \sigma^2 & & \\ \rho\sigma^2 & \sigma^2 & \\ \rho\sigma^2 & \rho\sigma^2 & \sigma^2 \end{bmatrix}$$

However it is important to recognize that it is widely known that when variables are measured over time, they usually have a very predictable form that does not particularly resemble compound symmetry. Specifically, data measured at adjacent time points (t=1 and t=2) are usually highly correlated, compared to the correlation between variables measured further separated in time (t=1 and t=3).

$$\text{e.g.,}\quad \boldsymbol{\Sigma} = \begin{array}{c} \begin{matrix} t_1 & t_2 & t_3 & t_4 \end{matrix} \\ \begin{bmatrix} 1.0 & .9 & .8 & .7 \\ & 1.0 & .9 & .8 \\ & & 1.0 & .9 \\ sym. & & & 1.0 \end{bmatrix} \end{array}$$

The variances might be roughly equal, but notice that the upper triangle isn't constant. The strengths of relationships fade with greater durations. For example, educational psychologists can routinely use performance in 3rd grade to predict kids' performance in 4th grade, but to use 3rd grade to predict 5th or 6th is more of a guess. Similarly, meteorologists can predict tomorrow's weather as a function of today's, but to use today's weather to predict next week's weather, starts getting dicey. The point is that for within-subjects designs or repeated measures data, the covariances are more likely to resemble that fading pattern over time, so the assumption of homogeneity of treatment differences covariances won't likely hold, in turn leading us to reject H_0 more often than we should.

MANOVA to the rescue! MANOVA is an alternative method of the analysis of repeated measures because it requires less restrictive assumptions. We'll still take difference scores as the explicit means of comparing a subject's performance in 1 condition to the performance in another: $d_1 = a_2 - a_1,\ d_2 = a_3 - a_2, \ldots\quad d_{p-1} = a_p - a_{p-1}$. However, in a MANOVA, we haven't ever been restrictive as to the form of $\boldsymbol{\Sigma}$. It can take on any form, so if the covariances happen to be equal, that's fine, but if they show the fading phenomenon, or any other inequalities, then it's still fine—MANOVA doesn't care. The variances need not even be equal. MANOVA is merely expecting some general covariance matrix:

$$\boldsymbol{\Sigma} = \begin{bmatrix} \sigma_1^2 & & & sym. \\ \sigma_{21} & \sigma_2^2 & & \\ \sigma_{31} & & \sigma_3^2 & \\ \cdots & & & \\ \sigma_{p-1,1} & & \cdots & \sigma_{p-1}^2 \end{bmatrix}$$

As a brief aside, note that it is still the case that if there is more than 1 group, we need to assume: $\boldsymbol{\Sigma}_1 = \boldsymbol{\Sigma}_2 = \cdots = \boldsymbol{\Sigma}$. For example, if we use a mixed design in which factor A is a within-subjects factor and B is a between-subjects factor, B×(A×S), and say a=4 and b=2. The 4 levels of the within-subjects factor would yield 3 difference scores, and their variances and covariances can take any form in MANOVA. However, we would still need to assume equality of covariance matrices across groups, here the groups are the different samples that have been exposed to different levels of B, the between-subjects factor. That is, we'd still need to assume the homogeneity of covariance matrices between groups, but we need not assume the homogeneity of covariances within groups (between variables).

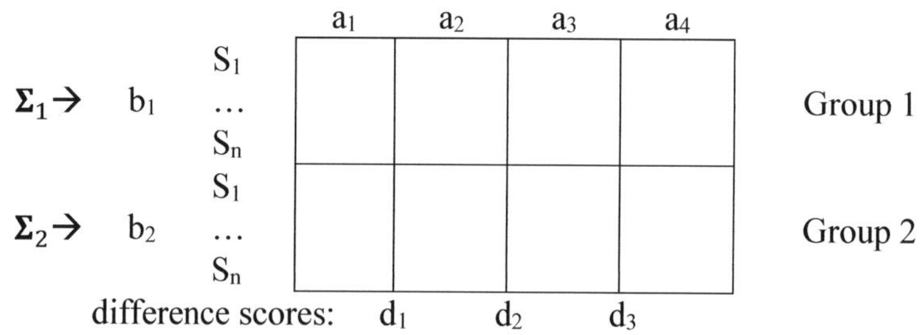

			a_1	a_2	a_3	a_4	
		S_1					
$\Sigma_1 \rightarrow$	b_1	...					Group 1
		S_n					
		S_1					
$\Sigma_2 \rightarrow$	b_2	...					Group 2
		S_n					
difference scores:			d_1	d_2	d_3		

$$\boldsymbol{\Sigma}_1 = \boldsymbol{\Sigma}_2 = \boldsymbol{\Sigma} = \begin{bmatrix} \sigma_1^2 & & \\ \sigma_{21} & \sigma_2^2 & \\ \sigma_{31} & \sigma_{32} & \sigma_3^2 \end{bmatrix}$$

How to Do It

If we'd like to use the MANOVA model to analyze within-subjects or repeated measures data, let's see how it's done. This first example is a simple and typical scenario. Say the repeated measures are of the type where the same variable is measured at several points in time and we'd like to compare the adjacent time points to evaluate growth or progress. We would create difference scores of the "profile" type—they compare pairs of adjacent time points, and slide from times 1 and 2, to 2 and 3, etc. The means in each of the p time points are in the p×1 vector $\bar{\boldsymbol{x}}$, the matrix $\boldsymbol{D}$ has coefficients to create the difference scores resulting in the (p-1)×1 vector $\bar{\boldsymbol{x}}_{diffs}$:

$$\underset{\boldsymbol{D}_{(p-1)\times p}}{\begin{bmatrix} 1 & -1 & 0 & 0 & \dots & 0 & 0 \\ 0 & 1 & -1 & 0 & \dots & 0 & 0 \\ & & & \dots & & & \\ & & & \dots & & & \\ 0 & 0 & 0 & 0 & \dots & 1 & -1 \end{bmatrix}} \underset{\bar{\boldsymbol{x}}}{\begin{bmatrix} \bar{x}_1 \\ \bar{x}_2 \\ \bar{x}_3 \\ \dots \\ \bar{x}_p \end{bmatrix}} = \underset{\bar{\boldsymbol{x}}_{diffs}}{\begin{bmatrix} \bar{x}_1 - \bar{x}_2 \\ \bar{x}_2 - \bar{x}_3 \\ \dots \\ \bar{x}_{p-1} - \bar{x}_p \end{bmatrix}}$$

Any linear combination of variables, as defined in the $\boldsymbol{D}$ matrix, will affect the covariances as well, so we compute: $T^2 = n(\boldsymbol{D}\bar{\boldsymbol{x}})'(\boldsymbol{DSD}')^{-1}(\boldsymbol{D}\bar{\boldsymbol{x}})$, and the F-test is: $F = \frac{n-p+1}{(n-1)(p-1)} T^2$ on (p-1) and (n-p+1) df.

Of course, depending on what the repeated measure's structure is and what the variables mean to us, we might transform **X** using some other matrix for $\boldsymbol{D}$. For example, with the 4 time points just discussed, rather than examining change in adjacent time points, 1 to 2, 2 to 3, etc., we might compare the performance in each condition to the first (or last, for that matter):

$$\underset{(p-1)\times p}{\begin{bmatrix} 1 & -1 & 0 & 0 \\ 1 & 0 & -1 & 0 \\ 1 & 0 & 0 & -1 \end{bmatrix}} \underset{p\times 1}{\begin{bmatrix} \bar{x}_1 \\ \bar{x}_2 \\ \bar{x}_3 \\ \bar{x}_4 \end{bmatrix}} = \underset{(p-1)\times 1}{\begin{bmatrix} \bar{x}_1 \ vs. \ \bar{x}_2 \\ \bar{x}_1 \ vs. \ \bar{x}_3 \\ \bar{x}_1 \ vs. \ \bar{x}_4 \end{bmatrix}}$$

As another example, maybe we have 3 time points, and a_1 a_2 a_3 represent 1 measure taken at p points in time, and maybe we think there might be some non-monotonic effect (a u-shape or an inverted u-shape). Then we would fit a (p-1) degree polynomial:

$$\begin{bmatrix} -1 & 0 & 1 \\ 1 & -2 & 1 \end{bmatrix} \begin{bmatrix} \bar{x}_1 \\ \bar{x}_2 \\ \bar{x}_3 \end{bmatrix} = \begin{bmatrix} linear\ component \\ quadratic\ component \end{bmatrix}$$

(p-1)×1 (p-1)×p p×1

Or perhaps there are 4 experimental conditions, or measures of subjects on 1 attribute made for each of 4 brands, a_1 a_2 a_3 a_4 , and we're interested in comparing the score in each condition (or brand) to the average over all the remaining conditions (or brands):

$$\begin{bmatrix} 1 & -.33 & -.33 & -.33 \\ -.33 & 1 & -.33 & -.33 \\ -.33 & -.33 & 1 & -.33 \end{bmatrix} \begin{bmatrix} \bar{x}_1 \\ \bar{x}_2 \\ \bar{x}_3 \\ \bar{x}_4 \end{bmatrix} = \begin{bmatrix} \bar{x}_1\ vs.rest \\ \bar{x}_2\ vs.rest \\ \bar{x}_3\ vs.rest \end{bmatrix}$$

(p-1)×p p×1 (p-1)×1

As yet another example, a "famous" set of transformations for repeated measures is called Helmert, where the data point in 1 condition is compared to all those that follow, but none taken previously:[32]

$$\begin{bmatrix} 1 & -.25 & -.25 & -.25 & -.25 \\ 0 & 1 & -.33 & -.33 & -.33 \\ 0 & 0 & 1 & -.5 & -.5 \\ 0 & 0 & 0 & 1 & -1 \end{bmatrix} \begin{bmatrix} \bar{x}_1 \\ \bar{x}_2 \\ \bar{x}_3 \\ \bar{x}_4 \\ \bar{x}_5 \end{bmatrix} = \begin{bmatrix} \bar{x}_1\ vs.rest \\ \bar{x}_2\ vs.measures\ taken\ later \\ \bar{x}_3\ vs.measures\ taken\ later \\ \bar{x}_4\ vs.measures\ taken\ later \end{bmatrix}$$

(p-1)×p p×1 (p-1)×1

As a final example, if we run an experimental design with more than 1 repeated measures factor, we might define the usual type of contrasts to get at the main effects and interaction terms. Say we run a 2-factor within-subjects experiment (A×B×S), where a=b=2:

	a_1		a_2	
	b_1	b_2	b_1	b_2
S_1				
S_2				
…				
S_n				

As usual, in each condition, we'd obtain a mean, and those means are now collected into vector format:

$$\bar{x}' = [\bar{x}_{a1b1} \quad \bar{x}_{a1b2} \quad \bar{x}_{a2b1} \quad \bar{x}_{a2b2}]$$

We would use linear combinations that contrasted the a1 vs. a2 cells to examine the main effect for factor A, and so on:

---

[32] Each of these alternative forms of contrasts may be obtained from SAS, using the keyword at the end of the "repeated" statement: profile, contrast, polynomial, mean, helmert.

$$\begin{bmatrix} .5 & .5 & -.5 & -.5 \\ .5 & -.5 & .5 & -.5 \\ .5 & -.5 & -.5 & .5 \end{bmatrix} \begin{bmatrix} \bar{x}_{11} \\ \bar{x}_{12} \\ \bar{x}_{21} \\ \bar{x}_{22} \end{bmatrix} = \begin{bmatrix} \frac{\bar{x}_{11}+\bar{x}_{12}}{2} - \frac{\bar{x}_{21}+\bar{x}_{22}}{2}; A\ main\ effect \\ \frac{\bar{x}_{11}+\bar{x}_{21}}{2} - \frac{\bar{x}_{12}+\bar{x}_{22}}{2}; B\ main\ effect \\ \frac{\bar{x}_{11}+\bar{x}_{22}}{2} - \frac{\bar{x}_{12}+\bar{x}_{21}}{2}; A \times B\ interaction \end{bmatrix}$$

$$(p-1)\times p \qquad p\times 1 \qquad (p-1)\times 1$$

The appendix to this chapter shows the SAS input syntax, and excerpts from the output, of the analysis of some data from a mixed design (2 within-subjects factors and 1 between-subjects factor). The analysis is shown via the univariate ANOVA approach alongside the MANOVA approach. In the example, the approaches differ, and when that happens, greater weight should be placed on the MANOVA results, because the difference is likely attributable to the data not conforming to the univariate assumptions.

Profile Analysis

Sometimes the analysis of repeated measures data using the MANOVA model is conducted via a "profile analysis." The model takes on a particular form which has its appeal but if it doesn't fit the research questions at hand, obviously, just use the MANOVA approach that has already been presented.

The idea underlying a profile analysis is that each of several groups may be described by data collected at multiple points in time. We'll draw the profile line for each group and compare them.

For example, in the design A×(B×S) (recall, A is the between-groups factor, and B the within-groups factor), measures at multiple points in time would be b=1, 2, …, b, and those data would be drawn as a profile, one for each group a=1, 2, …, a.

Here is a tiny example (from Harris, 1985, p.139). The dataset is to the left, the means are to the right, and the profiles are drawn below.

Group	x_1	x_2	x_3
1	3	4	11
1	9	8	10
1	3	9	6
2	10	9	11
2	11	9	10
2	6	9	12

Means:

	x_1	x_2	x_3	
Group 1	5	7	9	7
Group 2	9	9	11	9.667
	7	8	10	8.333

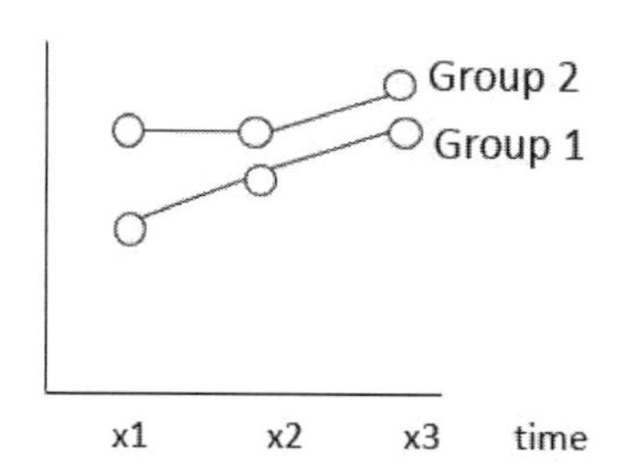

In a profile analysis, there are 3 questions or hypotheses:

1) Are the profiles at different levels (i.e., is there a "group" main effect)?
2) Are the profiles flat (i.e., is there a "time" or repeated measures variable main effect)?
3) Are the profiles parallel (i.e., is there an interaction between the groups and time points or multiple measures)?

The results are easy to obtain. In SAS, we would say: proc glm; class group; model x1 x2 x3 = group; repeated x 3 profile; For this example, those commands yield results: 1) on the group effect, $F_{1,4} = 6.40, p = .0647$, 2) on the effect over the multiple measures x_1-x_3, $F_{2,3} = 1.56, p = .3425$, and 3) for the interaction, $F_{2,3} = 0.28, p = .7745$. Well, all n.s., but what power would we expect of an itty bitty dataset.

SUMMARY

This chapter applied the MANOVA model to within-subjects or repeated measures data. The univariate approach to such data, as presented in Chapter 8, is perfectly acceptable, but that model carries restrictive assumptions and is not very robust to their violations. The MANOVA model approach to the same data requires less restrictive assumptions and therefore will do no worse than, and usually a superior job to the ANOVA model. If, out of curiosity, we were to run an ANOVA and a MANOVA on the same repeated measures dataset, and the results differ, we should trust the results produced by the MANOVA model more.

Go multivariate!

REFERENCES

1. Greenhouse, Samuel W. and Seymour Geisser (1959), "On Methods in the Analysis of Profile Data," *Psychometrika*, 24 (2), 95-112.
2. Greenwald, Anthony G. (1976), "Within-Subjects Designs: To Use or Not To Use?," *Psychological Bulletin*, 83 (2), 314-320.
3. Harris, Richard J. (1985), *A Primer of Multivariate Statistics*, 2nd ed., Academic Press.
4. O'Brien, Ralph G. and Mary Kister Kaiser (1985), "MANOVA Method for Analyzing Repeated Measures Designs: An Extensive Primer," *Psychological Bulletin*, 97 (2), 316-333.
5. Poor, David D. (1973), "Analysis of Variance for Repeated Measures Designs: Two Approaches," *Psychological Bulletin*, 80 (3), 204-209.
6. Algina, James and H. J. Keselman (1997), "Detecting Repeated Measures Effects with Univariate and Multivariate Statistics," *Psychological Methods*, 2 (2), 208-218.
7. Greer, Tammy and William P. Dunlap (1997), "Analysis of Variance with Ipsative Measures," *Psychological Methods*, 2 (2), 200-207.

APPENDIX

```
*This SAS job shows the multivariate approach to analyzing within-subjects or repeated measures data;
*First, compare these commands and the structure of the data (from Poor, 1973) to how we analyzed within-subjects data using the univariate ANOVA model (in Chapter 8). There, the data were strung out in vector form, here there is 1 line per subject.;
data rep2; input s a b1c1 b1c2 b1c3 b2c1 b2c2 b2c3 b3c1 b3c2 b3c3; datalines;
1 1 45 53 60 40 52 57 28 37 46
2 1 35 41 50 30 37 47 25 32 41
3 1 60 65 75 58 54 70 40 47 50
4 2 50 48 61 25 34 51 16 23 35
5 2 42 45 55 30 37 43 22 27 37
6 2 56 60 77 40 39 57 31 29 46
proc print data=rep2; proc glm data=rep2; class a;
model b1c1--b3c3 = a;
repeated b 3, c 3; run;
```

Two notes: 1) in the “repeated” statement, it says that b has 3 levels and c had 3 levels. Factor b is listed first so it is the outer loop, and c is the inner loop whose subscripts change fastest. 2) for the “model” statement, if you’re typing in Word, do not let the double dash -- get transformed into a single dash – .

SAS helpfully provides feedback as to how it interpreted our “repeated” code. We’ll look at the pattern in this table to make sure SAS read in the design as we intended.

The GLM Procedure

Repeated Measures Level Information

Dependent Variable	b1c1	b1c2	b1c3	b2c1	b2c2	b2c3	b3c1	b3c2
Level of b	1	1	1	2	2	2	3	3
Level of c	1	2	3	1	2	3	1	2

The MANOVA begins, presenting results on the within-subjects factors, B and C, and any effects that involve them. We’d report a significant effect for factor B ($F(2,3) = 28.15$, $p=.0114$), a borderline effect for A×B ($F(2,3) = 8.11$, $p=.0617$), etc.:

MANOVA Test Criteria and Exact F Statistics for the Hypothesis of no b Effect

Statistic	Value	F Value	Num DF	Den DF	Pr > F
Wilks' Lambda	0.05059831	28.15	2	3	0.0114
Pillai's Trace	0.94940169	28.15	2	3	0.0114
Hotelling-Lawley Trace	18.76350591	28.15	2	3	0.0114
Roy's Greatest Root	18.76350591	28.15	2	3	0.0114

MANOVA Test Criteria and Exact F Statistics for the Hypothesis of no b*a Effect

Statistic	Value	F Value	Num DF	Den DF	Pr > F
Wilks' Lambda	0.15607081	8.11	2	3	0.0617
Pillai's Trace	0.84392919	8.11	2	3	0.0617
Hotelling-Lawley Trace	5.40734806	8.11	2	3	0.0617
Roy's Greatest Root	5.40734806	8.11	2	3	0.0617

MANOVA Test Criteria and Exact F Statistics for the Hypothesis of no c Effect

Statistic	Value	F Value	Num DF	Den DF	Pr > F
Wilks' Lambda	0.01613663	91.46	2	3	0.0020
Pillai's Trace	0.98386337	91.46	2	3	0.0020
Hotelling-Lawley Trace	60.97082170	91.46	2	3	0.0020
Roy's Greatest Root	60.97082170	91.46	2	3	0.0020

MANOVA Test Criteria and Exact F Statistics for the Hypothesis of no c*a Effect

Statistic	Value	F Value	Num DF	Den DF	Pr > F
Wilks' Lambda	0.56498290	1.15	2	3	0.4247
Pillai's Trace	0.43501710	1.15	2	3	0.4247
Hotelling-Lawley Trace	0.76996506	1.15	2	3	0.4247
Roy's Greatest Root	0.76996506	1.15	2	3	0.4247

MANOVA Test Criteria and Exact F Statistics for the Hypothesis of no b*c Effect

Statistic	Value	F Value	Num DF	Den DF	Pr > F
Wilks' Lambda	0.0007537	331.45	4	1	0.0412
Pillai's Trace	0.9992463	331.45	4	1	0.0412
Hotelling-Lawley Trace	1325.7800000	331.45	4	1	0.0412
Roy's Greatest Root	1325.7800000	331.45	4	1	0.0412

MANOVA Test Criteria and Exact F Statistics for the Hypothesis of no b*c*a Effect

Statistic	Value	F Value	Num DF	Den DF	Pr > F
Wilks' Lambda	0.0004295	581.88	4	1	0.0311
Pillai's Trace	0.9995705	581.88	4	1	0.0311
Hotelling-Lawley Trace	2327.5000000	581.88	4	1	0.0311
Roy's Greatest Root	2327.5000000	581.88	4	1	0.0311

SAS then presents results for the between-subjects effects, here, just factor A:

Tests of Hypotheses for Between Subjects Effects

Source	DF	Type III SS	Mean Square	F Value	Pr > F
a	1	468.166667	468.166667	0.75	0.4348
Error	4	2491.111111	622.777778		

Then, interestingly, SAS provides the within-subjects results again, but here, as a function of using the univariate ANOVA modeling approach (that's handy, makes our comparisons easier, if we wish to do so). By the way, these univariate results are indeed equal to those presented, via the univariate ANOVA model, in the appendix to Chapter 8:

Univariate Tests of Hypotheses for Within Subject Effects

Source	DF	Type III SS	Mean Square	F Value	Pr > F	Adj Pr > F G - G	Adj Pr > F H - F
b	2	3722.333333	1861.166667	63.39	<.0001	0.0003	<.0001
b*a	2	333.000000	166.500000	5.67	0.0293	0.0569	0.0293
Error(b)	8	234.888889	29.361111				

Greenhouse-Geisser Epsilon 0.6476

Huynh-Feldt Epsilon 1.0668

Source	DF	Type III SS	Mean Square	F Value	Pr > F	Adj Pr > F G - G	Adj Pr > F H - F
c	2	2370.333333	1185.166667	89.82	<.0001	<.0001	<.0001
c*a	2	50.333333	25.166667	1.91	0.2102	0.2152	0.2102
Error(c)	8	105.555556	13.194444				

Greenhouse-Geisser Epsilon 0.9171
Huynh-Feldt Epsilon 2.0788

Source	DF	Type III SS	Mean Square	F Value	Pr > F	Adj Pr > F G - G	Adj Pr > F H - F
b*c	4	10.6666667	2.6666667	0.34	0.8499	0.7295	0.8499
b*c*a	4	11.3333333	2.8333333	0.36	0.8357	0.7156	0.8357
Error(b*c)	16	127.1111111	7.9444444				

Greenhouse-Geisser Epsilon 0.5134
Huynh-Feldt Epsilon 1.3258

We have 2 analyses now on the same data. The results (F's and p-values) are presented side-by-side in the following table:

Univariate analysis (from before) effect	F	df's	p-value	Multivariate analysis (now, new) effect	F	df's	p-value
A	.75	1,4	.435	A	.75	1,4	.435
B	63.39	2,8	.000*	B	28.15	2,3	.011*
C	89.82	2,8	.000*	C	91.46	2,3	.002*
A×B	5.67	2,8	.029*	A×B	8.11	2,3	.062 #
A×C	1.91	2,8	.210	A×C	1.15	2,3	.425
B×C	.34	4,16	.850	B×C	331.44	4,1	.041
A×B×C	.36	4,16	.836	A×B×C	581.87	4,1	.031

Comparing these sets of results, there are several things to notice:

1) There is some agreement and some disagreement between the 2 approaches. Both show significant effects for factors B and C. The ANOVA also shows an A×B interaction, which we lose in the MANOVA (it goes to "borderline"). On the other hand, in the MANOVA, we gain the B×C and A×B×C intereactions, neither of which were remotely close to significance in the ANOVA.
2) The F-test for the main effect of factor A is the same in both analyses because it's the between-subjects factor. (The results of A in interactions with either B or C, the within-subjects factors, will vary due to the effects of B and C being within.)
3) Given the qualitatively different results across the univariate and multivariate approaches to repeated measures (for A×B, B×C, and A×B×C), which should we believe? Usually (99% of time) we should tend toward interpreting and reporting the MANOVA over the ANOVA (remember, the repeated measures ANOVA has squirrely assumptions, which, if violated, the model is not robust to, etc.). However, here, for this tiny, "pretend" data set, I would worry that the 1 df for error (in the multivariate run) suggests instability.
4) Note that that implies, in this example, even if we had been hoping for an A×B interaction for theoretical reasons, and we didn't even care about the B×C and A×B×C interactions, we should still present the MANOVA results, in which our favorite interaction, A×B, is n.s. Tough noogies.

The bottom line, which of these analyses should you report? It's your choice. Probably (again 99% of the time), the MANOVA is the appropriate model. In any case, all results must come from the same model (all the univariate results or all the multivariate results; we cannot pick and choose).

Weird tidbit: when you run a repeated measures analysis the univariate way, the computer can run out of memory if the design is too large (e.g., if the number of subjects is big). When the repeated measures analysis is run the multivariate way, it doesn't choke (I've never had it not run). Obviously, the MANOVA program is based on a more efficient algorithm. In any event, when running the MANOVA model, in addition to the multivariate results, it spits out the results from a univariate analysis, so they can be obtained that way.

PROLOGUE

My hope, dear reader, is that you've found this book useful.
My aim had been to present the basics for readers new to ANOVA, as well as more advanced material for readers seeking greater breadth and depth.
I hope the book has served and will serve you well.

Do good. Make your professors proud.

Go in peace.

HOMEWORK SOLUTIONS

to check your understanding and progress

Homework from Chapter 2

Problem 1.

$\bar{x}_1 = 5, \quad \bar{x}_2 = 6, \quad \bar{x}_3 = 10, \quad \bar{x} = 7$

i	j	$(\bar{x}_i - \bar{x})(x_{ij} - \bar{x}_i)$	
1	1	(5-7)(4-5)	= 2
1	2	(5-7)(5-5)	= 0
1	3	(5-7)(6-5)	= -2
1	4	(5-7)(5-5)	= 0
2	1	(6-7)(5-6)	= 1
2	2	(6-7)(6-6)	= 0
2	3	(6-7)(6-6)	= 0
2	4	(6-7)(7-6)	= -1
3	1	(10-7)(9-10)	= -3
3	2	(10-7)(10-10)	= 0
3	3	(10-7)(11-10)	= 3
3	4	(10-7)(10-10)	= 0
			sum: 0

→

$$2\sum_{i=1}^{a}\sum_{j=1}^{n}(\bar{x}_i - \bar{x})(x_{ij} - \bar{x}_i) = 0$$

Problem 2.

a) Treatment means:
$\bar{x}_{A1} = \frac{30}{10} = 3, \quad \bar{x}_{A2} = 4, \quad \bar{x}_{A3} = 3, \quad \bar{x}_{A4} = 7, \quad \bar{x}_{A5} = 5,$ grand mean = 4.4

b) ANOVA Table:

For SS_A, $(3-4.4)^2 + \ldots(4-4.4)^2 + \ldots + (3-4.4)^2 + \ldots + (7-4.4)^2 + \ldots + (5-4.4)^2 = 112.0$
For $SS_{S(A)}$, $(8-3)^2 + \ldots + (3-4)^2 + \ldots + (7-3)^2 + \ldots + (5-7)^2 + \ldots + (8-5)^2 = 262.0$
For SS_{total}, $(8-4.4)^2 + \ldots + (6-4.4)^2 = 374.0$

Source	SS	df	MS	F	p-value
A	112.0	5-1=4	28.000	4.81	0.0026*
S(A)	262.0	a(n-1)=5(9)=45	5.822		
Total	374.0	an-1=49			

*p-value is < 0.05 so yes, the factor is significant, i.e., the groups' means are statistically different (we'd have to do "contrasts" to find out which ones specifically differ from which other ones, more in Chapter 5). The critical F-value may be obtained in excel using =F.INV.RT(.05,4,45), which yields 2.579, thus our F (4.81) exceeds this critical value. You may report in a paper $F_{4,45} = 4.81$, $p < 0.05$. To obtain the more precise p-value in the ANOVA table above, use excel's =F.DIST.RT(4.81,4,45), which yields p = 0.002572.

Problem 3.

a) Group means:

$$\bar{x}_{A1} = \frac{17}{5} = 3.4, \qquad \bar{x}_{A2} = 4.0, \qquad \bar{x}_{A3} = 5.2$$
$$\bar{x} = 4.2 \text{ grand mean}$$

b) ANOVA Table:

For SS_A, $(3.4\text{-}4.2)^2 + (3.4\text{-}4.2)^2 + \ldots(4\text{-}4.2)^2 + \ldots + (5.2\text{-}4.2)^2 = 8.40$
For $SS_{S(A)}$, $(6\text{-}3.4)^2 + (1\text{-}3.4)^2 + \ldots + (7\text{-}5.2)^2 = 40.00$
For SS_{total}, $(6\text{-}4.2)^2 + (1\text{-}4.2)^2 + \ldots + (7\text{-}4.2)^2 = 48.40$

Source	SS	df	MS	F	p-value
A	8.40	3-1=2	4.20	1.26	0.3186*
S(A)	40.00	a(n-1)=3(4)=12	3.33		
Total	48.40	an-1=14			

c) F if not significant. Using Excel's =F.INV.RT(.05,2,12) we learn that the critical F-value is 3.885 and our observed F-value falls short (we cannot reject the null). Using Excel's function =F.DIST.RT(F,df1,df2) we see that the p-value is 0.3186, well above .05. Whatever the group/treatment manipulation was, it was ineffective. None of the 3 group means are statistically different from each other.

Problem 4.

a) F(2,30) at $\alpha = .05$ → 3.316
b) F(3,100) at $\alpha = .01$ → 3.984
c) a = 5, n = 10, $\alpha = .05$, df1=4, df2=5(10-1)=45 → 2.579
d) a = 3, n = 7, $\alpha = .05$, df1=2, df2=3(7-1)=18 → 3.555

To get "a" (above), in Excel, use =F.INV.RT(.05,2,30). (Or you can look up the value in tables in the back of stats books.) Similarly for "b".

For "c" and "d," need to first figure that df will be a-1 and a(n-1).

More often, in Excel, we'd have an F-test and we'd want to get its p-value. Then use =F.DIST.RT(F,df1,df2).

Homework from Chapter 3

Problem 1.

A two-way interaction is the joint or combined effect of two independent variables on the dependent variable. The effect is distinct from (or above and beyond) the two variables' "main effects" on the dependent variable. If an interaction is significant, it means that the effect of independent variable A (or B) on the dependent variable differs depending on the level of the other independent variable B (or A). It means that the effect on the dependent variable depends on the levels of both independent variables, together. Note it is not the effect of one independent variable on the dependent variable (that's a main effect). It's not the effect of one independent variable on the other independent variable (effects are on the dependent variable, one independent variable doesn't affect another).

Problem 2.

Here are the 5 plots:

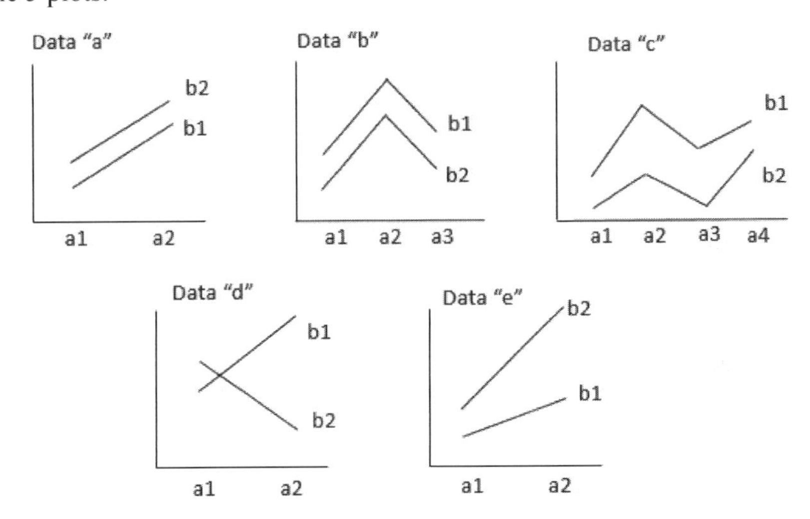

The plots make it clearer faster than the table of means that, in dataset:

a) There is an A main effect and small B main effect.
b) A and B main effects (lines are parallel, so no interaction).
c) A and B main effects and A×B interaction.
d) B main effect and A×B interaction.
e) A and B main effects and and A×B interaction.

Problem 3.

Source	df	SS	MS	F	p-value
A	2	68.074	34.037	54.06	<0.0001
B	2	47.185	23.593	37.47	<0.0001
A×B	4	38.370	9.593	15.24	<0.0001
error	18	11.333	0.630		
total	26	164.963			

All effects are significant (all p-values are <0.05). The means for the main effect for A follow, then those for B, then the interaction. (We'd have to do "contrasts" to see the nature of the mean differences, e.g., it's likely that a2>a3 and a3>a1):

$$\bar{x}_{a1} = 3.000, \quad \bar{x}_{a2} = 6.889, \quad \bar{x}_{a3} = 5.000$$
$$\bar{x}_{b1} = 3.444, \quad \bar{x}_{b2} = 4.778, \quad \bar{x}_{b3} = 6.667$$

	b1	b2	b3
a1	2.333	4.000	2.667
a2	5.000	7.000	8.667
a3	3.000	3.333	8.667

Here's the SAS input:

```
Data HW2; input A B Y; datalines;
a b id  x
1 1 1 2
1 1 2 2
1 1 3 3
1 2 1 5
1 2 2 4
1 2 3 3
1 3 1 3
1 3 2 3
1 3 3 2
2 1 1 6
2 1 2 5
2 1 3 4
2 2 1 7
2 2 2 6
2 2 3 8
2 3 1 9
2 3 2 8
2 3 3 9
3 1 1 2
3 1 2 3
3 1 3 4
3 2 1 4
3 2 2 3
```

```
3 2 3 3
3 3 1 8
3 3 2 9
3 3 3 9
;
proc print data=HW2;
proc glm data=HW2;
class a b; model y = a b a*b / ss3; means a b a*b; run;
```

Problem 4.

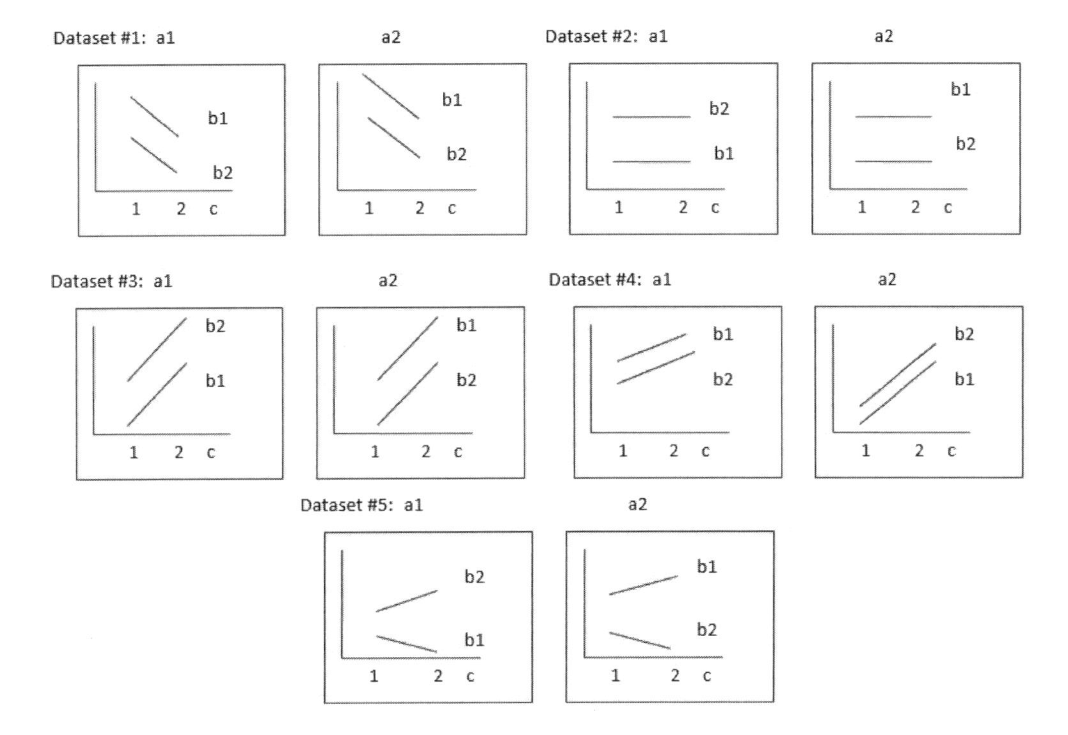

The plots make it clear that, in dataset:

1) A, B, and C main effects.
2) A×B interaction.
3) A×B interaction and C main effect.
4) A×B interaction and A×C interaction, C main effect. Maybe A main effect.
5) 3-way interaction, A×B interaction.

Problem 5.

Source	df	SS	MS	F	p-value
A	2	151.625	75.812	47.88	<.0001*
B	1	65.333	65.333	41.26	<.0001*
C	1	0.083	0.083	0.05	0.8198
A×B	2	31.542	15.771	9.96	0.0004*
A×C	2	19.542	9.771	6.17	0.0050*
B×C	1	0.750	0.750	0.47	0.4957
A×B×C	2	51.125	25.562	16.14	<.0001*
error	36	57.000	1.583		
total	47	377.000			*p<.05

We'd report that the main effects for A and B were significant, as were the A×B, A×C and A×B×C interactions. We'd interpret the means for those effects (they follow).

$$\bar{x}_{a1} = 5.000, \quad \bar{x}_{a2} = 6.062, \quad \bar{x}_{a3} = 9.187$$
$$\bar{x}_{b1} = 7.917, \quad \bar{x}_{b2} = 5.583$$

A×B means (should plot these):

	b1	b2
a1	5.625	4.375
a2	6.625	5.500
a3	11.500	6.875

A×C (plot these also):

	c1	c2
a1	4.250	5.750
a2	6.125	6.000
a3	10.000	8.375

A×B×C (and plot these):

	b1		b2	
	c1	c2	c1	c2
a1	5.50	5.75	3.00	5.75
a2	7.75	5.50	4.50	6.50
a3	11.00	12.00	9.00	4.75

Here's the SAS input:

```
data HW2prob5; input a b c x; datalines;
1 1 1  7
1 1 1  4
1 1 1  5
1 1 1  6
1 1 2  7
1 1 2  5
```

```
1 1 2  5
1 1 2  6
1 2 1  2
1 2 1  4
1 2 1  3
1 2 1  3
1 2 2  7
1 2 2  6
1 2 2  5
1 2 2  5
2 1 1 10
2 1 1  7
2 1 1  6
2 1 1  8
2 1 2  6
2 1 2  5
2 1 2  5
2 1 2  6
2 2 1  4
2 2 1  6
2 2 1  3
2 2 1  5
2 2 2  5
2 2 2  8
2 2 2  7
2 2 2  6
3 1 1 13
3 1 1 10
3 1 1 13
3 1 1  8
3 1 2 12
3 1 2 13
3 1 2 11
3 1 2 12
3 2 1  9
3 2 1  8
3 2 1  9
3 2 1 10
3 2 2  4
3 2 2  6
3 2 2  4
3 2 2  5
proc glm; class a b c; model x = a b c a*b a*c b*c a*b*c / ss3;
means a|b|c;
run;
```

Problem 6.

4 factor factorial: factors A, B, C, and D with #levels a, b, c, d:

$$Y_{ijk\ell m} = \mu + \alpha_i + \beta_j + \gamma_k + \delta_\ell + \alpha\beta_{ij} + \alpha\gamma_{ik} + \alpha\delta_{i\ell} + \beta\gamma_{jk} + \beta\delta_{j\ell} + \gamma\delta_{k\ell}$$
$$+\alpha\beta\gamma_{ijk} + \alpha\beta\delta_{ij\ell} + \beta\gamma\delta_{jk\ell} + \alpha\beta\gamma\delta_{ijk\ell} + \epsilon_{m(ijk\ell)}$$

ANOVA table

Source	df	SS	MS	F
A	(a-1)		MS/df	MS/MS$_{S(ABCD)}$
B	(b-1)			
C	(c-1)			
D	(d-1)			
A×B	(a-1)(b-1)			
A×C	(a-1)(c-1)			
A×D	(a-1)(d-1)			
B×C	(b-1)(c-1)			
B×D	(b-1)(d-1)			
C×D	(c-1)(d-1)			
A×B×C	(a-1)(b-1)(c-1)			
A×B×D	(a-1)(b-1)(d-1)			
A×C×D	(a-1)(c-1)(d-1)			
B×C×D	(b-1)(c-1)(d-1)			
A×B×C×D	(a-1)(b-1)(c-1)(d-1)			
S(ABCD)	abcd(n-1)			
Total	abcdn-1			

Definitional formula for SS$_{ABCD}$?

Recall that for a 2-way interaction, the effect is the cell mean that depends on those 2 factors and we subtract off all "lower order" effects (in this case, 2 main effects and the grand mean):

$$(\alpha\beta)_{ij} = \mu_{ij} - \alpha_i - \beta_j - \mu$$
$$= \mu_{ij} - (\mu_i - \mu) - (\mu_j - \mu) - \mu$$

which simplifies to:

$$(\alpha\beta)_{ij} = \mu_{ij} - \mu_i - \mu_j + \mu$$

And the same thing (conceptually) happened with a 3-way interaction:

$$(\alpha\beta\gamma)_{ijk} = \mu_{ijk} - (\alpha\beta)_{ij} - (\alpha\gamma)_{ik} - (\beta\gamma)_{jk} - \alpha_i - \beta_j - \gamma_k - \mu$$
$$= \mu_{ijk} - (\mu_{ij} - \mu_i - \mu_j + \mu) - (\mu_{ik} - \mu_i - \mu_k + \mu) - (\mu_{jk} - \mu_j - \mu_k + \mu)$$
$$-(\mu_i - \mu) - (\mu_j - \mu) - (\mu_k - \mu) - \mu$$
$$= \mu_{ijk} - \mu_{ij} - \mu_{ik} - \mu_{jk} + \mu_i + \mu_j + \mu_k - \mu$$

Well, guess what, same thing happens with a 4-way (or higher-order) interaction, A×B×C×D:

$$(\alpha\beta\gamma\delta)_{ijk\ell} = \mu_{ijk\ell} - (\alpha\beta\gamma)_{ijk} - (\alpha\beta\delta)_{ij\ell} - (\alpha\gamma\delta)_{ik\ell} - (\beta\gamma\delta)_{jk\ell}$$
$$-(\alpha\beta)_{ij} - (\alpha\gamma)_{ik} - (\alpha\delta)_{i\ell} - (\beta\gamma)_{jk} - (\beta\delta)_{j\ell} - (\gamma\delta)_{k\ell}$$
$$-\alpha_i - \beta_j - \gamma_k - \delta_\ell - \mu$$

$$= \mu_{ijk\ell} - (\mu_{ijk} - \mu_{ij} - \mu_{ik} - \mu_{jk} + \mu_i + \mu_j + \mu_k - \mu)$$
$$-(\mu_{ij\ell} - \mu_{ij} - \mu_{i\ell} - \mu_{j\ell} + \mu_i + \mu_j + \mu_\ell - \mu)$$
$$-(\mu_{ik\ell} - \mu_{ik} - \mu_{i\ell} - \mu_{k\ell} + \mu_i + \mu_k + \mu_\ell - \mu)$$
$$-(\mu_{jk\ell} - \mu_{jk} - \mu_{j\ell} - \mu_{k\ell} + \mu_j + \mu_k + \mu_\ell - \mu)$$
$$-(\mu_{ij} - \mu_i - \mu_j + \mu) - (\mu_{ik} - \mu_i - \mu_k + \mu) - (\mu_{i\ell} - \mu_i - \mu_\ell + \mu)$$
$$-(\mu_{jk} - \mu_j - \mu_k + \mu) - (\mu_{j\ell} - \mu_j - \mu_\ell + \mu) - (\mu_{k\ell} - \mu_k - \mu_\ell + \mu)$$
$$-(\mu_i - \mu) - (\mu_j - \mu) - (\mu_k - \mu) - (\mu_\ell - \mu) - \mu$$

$$= \mu_{ijk\ell} - \mu_{ijk} - \mu_{ij\ell} - \mu_{ik\ell} - \mu_{jk\ell}$$
$$+\mu_{ij} + \mu_{ik} + \mu_{i\ell} + \mu_{jk} + \mu_{j\ell} + \mu_{k\ell}$$
$$-\mu_i - \mu_j - \mu_k - \mu_\ell + \mu$$

So…the definitional formula for SS_{ABCD} follows:

$$SS_{ABCD} = \sum_{i=1}^{a}\sum_{j=1}^{b}\sum_{k=1}^{c}\sum_{\ell=1}^{d}\big(\bar{y}_{ijk\ell} - \bar{y}_{ijk} - \bar{y}_{ij\ell} - \bar{y}_{ik\ell} - \bar{y}_{jk\ell} + \bar{y}_{ij} + \bar{y}_{ik} + \bar{y}_{i\ell} + \bar{y}_{jk} + \bar{y}_{j\ell}$$
$$+ \bar{y}_{k\ell} - \bar{y}_i - \bar{y}_j - \bar{y}_k - \bar{y}_\ell + \bar{y}\big)^2$$

Problem 7.

The completed ANOVA table follows:

source	SS	df	MS	F
A	65.10	1	65.10	21.70
B	9.45	1	9.45	3.15
A×B	13.50	1	13.50	4.50
error	60.00	20	3.00	
total	148.05	23		

Because the F-tests for the price main effect and for the interaction are >4.35 (the critical value), these are the effects that are present in the data (i.e., statistically significant). Thus, we would reject hypotheses i and iii. Interpretations? The mean for “no price increase” is significantly greater than the mean for “price increase”, but this is tempered by the presence of a significant interaction. The effect that price had on the ratings depended also on package color (and vice versa). In particular, the combination of “no price increase” and “red” package seemed especially effective in resulting in higher ratings, and the combination of “price increase” and “blue” seemed to be especially bad. If you had to choose a package color, which would you choose? The appropriate answer would be: it does not matter. Statistically, the package color yielded no differences. However, the mean for the blue packaging was slightly higher, even if insignificant, so you might go with blue.

Homework from Chapter 5

Problem 1.

		control a1	No celeb Cognitive a2	No celeb Emotional a3	Celebrity Cognitive a4	Celebrity Emotional a5
Control vs. exp'l	c1	1	-.25	-.25	-.25	-.25
Celeb vs. none	c2	0	.5	.5	-.5	-.5
Cogn vs. emot no celeb	c3	0	1	-1	0	0
Cogn vs. emot w celeb	c4	0	0	0	1	-1

Testing whether the 4 comparisons are mutually orthogonal:

c_1 & c_2: (1)(0) +(-.25)(.5) +(-.25)(.5) +(-.25)(-.5) +(-.25)(-.5) =0

c_1 & c_3: (1)(0) +(-.25)(1) +(-.25)(-1) +(-.25)(0) +(-.25)(0) =0

c_1 & c_4: (1)(0) +(-.25)(0) +(-.25)(0) +(-.25)(1) +(-.25)(-1) =0

c_2 & c_3: (0)(0) +(.5)(1) +(.5)(-1) +(-.5)(0) +(-.5)(0) =0

c_2 & c_4: (0)(0) +(.5)(0) +(.5)(0) +(-.5)(1) +(-.5)(-1) =0

c_3 & c_4: (0)(0) +(1)(0) +(-1)(0) +(0)(1) +(0)(-1) =0

Problem 2.

a=5 trend coefficients:

a=5 ci's	$\bar{y}_1$	$\bar{y}_2$	$\bar{y}_3$	$\bar{y}_4$	$\bar{y}_5$
linear	-2	-1	0	1	2
quad	2	-1	-2	-1	2
cubic	-1	2	0	-2	1
quartic	1	-4	6	-4	1

ψ_1 & ψ_2: (-2)(2) + (-1)(-1) + (0)(-2) + (1)(-1) + (2)(2) = 0

ψ_1 & ψ_3: (-2)(-1) + (-1)(2) + (0)(0) + (1)(-2) + (2)(1) = 0

ψ_1 & ψ_4: (-2)(1) + (-1)(-4) + (0)(6) + (1)(-4) + (2)(1) = 0

ψ_2 & ψ_3: (2)(-1) + (-1)(2) + (-2)(0) + (-1)(-2) + (2)(1) = 0

ψ_2 & ψ_4: (2)(1) + (-1)(-4) + (-2)(6) + (-1)(-4) + (2)(1) = 0

ψ_3 & ψ_4: (-1)(1) + (2)(-4) + (0)(6) + (-2)(-4) + (1)(1) = 0

→Yes, since ALL pairs of contrasts are orthogonal, the set is "mutually orthogonal."

Problem 3.

modality:	paper	paper	tablet	tablet	tablet
order:	varied	varied	same	varied	varied
array:	column	scattered	-	-	-
rate:			3 sec.	3 sec.	1 sec.

ci's	$\bar{y}_1$	$\bar{y}_2$	$\bar{y}_3$	$\bar{y}_4$	$\bar{y}_5$
ψ_1	1	1	0	-1	-1
ψ_2	1	-1	0	0	0
ψ_3	0	0	0	1	-1
ψ_4	0	0	1	-1	0
ψ_5	-1	-1	4	-1	-1

ψ_5 is just one possibility. It is orthogonal to ψ_1, ψ_2, ψ_3. (The set of 5 contrasts is not "mutually orthogonal" because ψ_4 is not orthogonal to ψ_1 or ψ_3.)

Interpret as contrast of all groups where words are presented in varied order (groups 1, 2, 4, & 5) to group 3, where words are presented in constant order.

Homework from Chapter 6

Problem 1.

TM_i = factor "test market (city)"	a = 3, randomly selected
Ad_j = factor "ad"	b = 5, fixed
S(TM×A) = "monthly measure"	n = 2, random

	3	5	2	
	R	F	R	
source	i	j	k	EMS
TM_i	1	5	2	$10\sigma^2_{Testmarket} + \sigma^2_\epsilon$
Ad_j	3	0	2	$6\theta^2_{Ad} + 2\sigma^2_{TM\times Ad} + \sigma^2_\epsilon$
$(TM \times Ad)_{ij}$	1	0	2	$2\sigma^2_{TM\times Ad} + \sigma^2_\epsilon$
$\epsilon_{k(ij)}$	1	1	1	σ^2_ϵ

1) Test for main effect of test markets (cities):
   $F = MS_{TM}/MS_{S(TMA)}$ = 50/15 = 3.333
   compare to critical value* $F_{.05,2,15}$ = 3.682
   or p-value# = .063

2) Test for main effect of ads:
   $F = MS_{Ad}/MS_{TM\times A}$ = 30/20 = 1.500
   critical value* $F_{.05,4,8}$ = 3.838
   p-value# = .289

3) Test for interaction:
   $F = MS_{TM\times A}/MS_{S(TM\times A)}$ = 20/15 = 1.333
   critical value* $F_{.05,8,15}$ = 2.641
   p-value# = .300

   *Got those critical values in Excel using =F.INV.RT(.05,df1,df2).
   #Got the p-values in Excel using =F.DIST.RT(F,df1,df2).

So for all that fuss, nothing was significant. Not really the point of a HW exercise though, oui? ☺

Problem 2. factors A and B are fixed, factors C and D are random

effect	a F i	b F j	c R k	d R ℓ	n R m	EMS
α_i or A_i	0	b	c	d	n	$bcdn\theta_A^2 + bdn\sigma_{AC}^2 + bcn\sigma_{AD}^2 + bn\sigma_{ACD}^2 + \sigma_\epsilon^2$
B_j	a	0	c	d	n	$acdn\theta_B^2 + adn\sigma_{BC}^2 + acn\sigma_{BD}^2 + an\sigma_{BCD}^2 + \sigma_\epsilon^2$
C_k	a	b	1	d	n	$abdn\sigma_C^2 + abn\sigma_{CD}^2 + \sigma_\epsilon^2$
D_ℓ	a	b	c	1	n	$abcn\sigma_D^2 + abn\sigma_{CD}^2 + \sigma_\epsilon^2$
AB_{ij}	0	0	c	d	n	$cdn\theta_{AB}^2 + dn\sigma_{ABC}^2 + cn\sigma_{ABD}^2 + n\sigma_{ABCD}^2 + \sigma_\epsilon^2$
AC_{ik}	0	b	1	d	n	$bdn\sigma_{AC}^2 + bn\sigma_{ACD}^2 + \sigma_\epsilon^2$
$AD_{i\ell}$	0	b	c	1	n	$bcn\sigma_{AD}^2 + bn\sigma_{ACD}^2 + \sigma_\epsilon^2$
BC_{jk}	a	0	1	d	n	$adn\sigma_{BC}^2 + an\sigma_{BCD}^2 + \sigma_\epsilon^2$
$BD_{j\ell}$	a	0	c	1	n	$acn\sigma_{BD}^2 + an\sigma_{BCD}^2 + \sigma_\epsilon^2$
$CD_{k\ell}$	a	b	1	1	n	$abn\sigma_{CD}^2 + \sigma_\epsilon^2$
ABC_{ijk}	0	0	1	d	n	$dn\sigma_{ABC}^2 + n\sigma_{ABCD}^2 + \sigma_\epsilon^2$
$ABD_{ij\ell}$	0	0	c	1	n	$cn\sigma_{ABD}^2 + n\sigma_{ABCD}^2 + \sigma_\epsilon^2$
$ACD_{ik\ell}$	0	b	1	1	n	$bn\sigma_{ACD}^2 + \sigma_\epsilon^2$
$BCD_{jk\ell}$	0	1	1	1	n	$an\sigma_{BCD}^2 + \sigma_\epsilon^2$
$ABCD_{ijk\ell}$	0	0	1	1	n	$n\sigma_{BCD}^2 + \sigma_\epsilon^2$
$\epsilon_{m(ijk\ell)}$	1	1	1	1	1	σ_ϵ^2

F-tests:
MS_A/? → F' (will need a pseudo F test, no natural error term)
MS_B/? → F' (need a pseudo F)
MS_C/MD_{CD} and MS_D/MS_{CD}
MS_{AB}/? → F' (need a pseudo F)
MS_{AC}/MS_{ACD} and MS_{AD}/MS_{ACD}
MS_{BC}/MS_{BCD} and MS_{BD}/MS_{BCD}
$MS_{CD}/MS_{S(ABCD)}$
MS_{ABC}/MS_{ABCD} and MS_{ABD}/MS_{ABCD}
$MS_{ACD}/MS_{S(ABCD)}$ and $MS_{BCD}/MS_{S(ABCD)}$ and $MS_{ABCD}/MS_{S(ABCD)}$

To test A, set up a pseudo F-test: MS_A vs. ($MS_{AC} + MS_{AD} - MS_{ACD}$);
to test B, compare MS_B vs. ($MS_{BC} + MS_{BD} - MS_{BCD}$);
and to test A×B, compare MS_{AB} vs. ($MS_{ABC} + MS_{ABD} - MS_{ABCD}$).

Problem 3. SAS proc glm statements:

```
*for problem 1;
proc glm data = one;
class tm ad;   model  x = tm  ad  tm*ad;
```

```
*Tests for tm and tm*ad will be correct since SAS will use s(tmad) as the error term, but SAS
will also incorrectly use s(tmad) as the error term to test ad, so need to include explicit test
statement for ad. It follows;
test h=ad e=tm*ad;
means  tm ad tm*ad; run;

*for problem 2;
proc glm data = two;
class  a  b  c  d;
model  x = a b c d   a*b  a*c  a*d  b*c  b*d  c*d
          a*b*c  a*b*d  a*c*d  b*c*d  a*b*c*d;
test h = c d              e = c*d;
test h = a*c a*d          e = a*c*d;
test h = b*c b*d          e = b*c*d;
test h = a*b*c a*b*d      e = a*b*c*d;
*The tests for c*d, a*c*d, b*c*d, a*b*c*d will already be properly computed against the
          s(abcd) term.  Will need to compute pseudo f's by hand for a, b, and a*b;
means  a  b  c  d  a*b  a*c  a*d  b*c  b*d  c*d
        a*b*c  a*b*d  a*c*d  b*c*d  a*b*c*d; *or shorter, say; means a|b|c|d; run;
```

Problem 4 extra credit.

	a R i	b R j	c R k	n R ℓ	EMS
α_i	1	b	c	n	$bcn\sigma_A^2 + cn\sigma_{AB}^2 + bn\sigma_{AC}^2 + n\sigma_{ABC}^2 + +\sigma_\epsilon^2$
β_j	a	1	c	n	$acn\sigma_B^2 + cn\sigma_{AB}^2 + an\sigma_{BC}^2 + n\sigma_{ABC}^2 + \sigma_\epsilon^2$
γ_k	a	b	1	n	$abn\sigma_C^2 + bn\sigma_{AC}^2 + an\sigma_{BC}^2 + n\sigma_{ABC}^2 + \sigma_\epsilon^2$
$(\alpha\beta)_{ij}$	1	1	c	n	$cn\sigma_{AB}^2 + n\sigma_{ABC}^2 + \sigma_\epsilon^2$
$(\alpha\gamma)_{ik}$	1	b	1	n	$bn\sigma_{AC}^2 + n\sigma_{ABC}^2 + \sigma_\epsilon^2$
$(\beta\gamma)_{jk}$	a	1	1	n	$an\sigma_{BC}^2 + n\sigma_{ABC}^2 + \sigma_\epsilon^2$
$(\alpha\beta\gamma)_{ijk}$	1	1	1	n	$n\sigma_{ABC}^2 + \sigma_\epsilon^2$
$\epsilon_{l(ijk)}$	1	1	1	1	σ_ϵ^2

F-tests:

A	$F' = MS_A/(MS_{AB} + MS_{AC} - MS_{ABC})$
B	$F' = MS_B/(MS_{AB} + MS_{BC} - MS_{ABC})$
C	$F' = MS_C/(MS_{AC} + MS_{BC} - MS_{ABC})$
A×B	$F = MS_{AB}/MS_{ABC}$
A×C	$F = MS_{AC}/MS_{ABC}$
B×C	$F = MS_{BC}/MS_{ABC}$
A×B×C	$F = MS_{ABC}/MS_{S(ABC)}$

Homework from Chapter 8

Problem 1. The structure of the ANOVA table for the design A×(B×C×D×S)

Source	df	MS error term in F-ration
A	(a-1) = 5-1	S/A
B	(b-1) = 4-1	B×S(A)
C	(c-1) = 2-1	C×S(A)
D	(d-1) = 2-1	D×S(A)
A×B	(a-1)(b-1)	B×S(A)
A×C	(a-1)(c-1)	C×S(A)
A×D	(a-1)(d-1)	D×S(A)
B×C	(b-1)(c-1)	B×C×S(A)
B×D	(b-1)(d-1)	B×D×S(A)
C×D	(c-1)(d-1)	C×D×S(A)
A×B×C	(a-1)(b-1)(c-1)	B×C×S(A)
A×B×D	(a-1)(b-1)(d-1)	B×D×S(A)
A×C×D	(a-1)(c-1)(d-1)	C×D×S(A)
B×C×D	(b-1)(c-1)(d-1)	B×C×D×S(A)
A×B×C×D	(a-1)(b-1)(c-1)(d-1)	B×C×D×S(A)
S(A)	a(n-1) = 5(16-1)	-
B×S(A)	a(b-1)(n-1)	-
C×S(A)	a(c-1)(n-1)	-
D×S(A)	a(d-1)(n-1)	-
B×C×S(A)	a(b-1)(c-1)(n-1)	-
B×D×S(A)	a(b-1)(d-1)(n-1)	-
C×D×S(A)	a(c-1)(d-1)(n-1)	-
B×C×D×S(A)	a(b-1)(c-1)(d-1)(n-1)	-

Homework from Chapter 12

Problem 1. Table of numbers; allows us to write math and stats equations more succinctly. It is a little like learning a foreign language in that it gets easier the more times you're exposed to the material.

Problem 2.

$$A+B=\begin{bmatrix}1&2&1\\0&3&-1\\2&1&4\end{bmatrix}+\begin{bmatrix}3&2&1\\2&3&1\\-1&2&3\end{bmatrix}=\begin{bmatrix}4&4&2\\2&6&0\\1&3&7\end{bmatrix}$$

$$A-2B=\begin{bmatrix}1&2&1\\0&3&-1\\2&1&4\end{bmatrix}-2\begin{bmatrix}3&2&1\\2&3&1\\-1&2&3\end{bmatrix}$$
$$=\begin{bmatrix}1&2&1\\0&3&-1\\2&1&4\end{bmatrix}-\begin{bmatrix}6&4&2\\4&6&2\\-2&4&6\end{bmatrix}$$
$$=\begin{bmatrix}-5&-2&-1\\-4&-3&-3\\0&-3&-2\end{bmatrix}$$

$$A'+B=\begin{bmatrix}1&2&1\\0&3&-1\\2&1&4\end{bmatrix}'+\begin{bmatrix}3&2&1\\2&3&1\\-1&2&3\end{bmatrix}$$
$$=\begin{bmatrix}1&0&2\\2&3&1\\1&-1&4\end{bmatrix}+\begin{bmatrix}3&2&1\\2&3&1\\-1&2&3\end{bmatrix}$$
$$=\begin{bmatrix}4&2&3\\4&6&2\\0&1&7\end{bmatrix}$$

$A+C$ is not defined; A is 3×3 and C is 3×2, so they're not conformable.

$$(A+B)'=\begin{bmatrix}4&4&2\\2&6&0\\1&3&7\end{bmatrix}'=\begin{bmatrix}4&2&1\\4&6&3\\2&0&7\end{bmatrix}$$

$$(2A'-B)'=\left(2\begin{bmatrix}1&0&2\\2&3&1\\1&-1&4\end{bmatrix}-\begin{bmatrix}3&2&1\\2&3&1\\-1&2&3\end{bmatrix}\right)'$$
$$=\left(\begin{bmatrix}2&0&4\\4&6&2\\2&-2&8\end{bmatrix}-\begin{bmatrix}3&2&1\\2&3&1\\-1&2&3\end{bmatrix}\right)'$$
$$=\begin{bmatrix}-1&-2&3\\2&3&1\\3&-4&5\end{bmatrix}'$$
$$=\begin{bmatrix}-1&2&3\\-2&3&-4\\3&1&5\end{bmatrix}$$

Problem 3.

$$PQ = \begin{bmatrix} 4 & 2 & -1 \\ 2 & 3 & -1 \\ -1 & -1 & 1 \end{bmatrix} \begin{bmatrix} 1 & 2 & 3 & 4 \\ 1 & 1 & 1 & 0 \\ 1 & 2 & 2 & 1 \end{bmatrix} = \begin{bmatrix} 5 & 8 & 12 & 15 \\ 4 & 5 & 7 & 7 \\ -1 & -1 & -2 & -3 \end{bmatrix}$$

$$PQR = \begin{bmatrix} 5 & 8 & 12 & 15 \\ 4 & 5 & 7 & 7 \\ -1 & -1 & -2 & -3 \end{bmatrix} \begin{bmatrix} 1 & -2 \\ 0 & -2 \\ 3 & -1 \\ 2 & -1 \end{bmatrix} = \begin{bmatrix} 71 & -53 \\ 39 & -32 \\ -13 & 9 \end{bmatrix}$$

QR' is not defined; Q is 3×4 and R' is 2×4, so they're not conformable.

$$yx' = \begin{bmatrix} 0 \\ 1 \\ -1 \end{bmatrix} [1 \quad 2 \quad 3] = \begin{bmatrix} 0 & 0 & 0 \\ 1 & 2 & 3 \\ -1 & -2 & -3 \end{bmatrix}$$

$$x'y = [1 \quad 2 \quad 3] \begin{bmatrix} 0 \\ 1 \\ -1 \end{bmatrix} = -1$$

$$x'Px = [1 \quad 2 \quad 3] \begin{bmatrix} 4 & 2 & -1 \\ 2 & 3 & -1 \\ -1 & -1 & 1 \end{bmatrix} \begin{bmatrix} 1 \\ 2 \\ 3 \end{bmatrix} = [5 \quad 5 \quad 0] \begin{bmatrix} 1 \\ 2 \\ 3 \end{bmatrix} = 15$$

$$x'Py = [1 \quad 2 \quad 3] \begin{bmatrix} 4 & 2 & -1 \\ 2 & 3 & -1 \\ -1 & -1 & 1 \end{bmatrix} \begin{bmatrix} 0 \\ 1 \\ -1 \end{bmatrix} = [5 \quad 5 \quad 0] \begin{bmatrix} 0 \\ 1 \\ -1 \end{bmatrix} = 5$$

$$P(x+y) = \begin{bmatrix} 4 & 2 & -1 \\ 2 & 3 & -1 \\ -1 & -1 & 1 \end{bmatrix} \begin{bmatrix} 1 \\ 3 \\ 2 \end{bmatrix} = \begin{bmatrix} 8 \\ 9 \\ -2 \end{bmatrix}$$

Problem 4.

Columns sums of X:

$$j'X = \left[\sum_i x_{i1} \quad \sum_i x_{i2} \quad \cdots \quad \sum_i x_{in} \right]$$

Row sums of X:

$$Xj = \begin{bmatrix} \sum_j x_{1j} \\ \sum_j x_{2j} \\ \cdots \\ \sum_j x_{nj} \end{bmatrix}$$

$$WX = \begin{bmatrix} j'X \\ j'X \\ \dots \\ j'X \end{bmatrix} \qquad diag(WX) = \begin{bmatrix} \sum_i x_{i1} & 0 & \dots & 0 \\ 0 & \sum_i x_{i2} & \dots & 0 \\ \dots & & & \\ 0 & 0 & & \sum_i x_{in} \end{bmatrix}$$

$$XW = [Xj \quad Xj \quad \dots \quad Xj] \qquad diag(XW) = \begin{bmatrix} \sum_j x_{1j} & 0 & \dots & 0 \\ 0 & \sum_j x_{2j} & \dots & 0 \\ \dots & & & \\ 0 & 0 & & \sum_j x_{nj} \end{bmatrix}$$

$$\underset{1\times n}{} \qquad \underset{n\times 1}{}$$

$$j'j = [1 \quad 1 \quad \dots \quad 1] \begin{bmatrix} 1 \\ 1 \\ \dots \\ 1 \end{bmatrix} = n$$

n×1 vector of n's

$$Wj = \begin{bmatrix} 1 & 1 & \dots & 1 \\ 1 & 1 & \dots & 1 \\ \dots & & & \\ 1 & 1 & \dots & 1 \end{bmatrix} \begin{bmatrix} 1 \\ 1 \\ \dots \\ 1 \end{bmatrix} = \begin{bmatrix} n \\ n \\ \dots \\ n \end{bmatrix}$$

n×n matrix of n's

$$W^2 = \begin{bmatrix} 1 & 1 & \dots & 1 \\ 1 & 1 & \dots & 1 \\ \dots & & & \\ 1 & 1 & \dots & 1 \end{bmatrix} \begin{bmatrix} 1 & 1 & \dots & 1 \\ 1 & 1 & \dots & 1 \\ \dots & & & \\ 1 & 1 & \dots & 1 \end{bmatrix} = \begin{bmatrix} n & n & \dots & n \\ n & n & \dots & n \\ \dots & & & \\ n & n & \dots & n \end{bmatrix}$$

Problem 5.

A matrix X is orthogonal if $X'X = XX'$ is diagonal; orthonormal if $X'X = I$.

Problem 6.

A A'	C C'	E E'
(5×2)(2×5) = 5×5	(3×3)(3×3) = 3×3	(3×1)(1×3) = 3×3

$$\underset{(1\times4)}{B'}\underset{(4\times1)}{B} = 1\times1 \text{ (scalar)} \qquad \underset{(3\times5)}{D'}\underset{(5\times3)}{D} = 3\times3$$

A matrix "P" can premultiply A if it is (r×5)

$$\underset{(r\times5)}{P}\;\underset{(5\times2)}{A} = (r\times2)$$

A matrix "P" can postmultiply A if it is (5×c)

$$\underset{(5\times2)}{A}\;\underset{(2\times c)}{P} = (5\times c)$$

A matrix "P" can premultiply B if it is (r×4) (such as B')

$$\underset{(r\times4)}{P}\;\underset{(4\times1)}{B} = (r\times1)$$

A matrix "P" can postmultiply B if it is (1×c)

$$\underset{(4\times1)}{B}\;\underset{(1\times c)}{P} = (4\times c)$$

Given the covariance matrix S:

$$\begin{bmatrix} 9 & 10 & -12 \\ 10 & 25 & 12 \\ -12 & 12 & 16 \end{bmatrix}$$

$$D_S = \begin{bmatrix} 9.0 & 0 & 0 \\ 0 & 25.0 & 0 \\ 0 & 0 & 16.0 \end{bmatrix} \quad D_S^{1/2} = \begin{bmatrix} 3.0 & 0 & 0 \\ 0 & 5.0 & 0 \\ 0 & 0 & 4.0 \end{bmatrix} \quad D_S^{-1/2} = \begin{bmatrix} 1/3 & 0 & 0 \\ 0 & 1/5 & 0 \\ 0 & 0 & 1/4 \end{bmatrix}$$

$$R = D_S^{-1/2} S D_S^{-1/2} = \begin{bmatrix} 1/3 & 0 & 0 \\ 0 & 1/5 & 0 \\ 0 & 0 & 1/4 \end{bmatrix} \begin{bmatrix} 9 & 10 & -12 \\ 10 & 25 & 12 \\ -12 & 12 & 16 \end{bmatrix} \begin{bmatrix} 1/3 & 0 & 0 \\ 0 & 1/5 & 0 \\ 0 & 0 & 1/4 \end{bmatrix}$$

$$= \begin{bmatrix} 9/3 & 10/3 & -12/3 \\ 10/5 & 25/5 & 12/5 \\ -12/4 & 12/4 & 16/4 \end{bmatrix} \begin{bmatrix} 1/3 & 0 & 0 \\ 0 & 1/5 & 0 \\ 0 & 0 & 1/4 \end{bmatrix} = \begin{bmatrix} 9/9 & 10/15 & -12/12 \\ 10/15 & 25/25 & 12/20 \\ -12/12 & 12/20 & 16/4 \end{bmatrix}$$

$$R = \begin{bmatrix} 1.000 & 0.667 & -1.000 \\ & 1.000 & 0.600 \\ sym. & & 1.000 \end{bmatrix}$$

$$A'A = \begin{bmatrix} 1 & 0 & 6 & 5 \\ 2 & 3 & 3 & 4 \\ 3 & 2 & 44 & 22 \end{bmatrix} \begin{bmatrix} 1 & 2 & 3 \\ 0 & 3 & 2 \\ 6 & 3 & 44 \\ 5 & 4 & 22 \end{bmatrix} = \begin{bmatrix} 62 & 40 & 377 \\ 40 & 38 & 232 \\ 377 & 232 & 2433 \end{bmatrix}$$

$$tr(A'A) = 2533$$

$$AA' = \begin{bmatrix} 1 & 2 & 3 \\ 0 & 3 & 2 \\ 6 & 3 & 44 \\ 5 & 4 & 22 \end{bmatrix} \begin{bmatrix} 1 & 0 & 6 & 5 \\ 2 & 3 & 3 & 4 \\ 3 & 2 & 44 & 22 \end{bmatrix} = \begin{bmatrix} 14 & 12 & 144 & 79 \\ 12 & 13 & 97 & 56 \\ 144 & 97 & 1981 & 1010 \\ 79 & 56 & 1010 & 525 \end{bmatrix}$$

$$tr(AA') = 2533$$

$tr(A'A) = tr(AA')$ because each

$$tr = \sum_{i=1}^{r} \sum_{j=1}^{c} a_{ij}^2$$

Spelling that out a bit more:

General notation for this 4×3 A matrix, $A = \begin{bmatrix} a_{11} & a_{12} & a_{13} \\ a_{21} & a_{22} & a_{23} \\ a_{31} & a_{32} & a_{33} \\ a_{41} & a_{42} & a_{43} \end{bmatrix}$

$$A'A = \begin{bmatrix} a_{11} & a_{21} & a_{31} & a_{41} \\ a_{12} & a_{22} & a_{32} & a_{42} \\ a_{13} & a_{23} & a_{33} & a_{43} \end{bmatrix} \begin{bmatrix} a_{11} & a_{12} & a_{13} \\ a_{21} & a_{22} & a_{23} \\ a_{31} & a_{32} & a_{33} \\ a_{41} & a_{42} & a_{43} \end{bmatrix} = \begin{bmatrix} \sum_i a_{i1}^2 & na & na \\ na & \sum_i a_{i2}^2 & na \\ na & na & \sum_i a_{i3}^2 \end{bmatrix}$$

(Not worrying about the off-diagonals for the moment because we're going right to the trace which is a sum of only the diagonal elements. An off-diagonal element would be $\sum_i (a_{ij} a_{ij'})$.)

$$tr(A'A) = \sum diag = \sum_j \left(\sum_i a_{ij}^2 \right)$$

$$diag(AA') = \begin{bmatrix} a_{11} & a_{12} & a_{13} \\ a_{21} & a_{22} & a_{23} \\ a_{31} & a_{32} & a_{33} \\ a_{41} & a_{42} & a_{43} \end{bmatrix} \begin{bmatrix} a_{11} & a_{21} & a_{31} & a_{41} \\ a_{12} & a_{22} & a_{32} & a_{42} \\ a_{13} & a_{23} & a_{33} & a_{43} \end{bmatrix}$$

$$= \begin{bmatrix} \sum_j a_{1j}^2 & & & \\ & \sum_j a_{2j}^2 & & \\ & & \sum_j a_{3j}^2 & \\ & & & \sum_j a_{4j}^2 \end{bmatrix}$$

$$tr(AA') = \sum diag = \sum_i \left(\sum_j a_{ij}^2 \right)$$

With "c" being a data vector n×1, then (1'c)/n:

$$(1'c)/n = \left(\frac{1}{n}\right)[1 \quad 1 \quad \dots \quad 1]\begin{bmatrix} c_1 \\ c_2 \\ \dots \\ c_n \end{bmatrix} = \left(\frac{1}{n}\right)\sum_i c_i = \bar{c}$$

It's the mean!
In terms of 1'1:

$$(1'1) = [1 \quad 1 \quad \dots \quad 1]\begin{bmatrix} 1 \\ 1 \\ \dots \\ 1 \end{bmatrix} = 1 + 1 + \dots + 1 = \sum_{i=1}^{n} 1 = n$$

So mean of c:

$$(1'c)/n = (1'c)(1'1)^{-1}$$

Problem 7. The covariance matrix of X…

$$X = \begin{bmatrix} 1 & 7 & 5 \\ 2 & 4 & 9 \\ 3 & 6 & 2 \\ 6 & 5 & 0 \\ 8 & 3 & 9 \end{bmatrix}$$

$$\bar{x}' = \left(\frac{1}{n}\right)(1'X) = [4 \quad 5 \quad 5]$$

$$X_d = X - 1\bar{x} = \begin{bmatrix} 1 & 7 & 5 \\ 2 & 4 & 9 \\ 3 & 6 & 2 \\ 6 & 5 & 0 \\ 8 & 3 & 9 \end{bmatrix} - \begin{bmatrix} 4 & 5 & 5 \\ 4 & 5 & 5 \\ 4 & 5 & 5 \\ 4 & 5 & 5 \\ 4 & 5 & 5 \end{bmatrix} = \begin{bmatrix} -3 & 2 & 0 \\ -2 & -1 & 4 \\ -1 & 1 & -3 \\ 2 & 0 & -5 \\ 4 & -2 & 4 \end{bmatrix}$$

$$COV(X) = \left(\frac{1}{n-1}\right)X'_d X_d = \left(\frac{1}{5-1}\right)\begin{bmatrix} -3 & -2 & -1 & 2 & 4 \\ 2 & -1 & 1 & 0 & -2 \\ 0 & 4 & -3 & -5 & 4 \end{bmatrix}\begin{bmatrix} -3 & 2 & 0 \\ -2 & -1 & 4 \\ -1 & 1 & -3 \\ 2 & 0 & -5 \\ 4 & -2 & 4 \end{bmatrix}$$

$$= \left(\frac{1}{4}\right)\begin{bmatrix} 34 & -13 & 1 \\ -13 & 10 & -15 \\ 1 & -15 & 66 \end{bmatrix} = \begin{bmatrix} 8.50 & -3.25 & .25 \\ & 2.50 & -3.75 \\ sym. & & 16.50 \end{bmatrix}$$

Homework from Chapter 14

Problem 1.

1) $\boldsymbol{E} = \boldsymbol{H_{AB}}$ (derive EMS for B when A is random)

	a	b	c	n	
	R	F	F	R	
	i	j	k	ℓ	EMS
α_i	1	b	c	n	$bcn\sigma_A^2 + \sigma_\epsilon^2$
β_j	a	0	c	n	$acn\theta_B^2 + cn\sigma_{AB}^2 + \sigma_\epsilon^2$
γ_k	a	b	0	n	$abn\theta_C^2 + bn\sigma_{AC}^2 + \sigma_\epsilon^2$
$(\alpha\beta)_{ij}$	1	0	c	n	$cn\sigma_{AB}^2 + \sigma_\epsilon^2$
$(\alpha\gamma)_{ik}$	1	b	0	n	$bn\sigma_{AC}^2 + \sigma_\epsilon^2$
$(\beta\gamma)_{jk}$	a	0	0	n	$an\sigma_{BC}^2 + n\sigma_{ABC}^2 + \sigma_\epsilon^2$
$(\alpha\beta\gamma)_{ijk}$	1	0	0	n	$n\sigma_{ABC}^2 + \sigma_\epsilon^2$
$\epsilon_{l(ijk)}$	1	1	1	1	σ_ϵ^2

2) $\lambda_i = 0$ for $i \geq 5$

$$rank(HE^{-1}) = s = \min(p, df_{effect}) = \min(4,4) = 4$$

3) p = # dependent variables = 4

4) $df_{effect} = (b-1) = (5-1) = 4$

5) $df_{error} = (a-1)(b-1) = (7-1)(5-1) = 24$

6) &7) The 4 test statistics, Λ, V, R and T

$$\Lambda = \left(\frac{1}{1+5.20}\right)\left(\frac{1}{1+3.75}\right)\left(\frac{1}{1+2.00}\right)\left(\frac{1}{1+.75}\right) = .006468$$

$$q = \sqrt{\frac{4^2 4^2 - 4}{4^2 + 4^2 - 5}}, \quad m = 24 - \frac{4+1-4}{2} = 23.5$$

$$F = \frac{(1 - .0065^{1/3.055})[(23.5)(3.055) - .5(4)(4) + 1]}{.0065^{1/3.055}(4)(4)} = \frac{52.33}{3.07} = 17.05$$

On 16 and 64.79 (or 64) df, critical value is $F_{.05;16;64} = 1.84$.
According to this test, reject H_0.

For V:

$$V = \left(\frac{5.20}{1+5.20}\right)\left(\frac{3.75}{1+3.75}\right)\left(\frac{2.00}{1+2.00}\right)\left(\frac{.75}{1+.75}\right) = 2.723$$

$$F = \left(\frac{(24-4+4)(2.723)}{4(4-2.723)}\right) = \left(\frac{65.352}{5.108}\right) = 12.794$$

On 16 and 96 df, critical value is $F_{.05;15;60} = 1.84$
According to this test, reject H_0.

For R:

$$R = 5.20$$

(Don't have an easy F-approximation to calculate by hand.)

For T:

T = 5.20 + 3.75 + 2.00 + .75 = 11.70
a = 16 b = 36.27 (use 36 in F table)
c = .796B = 1.556
F = T/C = 11.70/.796 = 14.698 on 16 and 36 df
$F_{.05;15;30} = 2.01$ (critical value)

8) H_0: no main effect for B:

$$H_0\text{: } \mu_{B1} = \mu_{B2} = \mu_{B3} = \mu_{B4} = \mu_{B5}$$

Each μ_{Bj} is a p×1 (4×1) vector means for B = j

9) For Λ: F = 17.05 > 1.84) → reject H_0

For V: F = 12.79 >1.84) → reject H_0

For R: have to get the F from SAS

For T: F = 14.698 > 2.01) → reject H_0

∴ Can probably reject H_0

10) $df_{error} = 24$, almost = 30; quite big, and we're told assumptions hold, so probably all 4 test statistics will lead to valid statistical conclusions.

If we could distinguish among the 4 tests at all, if they seemed to lead to slightly different conclusions re rejecting the null hypothesis or not, we might not rely too much on R, since our "structure" looks diffuse (i.e., it's not the case that $\lambda_1 \gg \lambda_2 > \lambda_3 > \lambda_4$; they're all of comparable size (a concentrated structure would be more like: $\lambda_1 = 15$, $\lambda_2 = 3$, $\lambda_3 = 2$, $\lambda_4 = .5$).

Homework from Chapter 15

Problem 1. Start simply: Look at means and standard deviations and correlations, for each group, for each factor, and overall. Plot the data (by hand to get a feel for the data, and in Excel to make the plots prettier for possible inclusion in a paper).

What patterns do you see in the data at this point? What might you expect to find in further analyses?

Conduct an ANOVA on each variable. Include a derivation of the expected mean squares for this design to justify the F-tests you compute. Follow-up any significant tests of main effects and interaction with whatever tests of contrasts or simple effects that seem appropriate or interesting for these data.

Conduct a MANOVA on these data, and follow-up significant tests of main effects and interactions with appropriate contrasts or tests of simple effects.

Plot the direction of the discriminant axis. Does it look "reasonable"? Does it provide greater group distinction than the axis that could be drawn connecting the groups' centroids? Use the simple descriptives and the plots in your explanations.

Are the results of the ANOVAs and the MANOVA in agreement? Why or why not? If they are ever not in agreement for any of the tests you conducted, which of these (the ANOVAs, or the MANOVA) should you weight more heavily?

What have you learned about these data?

(Students were also instructed: Summarize all the results you wish to convey on paper (in sentences, tables, and/or figures). For any table or statistic that you cite, include the page number where I could find the value in your SAS output file, and in that output, circle the result(s) you referred me to. Turn in a copy of your SAS input and output files.)

Brief summary of HW7 SAS input and output files:

1) data are printed and plotted.

2) proc means and proc corr's result in the following info:

Each cell contains:
$\bar{x}_1$ $\bar{x}_2$
(s_{x1}) (s_{x2})
($r_{x1,x2}$)

	male	female	Drug marginals
Drug A	6.500 6.250 (1.915) (2.062) (r = 0.887)	7.500 8.250 (1.291) (2.062) (r = 0.188)	7.000 7.250 (1.604) (2.188) (r = 0.652)
Drug B	7.250 8.250 (1.258) (2.630) (r = 0.781)	7.750 8.750 (1.708) (3.096) (r = 0.678)	7.500 8.500 (1.414) (2.673) (r = 0.718)
Drug C	16.000 12.000 (3.916) (2.160) (r = 0.985)	13.500 8.500 (2.517) (2.887) (r = 0.964)	14.750 10.250 (3.327) (3.012) (r = 0.905)
Gender marginals	9.917 8.833 (5.089) (3.243) (r = 0.891)	9.583 8.500 (3.370) (2.468) (r = 0.366)	9.750 8.667 (4.225) (2.823) (r = 0.714)

- Drug C seems to be raising the means (and standard deviations...). We'll see if the differences are significant.
- Also note that the two dependent variables (X_1 and X_2) are fairly highly correlated in the entire table, except for females under the influence of drug A. Given the (nearly uniformly) high correlations between X_1 and X_2, we may consider simplifying our modeling to analyzing some combination of X_1 and X_2 (e.g., their mean).

Plots also had indicated drug C is likely to be statistically diff from drugs A and B.

3) We are considering both factors, sex and drug (and rock 'n roll) to be fixed, so we know testing for the two main effects and for the interaction, the denominator will be the S/gender*drug MS. But, for practice, we can derive the EMS to be sure.

model $ABS_{ijk} = \mu + \alpha_i + \beta_j + \alpha\beta_{ij} + \epsilon_{k(ij)}$

	a	b	n		
	F	F	R		
	i	j	k	EMS	F
α_i	0	b	n	$bn\theta_A^2 + \sigma_\epsilon^2$	$MS_A/MS_{S(AB)}$
β_j	a	0	n	$an\theta_B^2 + \sigma_\epsilon^2$	$MS_B/MS_{S(AB)}$
$\alpha\beta_{ij}$	0	0	n	$n\theta_{AB}^2 + \sigma_\epsilon^2$	$MS_{AB}/MS_{S(AB)}$
$\epsilon_{k(ij)}$	1	1	1	σ_ϵ^2	

4) Univariate ANOVAs

Dependent variable X_1:

source	df	SS	MS	F	p
gender	1	.667	.667	.13	.726
drug	2	301.000	150.500	28.67	.000*
gender×drug	2	14.333	7.167	1.37	.281
error	18	94.500	5.250		
total	23		410.500		

contrasts	df	SS	MS	F	p
drug 1&2 vs. 3	1	300.000	300.000	57.14	.000*
drug 1 vs. 2	1	1.000	1.000	.19	.668
drug 1 vs. 3	1	240.250	240.250	45.76	.000*
drug 2 vs. 3	1	210.250	210.250	40.05	.000*

(First two contrasts are orthogonal. Last three contrasts are all pairwise comparisons.)

Dependent variable X_2:

source	df	SS	MS	F	p
gender	1	.667	.667	.11	.749
drug	2	36.333	18.167	2.87	.083 marg'l
gender×drug	2	32.333	16.167	2.55	.106
error	18	114.000	6.333		
total	23	183.000			

Seems to be no gender effect and no gender×drug interaction. There seem to be a drug effect, where the significance is driven by drug C being quite diff from drugs A and B.

Multivariate ANOVA (p=2 dep vars)

effect	F	p
gender	F(2,17) = .06	.938
drug	F(4,34) = 12.20	.000
drug 1&2 vs 3	F(2,17) = 38.88	.000
drug 1 vs 2	F(2,17) = .56	.583
drug 1 vs 3	F(2,17) = 28.61	.000
drug 2 vs 3	F(2,17) = 29.99	.000
gender×drug	F(4,34) = 1.16	.346

Consistent with the ANOVA findings, there seems to be no gender effect and no gender by drug interaction. (Life is not always so easy—results from ANOVAs and MANOVA may sometimes lead to diff conclusions.) The drug main effect seems to be present, due primarily to the fact that drug C is diff from the first two drugs.

To verify some random set of df, pick say, the F test for the gender by drug interaction. It had 4 and 34 df. Is this correct? For the F approximation on, say Λ, the F has $p(df_{effect})$ df in the numerator and $(mq - (.5)p(df_{effect}) + 1)$ df in the denominator. $p(df_{effect}) = 2(2) = 4$, so that's ok.

The mq etc. term is more complicated:

$$m = (df_{error} - (p + 1 - df_{effect})/2) = (18 - (2 + 1 - 2)/2) = 17.5$$

$$q = \sqrt{\frac{p^2(df_{effect})^2 - 4}{p^2 + (df_{effect})^2 - 5}} = \sqrt{\frac{4(4) - 4}{4 + 4 - 5}} = \sqrt{\frac{12}{3}} = 2$$

$(mq - (.5)p(df_{effect}) + 1) = (17.5)(2) - (.5)(2)(2) + 1 = 34$, so that's ok.

Eigenvectors for the diff effects follow, along with a reminder as to whether the effect was significant or not:

effect	eigenvector	significant?
gender	[.072 .034]'	no
drug	[.148 -.077]'	yes
drug 1&2 vs. 3	[.148 -.077]'	yep
drug 1 vs 2	[-.061 .127]'	no
drug 1 vs 3	[.145 -.069]'	yes
drug 2 vs 3	[.150 -.084]'	yes
gender×drug	[-.003 .096]'	no

- Plot these vectors. Use the origin as one point and the coordinates listed in the column labelled "eigenvector" as the other point. The "discriminant axis" is in the direction pointing toward the eigenvector coordinates. See if you can "project" the data points onto these various new axes to see how and why the different effects are significant in the data.
- If you could draw increasingly larger encompassing confidence regions or ellipsoids around each of the sets of data points, and saw where they had begun to intersect with other ellipsoids, would the significant discriminant functions (i.e., eigenvectors) be in the direction perpendicular (i.e., orthogonal) to the lines of intersection? (answer better be yes).
- Note that the only two effects that have more than one nonzero eigenvalue are the drug main effect and the gender by drug interaction. Especially for testing the drug main effect, the first eigenvalue is much larger than the next eigenvalue. If the results of the 4 diff test stats (Λ, V, R, T (you remember)) had been different, perhaps we'd "weight" R more heavily in our minds when making decisions?

Note once again that while the ANOVAs and the MANOVA gave consistent results, that need not have been the case. With data such as these, we might "weight" the MANOVA results more heavily (i.e., believe them a bit more) than the ANOVA results (had they been inconsistent), given that the plot of the data indicates the groups are more clearly separable using both X_1 and X_2 than using either variable alone. That is, the marginal distributions aren't as separable as the bivariate plot.

The ANOVA and MANOVA results are probably similar to some extent because the two dep vars are so highly correlated. If the two vars are measuring the same thing, their joint analysis shouldn't be surprisingly diff from their separate analyses. The final analysis looks at a univariate ANOVA where the dep var is the mean of the original dep vars. (Though X_1 and X_2 are not entirely redundant. If they were, the ANOVA on X_2 would have been more similar to the ANOVA on X_1. Or vicey versie!)

SAS input would look something like this:

```
*1=male, 2=female;
Data abc;
input sex drug X1 X2 symb $;
datalines;
1 1     5  6   A
1 1     5  4   A
1 1     9  9   A
1 1     7  6   A
1 2     7  6   B
1 2     7  7   B
1 2     9 12   B
1 2     6  8   B
1 3    21 15   C
1 3    14 11   C
1 3    17 12   C
1 3    12 10   C
2 1     7 10   D
2 1     6  6   D
2 1     9  7   D
2 1     8 10   D
2 2    10 13   E
2 2     8  7   E
2 2     7  6   E
2 2     6  9   E
2 3    16 12   F
2 3    14  9   F
2 3    14  8   F
2 3    10  5   F
proc print data=abc;
proc plot; plot X2*x1 = symb;
proc means; var X1 X2; proc corr cov; var X1 X2;
proc sort; by sex; proc means; var X1 X2; by sex;
proc corr cov; var X1 X2; by sex;
proc sort; by drug; proc means; var X1 X2; by drug;
proc corr cov; var X1 X2; by drug;
proc sort; by sex drug; proc means; var X1 X2; by sex drug;
proc corr cov; var X1 X2; by sex drug;
proc glm data=abc; class sex drug;
  model X1 X2 = sex drug sex*drug / ss3;
  manova H = sex / printe printh summary;
  manova H = drug / printe printh summary;
  contrast 'drugs A and B vs. C' drug .5 .5 -1;
```

```
contrast 'drug A vs. B' drug 1 -1 0;
contrast 'drug A vs. C' drug 1 0 -1;
contrast 'drug B vs. C' drug 0 1 -1;
manova H = sex*drug / printe printh summary;
contrast 'simple1' sex 1 -1 sex*drug 1 0 0 -1 0 0;
contrast 'simple2' sex 1 -1 sex*drug 0 1 0 0 -1 0;
contrast 'simple3' sex 1 -1 sex*drug 0 0 1 0 0 -1;
means sex drug sex*drug / scheffe; run;
```

As you can see, I'm analyzing the beejeebers out of these data. This IS after all, part of your final. Treat these data as if they were your thesis or RA data—you'd want to know those data inside and out…

The SAS output from the above commands would look like this:

→Univariate descriptives on X_1 and X_2:

Variable	N	Mean	Std Dev	Minimum	Maximum
X1	24	9.7500000	4.2246688	5.0000000	21.0000000
X2	24	8.6666667	2.8232985	4.0000000	15.0000000

→Correlation between X_1 and X_2 (with a correlation this high, I might have simply taken a mean of X_1 and X_2, call it a "scale," and go forward with a univariate ANOVA):

	X1	X2
X1	1.00000	0.71446 <.0001
X2	0.71446 <.0001	1.00000

The SAS output continues, listing the means broken down by gender, then drug, then by both gender and drug, and same for the correlations. (That info was summarized above.)

→Then Proc Glm kicks in. First the univariate ANOVA for X_1:

Dependent Variable: X1

Source	DF	Sum of Squares	Mean Square	F Value	Pr > F
Model	5	316.0000000	63.2000000	12.04	<.0001
Error	18	94.5000000	5.2500000		
Corrected Total	23	410.5000000			

R-Square	Coeff Var	Root MSE	X1 Mean
0.769793	23.50039	2.291288	9.750000

Source	DF	Type III SS	Mean Square	F Value	Pr > F
SEX	1	0.6666667	0.6666667	0.13	0.7257
DRUG	2	301.0000000	150.5000000	28.67	<.0001
SEX*DRUG	2	14.3333333	7.1666667	1.37	0.2806

→Then for X_2:

```
Dependent Variable: X2
                                          Sum of
 Source                      DF          Squares     Mean Square    F Value    Pr > F
 Model                        5       69.3333333      13.8666667       2.19    0.1008
 Error                       18      114.0000000       6.3333333
 Corrected Total             23      183.3333333
               R-Square     Coeff Var      Root MSE       X2 Mean
               0.378182      29.03782      2.516611      8.666667
 Source                      DF      Type III SS     Mean Square    F Value    Pr > F
 SEX                          1       0.66666667      0.66666667       0.11    0.7493
 DRUG                         2      36.33333333     18.16666667       2.87    0.0829
 SEX*DRUG                     2      32.33333333     16.16666667       2.55    0.1057
```

Then the MANOVA for X_1 and X_2 modeled jointly. SAS produces the results organized by the H (effect), thus, →first for gender:

```
                       Multivariate Analysis of Variance
                            E = Error SSCP Matrix
                                  X1                 X2
                    X1          94.5               76.5
                    X2          76.5                114
                       H = Type III SSCP Matrix for SEX
                                  X1                X2
                    X1      0.6666666667      0.6666666667
                    X2      0.6666666667      0.6666666667

           Characteristic Roots and Vectors of: E Inverse * H, where
H = Type III SSCP Matrix for SEX, E = Error SSCP Matrix
           Characteristic                Characteristic Vector  V'EV=1
                     Root    Percent               X1             X2
               0.00751918     100.00       0.07175780     0.03444374
               0.00000000       0.00      -0.13423121     0.13423121

                  MANOVA Test Criteria and Exact F Statistics
                  for the Hypothesis of No Overall SEX Effect
H = Type III SSCP Matrix for SEX, E = Error SSCP Matrix
                               S=1    M=0     N=7.5
Statistic                         Value    F Value    Num DF    Den DF    Pr > F
Wilks' Lambda                0.99253694       0.06         2        17    0.9383
Pillai's Trace               0.00746306       0.06         2        17    0.9383
Hotelling-Lawley Trace       0.00751918       0.06         2        17    0.9383
Roy's Greatest Root          0.00751918       0.06         2        17    0.9383
```

→Next for drug:

```
                            E = Error SSCP Matrix
                                  X1                 X2
                    X1          94.5               76.5
                    X2          76.5                114
                       H = Type III SSCP Matrix for DRUG
                                   X1                X2
                    X1            301              97.5
                    X2           97.5      36.333333333
```

```
          Characteristic Roots and Vectors of: E Inverse * H, where
H = Type III SSCP Matrix for DRUG, E = Error SSCP Matrix
          Characteristic                  Characteristic Vector  V'EV=1
                    Root     Percent                 X1              X2
              4.57602675       98.63         0.14784109     -0.07693601
              0.06350991        1.37        -0.03619684      0.11526161

                MANOVA Test Criteria and F Approximations for
                   the Hypothesis of No Overall DRUG Effect
H = Type III SSCP Matrix for DRUG, E = Error SSCP Matrix
                             S=2    M=-0.5    N=7.5
Statistic                        Value    F Value    Num DF    Den DF    Pr > F
Wilks' Lambda               0.16862952      12.20         4        34    <.0001
Pillai's Trace              0.88037810       7.08         4        36    0.0003
Hotelling-Lawley Trace      4.63953666      19.40         4    19.407    <.0001
Roy's Greatest Root         4.57602675      41.18         2        18    <.0001
```

→Lastly for gender×drug:

```
                             E = Error SSCP Matrix
                                     X1                  X2
                    X1             94.5                76.5
                    X2             76.5                 114
                     H = Type III SSCP Matrix for SEX*DRUG
                                     X1                  X2
                    X1     14.333333333        21.333333333
                    X2     21.333333333        32.333333333

          Characteristic Roots and Vectors of: E Inverse * H, where
H = Type III SSCP Matrix for SEX*DRUG, E = Error SSCP Matrix
          Characteristic                  Characteristic Vector  V'EV=1
                    Root     Percent                 X1              X2
              0.28372273       97.94        -0.00284433      0.09555092
              0.00596889        2.06        -0.15218117      0.10037136

                MANOVA Test Criteria and F Approximations for
                  the Hypothesis of No Overall SEX*DRUG Effect
H = Type III SSCP Matrix for SEX*DRUG, E = Error SSCP Matrix
                             S=2    M=-0.5    N=7.5
Statistic                        Value    F Value    Num DF    Den DF    Pr > F
Wilks' Lambda               0.77436234       1.16         4        34    0.3459
Pillai's Trace              0.22694905       1.15         4        36    0.3481
Hotelling-Lawley Trace      0.28969161       1.21         4    19.407    0.3381
Roy's Greatest Root         0.28372273       2.55         2        18    0.1056
```

Then lots of MANOVA contrast info is printed in the output (I think I'd stick to the univariate contrast—run proc glm with the contrast statement but without the manova statement).